Cost-effective Energy Management

K.S. Bajaj, Ph.D., P.E.
Director of Systems Engineering
Electronic Data Systems Corp.

T. Singh, Ph.D.
Associate Professor
Wayne State University

Business News Publishing Co.
Troy, Michigan

Library of Congress Cataloging in Publication Data
Bajaj, K.S., 1942–
    Cost-effective energy management

    Bibliography: p.
    1. Buildings—Energy Conservation—
Cost effectiveness.
I. Singh, T. (Trilochan) l937–   II. Title
TJ163.5.B84B34   658.2'6   82-4243
ISBN 0-912524-22-7   AACR2

Printed in the United States of America

Contents

Chapter 1

ENERGY USE AND CONSERVATIONAL POTENTIAL

Efficient energy utilization is a national need. Never before in our history have we, as a nation, been more aware of the scarcity of this resource that is as much a part of the American way as apple pie and church on Sunday. Everything we do requires energy: driving to work, lighting, heating/cooling our buildings, manufacturing marketable goods from raw materials and the delivery of services.

In recent years there has been considerable talk of new and innovative methods of energy utilization. Considerable progress is being made in the design of energy efficient transportation systems. Extensive research is being directed towards finding alternate energy sources. The energy crisis is truly a test of American ingenuity and inventiveness. But we need more than new inventions; we need to examine our current energy usage system in the light of decreasing availability, increasing dependence on foreign fossil fuels and extremely high prices for imported energy. Figure 1-1 shows the average price of imported oil from 1970 through 1979, in terms of 1979 dollars. Prices more than tripled during the 1973–74 oil embargo, leveled off at a high plateau in the years 1974 through early 1978, when they once again aggressively advanced to new heights.

High price levels have inspired radical changes in energy consumption, most notably in gasoline consumption. In 1978, U.S. gasoline consumption was 7.4 million barrels per day. During 1979, consumption fell to 7.0 million barrels a day while cars on the road increased by 2½ percent. Americans may be driving some fewer miles, but the greatest decline in consumption is due more to fuel efficient cars and energy consciousness being shown by drivers.

Similar declines in per capita energy consumption are now in progress due to more efficient appliances and heating systems, energy conserving housing, changes in residential locations and design and imperceptible alterations in life styles. To keep this downward trend in force, Americans must persist in their new-found energy conserving techniques and thus conserve our limited stores of energy.

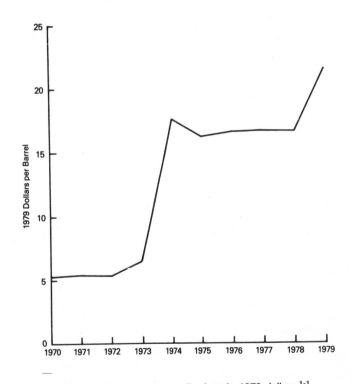

FIG. 1-1. Annual average oil prices in 1979 dollars.[1-1]

1

Nuclear Power

The utility industry in the United States has accepted nuclear power as a safe, reliable and economic means of meeting a large proportion of its requirements for new capacity. In 1979, nuclear power provided 11% of the utility industry's requirements and less than 3.5% of the nation's total energy needs. By 1990 it will provide utilities with almost 18% of their capacity and 6% of the nation's demand. However, the growth of nuclear power suffered a setback in 1979 due to the accident at Three Mile Island.

Present day power reactors utilize only a small fraction of the nuclear raw material. In view of the rapid increase in the projected demand for uranium, some concern has been expressed as to whether there will be sufficient material to meet these requirements. Based on known reserves, there is enough moderately priced uranium ($30 per pound of U_3O_8) to produce power for 78 years at the 1979 rate of U.S. nuclear power production. It is estimated that, by the year 1990, annual nuclear power production will reach about 8.0 quads thermal. At that rate of generation, the probable supply would last for about 61 years in conventional light water reactors. However, if the breeder reactor can be developed and put into commercial operation, the fuel supply could be extended some 70-fold, giving a life span of more than 4,000 years.

Geothermal

Among the often mentioned sources is geothermal energy that is derived from underground reservoirs containing either steam, hot water, hot brine, or hot dry rock. It is a source of hot water, for heating, and steam for conventional electric power generation. Estimates of the amount of power that can be economically generated from potential geothermal sites in the U.S. range from 0.5 to 3% of total electric consumption.

Wind Energy

Wind energy is another candidate, which is being given serious consideration. The amount of energy possessed by the wind over the United States is enormous, but it is quite dilute and varies in intensity with geographical region. As yet, it is not economically feasible to generate electricity from wind power.

Solar Energy

The solar energy available in the United States is in excess of 30,000 quads per year, sufficient to supply 200 times the total energy requirements for the year 1990. The difficulty in utilizing this energy is that it is dilute and intermittent. There are three major areas for solar energy utilization: heating and cooling, thermal electric, and photovoltaic. The feasibility of solar heating of houses has been demonstrated and it is economi-

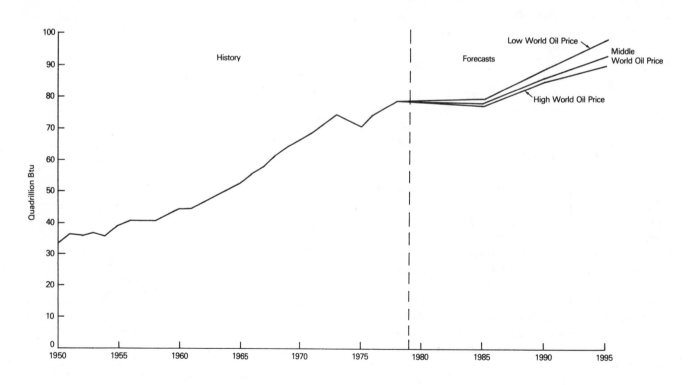

FIG. 1-2. Trends in U.S. energy consumption.[1-1]

TABLE 1-1. Summary and Forecast of U.S. Gross Energy Production (Quadrillion Btu)[1-1]

Source	1978	1990
Oil and Natural Gas Liquids[b]	20.4	20.2
Natural Gas[b]	19.8	17.0
Coal[b]	15.0	28.1
Nuclear Power[c]	3.0	8.0
Other[d]	3.0	3.6
Total	61.1	76.9

ENERGY SOURCES AND SUPPLY

Table 1-1 shows the domestic energy production in 1978 and projects production in 1990, based on an average 2.5% annual increase in Gross National Product. As the Table shows, future energy demand will be supplied by numerous sources.

Coal

In terms of energy equivalents, estimated recoverable reserves of coal comprise approximately 83% of all the fossil fuels in the U.S. However, all of this recoverable coal may not be economically mineable. Only 57% of the bituminous coal is currently being recovered from active mines.

Air pollution control has made the sulphur content of the coal an important factor in evaluating reserves. Approximately two thirds of the estimated reserves consist of low sulphur (1.0% or less) coal. However, more than half of this cleaner burning coal is composed of lower Btu coals such as lignite and sub-bituminous grades.

The states east of the Mississippi River contain slightly over 40% of the high Btu bituminous and anthracite coals containing 1.0% sulphur or less. Approximately 98% of total lignite reserves, which are largely low in sulphur, are located in North Dakota and Montana. Extra high voltage transmission of electricity, and developments in the technology of using low Btu coal as a practical and economical fuel, make it possible to have large power generating stations burn sub-bituminous coal or lignite near the source.

At the present rate of consumption, coal supplies will last close to 400 years. According to U.S Bureau of Mines forecasts, the rate of energy consumption by fuel burning electric generating plants will increase from the present 15.4 quads per year to about 26.4 quads per year by the year 2000. At this higher rate of consumption, our coal reserves will be exhausted in approximately 200 years. Should additional coal be used to produce synthetic fuels for transportation and other non-electric generation uses, U.S. reserves will last about 100 years.

Oil

About 46% of our energy needs are met by oil, making it our major fuel. Estimates of ultimate United States crude oil reserves are in the range of 500 billion barrels. At the present recovery efficiency, total recoverable reserves are about 175 billion barrels. Proved reserves represent the working inventory of the petroleum industry and have been kept at approximately 31 billion barrels. About 3 billion barrels of domestic crude oil is now being produced per year. To meet the ever rising demand new sources of oil are being explored.

It is estimated that shale oil could fulfill our petroleum needs, at the present rate of consumption, for about 75 years. Leaner shale, yielding 10 to 25 gallons per ton, could extend this severalfold. However, there are severe environmental and other problems associated with its recovery. First, the extraction of oil from shale requires an enormous amount of water, which is not readily available where shale is abundant. Secondly, the disposal of spent shale, in an environmentally acceptable manner, is a monumental task that involves great expense.

A number of pilot projects are presently underway. The largest, TOSCO, produces 700 barrels of oil a day from 1,000 tons of oil shale. The Department of Interior estimates that production could reach one million barrels per day by 1985.

Tar sands are less plentiful in the U.S. than oil shale and oil recovery is more difficult. Hence, no large extraction effort is anticipated.

Natural Gas

Natural gas is a mixture of methane and other low molecular weight hydrocarbons. Ordinarily, it has a negligible sulphur content. However, if the sulphur content is significant, the gas can be efficiently processed to reduce the sulphur compounds before it is marketed.

The proved recoverable reserves of gas in the U.S. are estimated at 300 trillion cubic feet, while the total potential supply is estimated to be 690 trillion feet, resulting in a total recoverable reserve of about 983 trillion feet.

Hydroelectric Power

Hydroelectric power does not require fuel for generation and, therefore, does not create any sulphur oxides. Hydroelectric generation presently accounts for 13% of the electric energy produced in the United States. Because of the widespread opposition to development of potential hydroelectric sites, it is unlikely that more than 6.0 quads of power will ever be exploited. It appears that the growth of hydroelectric power will not keep pace with fossil and nuclear fuels.

cally competitive with electric power in certain areas.

ERDA estimates that by 1990, about 1% of the national energy will be provided by the sun.

ENERGY DEMAND

Higher energy prices and Federal energy conservation programs have already led to a reduction in average energy consumption. Projections assume even greater future reductions in energy intensity by each sector of the economy, Table 1-2 and Figure 1-2.

TABLE 1-2. U.S. Energy Consumption and Forecast[1-1]

Year	Residential/ Commercial	Industrial	Transportation	Electricity Generation and Transmission Losses
1965	14.0	19.1	12.8	7.9
1973	18.8	24.0	18.9	14.0
1978	18.8	23.2	20.9	16.8
1990	18.3	26.9	18.6	23.0

Domestic energy consumption is projected to be only 11.2% greater in 1990 than in 1978. Furthermore, modest decreases in total consumption are forecast for the residential, commercial and transportation sectors. Overall energy consumption by the industrial sector is forecast to be greater in 1990 than it was in 1978, but energy use per unit of industrial output is expected to decline considerably. Electricity generating and transmission losses are projected to grow, reflecting the expected growth in electric demand.

Residential Sector

Most fossil fuel expended in the residential sector is used for heating, cooling, cooking and domestic hot water. Residental energy consumption is forecast to remain roughly constant through 1990 despite growing population and rising income levels. Table 1-3 shows that energy use per household is projected to decline by about 25% be-

TABLE 1-3. Residential Energy Consumption and Prices[1-1]

Year	Net Energy Consumed per Thousand Households (million Btu)	Average Residential Energy Price (1979 dollars per million Btu)
1960	140	4.02
1973	160	3.85
1978	150	5.19
1990	110	9.43

tween 1978 and 1990. By contrast, energy use per household increased by 14% between 1960 and 1973.

The projected decrease in residential energy is the result of higher average energy prices, projected to grow at about 5% annually, and the impact of energy conservation in buildings and equipment.

Commercial Sector

Fossil fuels are used in the commercial sector for heating and cooling buildings and for supplying hot water. A useful indicator of energy intensity is energy use per square foot of floor space and energy consumed per dollar of real GNP, Table 1-4. Both of these indicators increased slightly between 1960 and 1973, while both are forecast to decrease between 1978 and 1990 due to increased energy prices and energy conservation programs.

TABLE 1-4. Commercial Energy Consumption and Prices[1-1]

Year	Net Energy Consumed per Square Foot of Floorspace (million Btu)	Net Energy Consumed per Dollar of GNP (trillion Btu)	Average Commercial Energy Price (1979 dollars per million Btu)
1960	0.281	3.7	4.25
1973	0.283	3.8	3.77
1978	0.251	3.3	5.48
1990	0.189	2.4	9.52

Industrial Sector

Industrial use of energy, which decreased after the oil embargo, is projected to grow modestly between 1978 and 1990. Table 1-5 shows that the energy intensity of industrial production continues to decline, but at a slower pace. Between 1978 and 1990, energy intensity is projected to decline from 32.5 to 24.6 thousand Btu per dollar of *manufacturing value-added*. This decrease of 2.3% per year is in line with the pre-embargo decline of 2.2% per year and the 1973 to 1978 decline of 2.6% annually.

TABLE 1-5. Industrial Energy Intensity and Prices[1-1]

Year	Net Energy per Dollar of GNP (thousand Btu per dollar)	Net Energy per Dollar of Manufacturing Value-Added (thousand Btu per dollar)	Average Industrial Energy Price (1979 dollars per million Btu)
1960	13.0	49.6	1.64
1973	11.7	37.0	1.73
1978	10.0	32.5	2.98
1990	8.3	24.6	5.92

TABLE 1-6. Factors Affecting Gasoline Consumption[1-1]

Year	Automobiles per Capita	Average Fleet Miles per Gallon	Miles Traveled per Car (thousands)	Gasoline Price (1979 cents per gallon)	Automobile Gasoline Consumption (billion gallons)
1960	0.35	14.3	9.45	72.5	41.2
1973	0.51	13.3	9.77	61.1	78.0
1978	0.56	14.3	9.81	70.5	83.6
1990	0.59	22.3	9.50	158.5	57.7

Transportation Sector

All the energy used by the transportation sector is petroleum based and is largely consumed by automobiles. Table 1-6 shows the factors influencing gasoline consumption. Between 1978 and 1990, gasoline prices are expected to rise 7% annually. Vehicle miles travelled are projected to decline by about 3% annually and fleet efficiency is forecast to increase from about 14 miles per gallon to over 22 mpg. Total transportation energy use per capita is forecast to decrease less than 2% annually.

ELECTRICAL GENERATION

During the 1960's and through 1973, the demand for electricity rose at an average rate of about 5.6% It is estimated that electric energy demand through 1990 will increase 3.4% per annum.

Figure 1-3 compares the energy demand in 1978 with that forecast for 1990. Two large shifts are evident:

- Increased used of electricity in all sectors except transportation.
- Increased use of coal in the industrial sector.

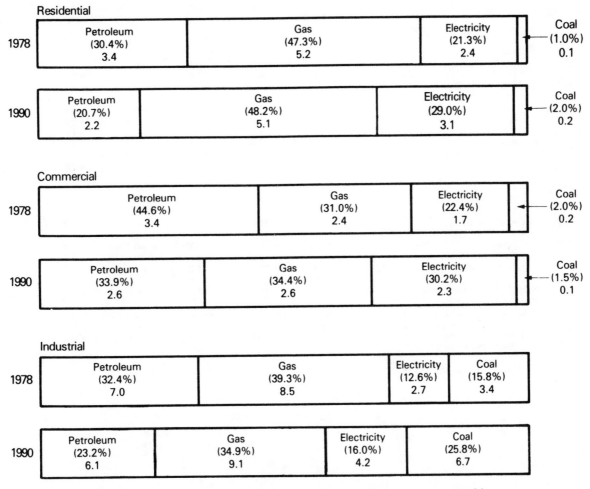

FIG. 1-3. Composition of net energy consumption by sector (quadrillion Btu).[1-1]

Both of these trends point towards decreased use of oil and increased use of coal.

ENERGY CONSERVATION AND ITS POTENTIAL

Table 1-7 presents the forecasts of energy production and consumption for 1990. Domestically produced energy is projected to increase almost 2% per year compared to a decline in the 5 years after the embargo. In 1990, coal production is projected to be almost double that of 1978. Oil and gas production are projected to continue to decline, while other sources continue to grow.

In summary, domestic supply is projected to increase more rapidly in the 1980's than is total consumption. As a result, imports are projected to decline. Although imports are projected to be lower, they will continue to be significant. As a result, import supply disruptions could still have very large economic impacts. More efficient use of existing resources, coupled with more domestic energy supply, will make the U.S. energy independent and will reduce our exposure to the whims of foreign governments.

Studies show that energy requirements in existing residential and commercial buildings can be reduced by 20%, and consumption in new buildings can be reduced by nearly 50%, or more if solar technology is used. Energy requirements for industrial plants can be reduced by 20%. It is estimated that, by using energy conservation strategies in all the areas of energy demand, total energy consumption of U.S.A. can be reduced by 33%.

TABLE 1-7. Summary of U.S. Production and Consumption[1-1]

	1973	1978	Midprice Case 1990	Average Annual Percent Growth Rate 1973–78	1978–90
Domestic Energy Production					
Oil	22.1	20.4	20.2	−1.6	−0.1
Gas	22.2	19.8	17.0	−2.3	−1.3
Coal	14.4	15.0	28.1	+0.8	+5.4
Other	3.8	6.0	11.6	+9.6	+5.6
Gross Domestic Energy					
Production	62.4	61.2	76.9	−0.4	+1.9
Oil Imports	13.0	17.1	11.1	+5.6	−3.5
Gas Imports	1.0	0.9	1.0	−2.1	+0.9
Coal Imports	−1.4	−1.1	−2.7	−4.7	+7.8
Net Imports	12.6	16.9	9.4	+6.0	−4.8
Gross Energy Supply	70.0	78.1	86.3	0.8	+0.8
Losses	−14.1	−16.6	−23.2	3.3	+2.8
Losses as a Percent of Gross Energy Supply	19	21	27	+2.0	+2.1
Net Available Energy[a]	61.5	61.2	63.8	−0.1	+0.3
Domestic Energy Consumption					
Residential	11.1	10.7	10.7	−0.7	0.0
Commercial	7.7	7.6	7.6	−0.1	0.0
Transportation	18.7	20.7	18.6	+2.1	−0.9
Industrial	24.0	26.2	26.9	+1.8	+0.2
Total Energy Consumption	61.5	61.2	63.8	−0.1	+0.3
Price of Imported Oil (1979 dollars per barrel)	6.50	15.50	41.00	+19.0	+8.4
Energy Price Index (billion 1979 dollars per quadrillion Btu)	0.79	1.84	5.11	+18.4	+8.9
Real GNP (billion 1979 dollars)	2,044	2,314	3,130	+2.5	+2.5

[a]After conversion and distribution losses.

Chapter 2

ASHRAE ENERGY STANDARDS

If buildings could speak out, they would say, "We do not waste energy, it's you!"

Buildings are not designed for energy use or conservation, rather they are designed to provide spaces where certain activities take place. Energy is primarily used to extend the range of humanly acceptable conditions within the building and, if the buildings were unoccupied, there would be no reason for using energy at any time. This obvious fact is generally ignored by studies designed to reduce energy used in buildings.

Energy used in buildings has increased dramatically. On average, the energy used per square foot of building has more than doubled in the past two decades, while the use of electricity has more than doubled and in some cases quadrupled.

Dr. Charles Lawrence, Public Utility Specialist for New York City, studied more than 80 post-World War II office buildings in New York, all serving similar users. There were great individual differences, with a ratio of more than 3.5 to 1 between high and low energy users. However, when buildings were grouped according to the year of completion, there was an average increase, in energy usage between 1950 and 1970, of two times. In addition, the energy invested in building construction has increased markedly due to changes in building materials, methods of building, methods of computation, and attitudes toward design.[2-1]

To halt this ever increasing energy use in buildings, the American Society of Heating, Refrigerating and Air-Conditioning Engineers has, over several years, created ASHRAE Standard 90. Actually, the Standard is comprised of three documents; 90A-1980, 90B-1975 and 90C-1977.[2-2] This Standard, or a similar one with modifications, has become the law governing new building construction in the majority of the 50 States. ASHRAE 90 specifies design requirements that will limit the use of energy and improve the performance and management of equipment in buildings, and provides a means of determining their impact on the depletion of energy resources.

ASHRAE has also released a set of 100 Series Standards advocating energy conservation in various types of existing buildings. ASHRAE Standards 90 and the 100 Series establish parameters for energy conservation in the building envelope; heating, ventilation, and air conditioning (HVAC) systems; HVAC equipment; service water heating; energy distribution system; and lighting. The main features of ASHRAE 90 will be discussed briefly, bearing in mind that they are not applicable if the energy usage in a building is less than 3.4 Btuh/ft^2 (10.8 W/m^2) of floor area.

HEATING CRITERIA

Walls   The combined thermal coefficient of transmission (U_o value) for the gross area of exterior wall of a heated building shall not exceed the values shown in Figures 2-1A and 2-1B. The U_o value is estimated as follows:

$$U_o = \frac{U_w A_w + U_g A_g + U_d A_d}{A_o}$$

where U = coefficient of thermal transmission

A = area

subscript w = walls

g = glass (fenestration)

d = doors

o = overall value

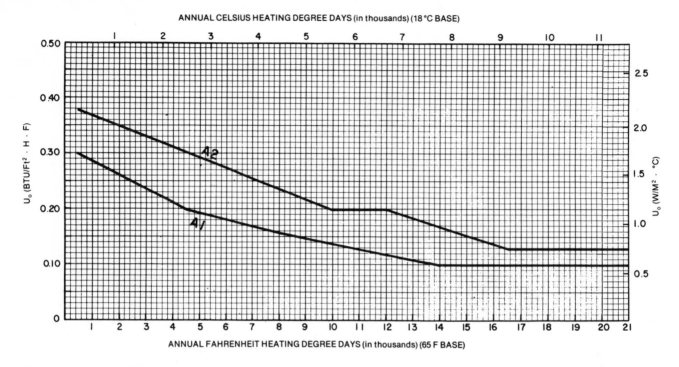

FIG. 2-1A. Maximum U_o for walls, heating. Curve A1 is for detached one- and two-family dwellings. Curve A2 is for other residential buildings, three stories or less.[2-2]

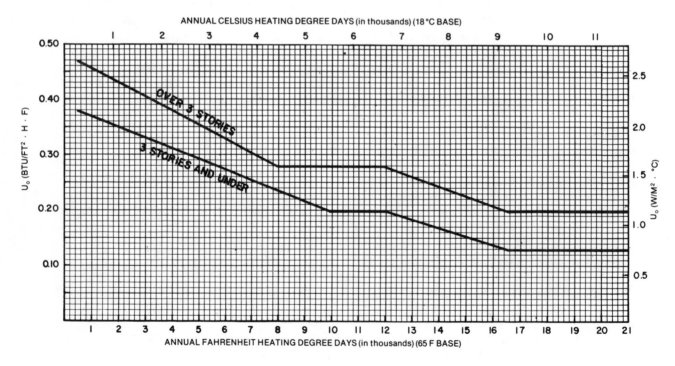

FIG. 2-1B. Maximum U_o for walls, heating, for non-residential buildings.[2-2]

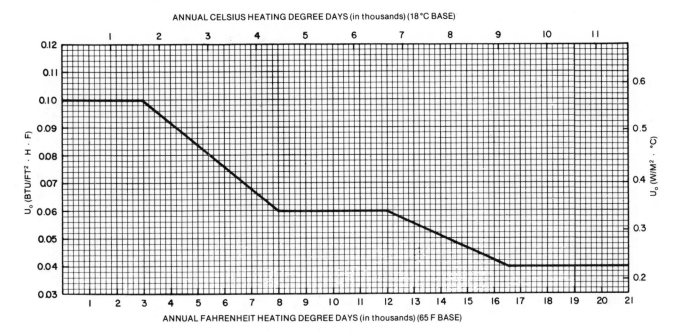

FIG. 2-2. Maximum U_o for roofs and ceilings in non-residential buildings.[2-2]

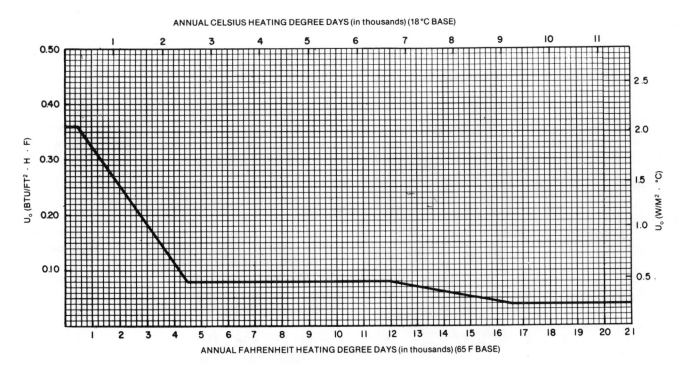

FIG. 2-3. Maximum U_o for floors over unheated spaces, all buildings.[2-2]

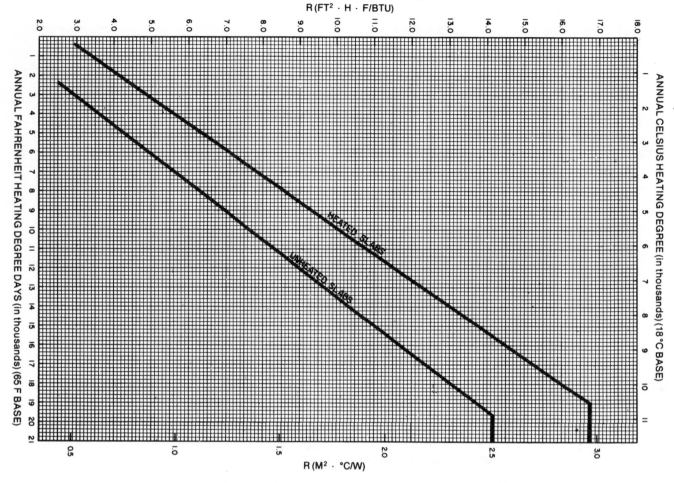

FIG. 2-4. Minimum R-values for heated and unheated slabs on grade.[2-2]

Where more than one type of wall, window, or door is used, the term *UA* shall be expanded into its subelements, as

$$U_{w1}A_{w1} + U_{w2}A_{w2} + \text{------}$$

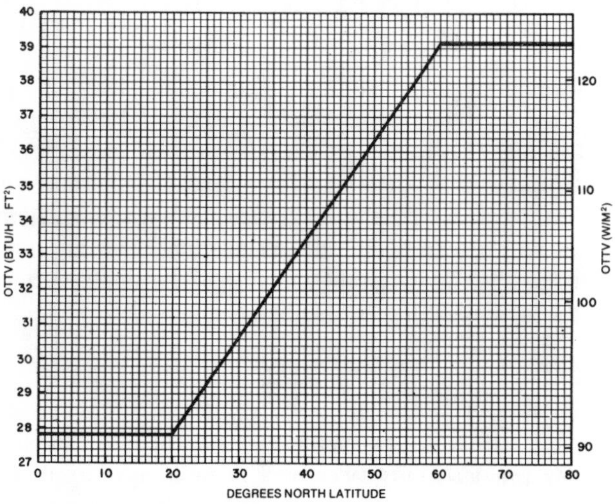

Roof/Ceiling   The overall thermal transmittance value, U_o, for the roof/ceiling of a heated building shall not exceed the values shown in Figure 2-2. The overall U value shall be computed as follows:

$$U_o = \frac{U_{roof}A_{roof} + U_{skylight}A_{skylight}}{\text{Gross Area of a roof/ceiling assembly}}$$

Floors   The U_o value for floors over heated and/or mechanically cooled spaces over unheated spaces shall not exceed those shown in Figure 2-3.

Slab on Grade   The thermal resistance (*R*) of the insulation around the perimeter of slab-on-grade floors shall be as shown in Figure 2-4.

COOLING CRITERIA

Walls   The overall thermal transfer value, *OTTV*, for the gross area of exterior walls of mechnically cooled

FIG. 2-5. OTTV, cooling, non-residential buildings.[2-2]

TABLE 2-1

Mass per unit wall area		TD_{eq}	
lbs/ft²	kg/m²	°F	°C
0-25	0-125	44	24.5
26-40	126-149	37	21.0
41-70	196-345	30	17.0
71 and above	346 and above	23	13.0

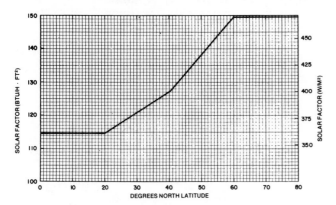

FIG. 2-6. Solar factor.[2-2]

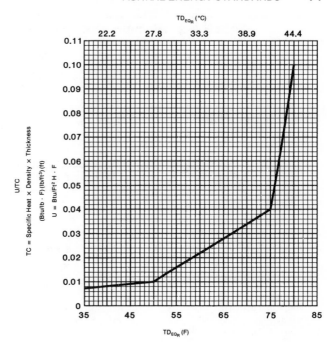

FIG. 2-7. Equivalent temperature difference.[2-2]

buildings, shall not exceed the value shown in Figure 2-5. The following equation is used to determine $OTTV_w$:

$$OTTV_W = \frac{U_w \times A_w \times TD_{eq} + A_f \times SF \times SC + U_f \times A_f \times \Delta t}{A_o}$$

where
U = Coefficient of thermal transmission
A = Area
TD_{eq} = Equivalent temperature difference (Table 2-1)
SF = Solar Factor (Figure 2-6)
SC = Shading coefficient, fenestration
Δt = Indoor and outdoor temperature difference
subscript w = wall
f = fenestration
o = overall

Roof/ceiling The combined thermal transmittance value, *OTTV*, for the roof/ceiling assembly shall not exceed that determined by the heating criteria. Further, the *OTTV* for the gross area of the roof assembly shall not exceed 8.5 Btuh/ft² (26.8 W/m²). The following equation is used to determine $OTTV_r$:

$$OTTV_r = \frac{U_r \times A_r \times TD_{eqr} + 138A_s \times SC_s + U_s \times A_s \times \Delta t}{A_o}$$

where U = Coefficient of thermal transmission

A = Area
TD_{eq_r} = Equivalent temperature difference (Figure 2-7)
SC = Shading coefficient, skylight
Δt = Indoor and outdoor temperature difference
subscript r = opaque roof
s = skylight
o = overall

AIR LEAKAGE

Windows Infiltration around windows shall not exceed 0.5 ft³/min per foot (0.77 m³/sec per meter) of sash crack.

Sliding doors Air leakage through swinging, revolving or sliding doors shall not exceed 11 ft³/min per lineal foot (17 m³/sec per lineal meter) of door crack.

Caulking Exterior joints around windows and door frames, between wall cavities and window and door frames, between wall and foundation, and all other openings in the exterior walls shall be caulked, gasketed, weatherstripped or otherwise sealed.

HVAC

HVAC systems Criteria for determining heating/cooling loads, design requirements and control requirements for comfort applications in buildings, insulation requirements for HVAC systems and duct construction

TABLE 2-2. Minimum COP (EER) for Electrically Driven Cooling Equipment[2-2]

| | Standard Rating Capacities | | | |
| | Under 19 kW (65,000 Btu/h) | | 19 kW (65,000 Btu/h) and Over | |
Effective Date	Air-Cooled	Evap. or Water Cooled	Air-Cooled	Evap. or Water Cooled
Beginning January 1, 1980	1.99 (6.8)	—	2.20 (7.5)	—
Beginning January 1, 1984	2.28 (7.8)	2.58 (8.8)	2.40 (8.2)[6]	2.69 (9.2)

TABLE 2-3. Standard Rating Conditions for Heat-Operated Cooling Equipment[2-2]

| | | Heat Source | |
| Standard Rating Conditions | | Direct Fired (Gas, Oil) | Indirect Fired (Steam, Hot Water) |
Air Conditioners[1]	Units	Temperatures	Temperatures
Entering Conditioned Air	°C (F)	26.7 (80) DB; 19.4 (67) WB	—
Entering Condenser Air	°C (F)	35.0 (95) DB; 23.9 (75) WB	—
Water Chillers[2]			
Leaving Chilled Water	°C (F)	7.2 (45)	6.7 (44)
Fouling Factor	m² · K/W (ft² · h · F/Btu)		.00009 (.0005)
Entering Chilled Water	°C (F)	Per Mfgr. Spec.	12.2 (54)
Entering Condenser Water	°C (F)	23.9 (75)	29.4 (85)
Fouling Factor	m² · K/W (ft² · h · F/Btu)		.00018 (.0005)
Leaving Condenser Water	°C (F)	35.0 (95)	—
Condenser Water Flow Rate	litre/W · min (gpm/ton)	—	Per Mfgr. Spec.

TABLE 2-4. Minimum COP for Heat-Operated Cooling Equipment[2-2]

| Heat Source | |
Direct Fired (Gas, Oil)	Indirect Fired (Steam, Hot Water)
0.48	0.68

have been established. Details are given in the ASHRAE 90 Standard.

HVAC equipment   Criteria for equipment and component performance, such as standard rating conditions, minimum energy efficiency ratio (EER) or coefficient of performance (COP) and manufacturer's documentation, have been established and are summarized in Tables 2-2

TABLE 2-5. Standard Rating Conditions for Electrically Operated Heat Pumps[2-2]

Conditions	Type	Air Source		Water Source
Air Entering Equipment	°C (F)	21.1 DB (70 DB)	21.1 DB (70 DB)	21.1 DB (70 DB)
Outdoor Unit Ambient	°C (F)	8.3 DB/6.1 WB (47 DB/43 WB)	−8.3 DB/−9.4 WB (17 DB/15 WB)	— —
Entering Water Temperature	°C (F)	— —	— —	15.6 (70)
Water Flow Rate	—	—	—	As used in cooling mode[3]

TABLE 2-6. Minimum COP for Electrically Driven Heat Pumps[2-2]

| | Air-Source | | Water-Source |
Heat Source Entering Temperature °C (F)	8.3 DB/6.1 WB (47 DB/43 WB)	−8.3 DB/−9.4 WB (17 DB/15 WB)	15.6 (70)
Beginning January 1, 1980	2.5	1.5	2.5
Beginning January 1, 1984	2.7	1.8	3.0

TABLE 2-7.[2-2]

ANNUAL FUEL AND ENERGY CALCULATION FORM 12-1

Line	Column	Building/Project Energy Req'ments		Fuel and Energy Supplied to Site								
		A1	A2	B1	B2	B3	B4	B5	B6	B7	B8	B9
	Function	Thermal 10^6 Btu	Electric 10^3 KWH	Coal 10^6 Btu	Gas 10^6 Btu	Light Oil 10^6 Btu	Heavy Oil 10^6 Btu	Elec. Win. 10^3 KWH	Elec. Sum. 10^3 KWH	Elec. Ann. 10^3 KWH	Other	Other
1	Heating											
2	Cooling											
3	Water Heating											
4	HVAC Auxiliaries											
5	Lighting											
6	Elevators											
7	Computers											
8	Cooking											
9	Process											
10	Other											
11	Other											
12	Other											
13	Total Carry Fwd. to Form 12-2											

ANNUAL FUEL AND ENERGY CALCULATION FORM 12-2

Line	Annual Fuel and Energy Calculation Form 12-2 Fuel and Energy Supplied to Site			C.O. Total From Form 12-1 Line 13	RUF From Supplier or From Tables	Fuel and Energy Resources Used on Site and Off Site To Meet Energy Requirements of Building/Project						
						C1	C2	C3	C4	C5	C6	C7
						S. Tons Coal	MCF Nat'l	BBL Crude Oil	Grams U-235	10^3 KWH Hydro	Other	Other
14	Fuel Oil, Light											
15	Fuel Oil, Heavy											
16	Gas	Nat'l	MCF									
		Oil	BBL									
17	Coal											
18	Elec. Winter											
		Coal	S. Tons									
		Gas	MCF									
		Oil	BBL									
		Nuc	Grams									
		Hydro	10^3 KWH									
		Other										
19	Elec. Summer											
		Coal	S. Tons									
		Gas	MCF									
		Oil	BBL									
		Nuc	Grams									
		Hydro	10^3 KWH									
		Other										
20	Elec. Annual											
		Coal	S. Tons									
		Gas	MCF									
		Oil	BBL									
		Nuc	Grams									
		Hydro	10^3 KWH									
		Other										
21	(Other)											
22	Total Resources											

through 2-6. Details are given in the ASHRAE 90 Standard.

WATER HEATING

ASHRAE 90 requires that hot water for domestic, sanitary and swimming pool purposes shall be generated and delivered in an energy-saving manner. It gives details on performance efficiency, standby losses, insulation requirements, temperature controls and pump operation.

ELECTRICAL DISTRIBUTION AND LIGHTING

Power factor   Utilization equipment rated greater than 1000W and lighting rated greater than 15W, with an inductive reactance component, shall have a power factor of not less than 85 percent under rated load conditions. Utilization equipment with a power factor of less than 85 percent shall be corrected to at least 90 percent under rated load conditions. Power factor correction devices shall be switched with the equipment, except where unsafe.

Service voltage   Where there is a choice, the service voltage producing the least energy loss should be selected.

Voltage drop   The maximum total voltage drop shall not exceed 3 percent in branch circuits or feeders, for a total of 5 percent to the furthest outlet based on steady state conditions.

Lighting switches   Switching shall be provided for each lighting circuit, so that partial lighting required by custodians or for complementary use with natural light may be operated selectively.

Power budget   The lighting power budget is the upper limit of power available to provide needed lighting in accordance with criteria and calculations specified in ASHRAE 90. The budget value shall be based on the intended use of the space within the building and on efficient energy utilization.

ALTERNATE DESIGNS

Alternate building, system, and equipment designs will be acceptable when they are shown to have equal or less energy consumption when compared to similar buildings meeting the criteria laid down in ASHRAE 90.

ANNUAL ENERGY USAGE

ASHRAE 90 specifies the procedures for reporting the calculated quantities of each form of energy delivered to the building site, to satisfy the building project needs, and for extending these quantities to resource quantities by application of the appropriate resource utilization factors (RUF). The forms used for recording and adjusting the annual fuel and energy requirements are identified as Table 2-7.

EXISTING BUILDINGS

The ASHRAE 100 Series Standards for energy conservation in existing buildings require that various components of existing buildings be modified, wherever feasible, to satisfy the criteria of ASHRAE 90. A proposed modification must be implemented if the annual energy cost savings are 25 percent or more of the cost of the modification.

Because of these Standards, significant savings are being achieved in the design of new buildings and the operation of existing buildings. A study by Arthur D. Little, Inc., for the Department of Energy, evaluated the impact of energy conservation on new construction if the ASHRAE 90 Standards were implemented nationwide. Table 2-8 shows their estimate of annual source energy requirements of new construction built over the period 1978 through 1990.

TABLE 2-8.[2-3]

Comparative Annual Source—Energy Requirements for New Construction Built Over the Period 1978 Through 1990[a]
(10^12 Btu in the Year 1980)

	Northeast				North Central				South				West				Total U.S.			
	Gas	Oil	Electric	Total	Gas	Oil	Electric	Total	Gas	Oil	Electric	Total	Gas	Oil	Electric	Total	Gas	Oil	Electric	Total
Base Case[b]	14	49	292	355	128	23	589	740	76	13	1,002	1,091	119	5	420	544	337	90	2,303	2,730
RUF Case[c]	102	7	170	279	232	14	351	597	212	6	664	882	153	2	351	506	699	29	1,536	2,264
Difference	+88	−42	−122	−76	+104	−9	−238	−143	+136	−7	−338	−209	+34	−3	−69	−38	+362	−61	−767	−466
Percent Change				*−21.4%*				*−19.3%*				*−19.2%*				*−7.0%*				*−17.1%*

[a] Estimated based on RUF values as given in Section 12 of ASHRAE Standard 90-75R; figures do not account for shifts in resource base.
[b] Assumes: a) Application of source energy analysis procedure such as Section 12 of ASHRAE Standard 90-75R; b) fuel type selection made based on such an analysis; and c) nationwide compliance and effective enforcement.

SOURCE: Arthur D. Little, Inc., estimates.

Chapter 3

HEAT LOSS AND GAIN
IN BUILDINGS

The heat loss through building surfaces such as walls, roof, and windows determines the heating load for a particular building. The heat gain through these surfaces, including radiant heat from the sun, determines the cooling load. These heat losses and gains occur as heat is transmitted by conduction, convection and radiation through the exterior surfaces of the building and by outside air infiltrating the building through and around windows and doors and through minute openings in the building surface.

We shall review the rate of heat transfer through the walls, roof and windows of a typical building in order to determine the design heating and cooling load for one hour of operation and then determine the seasonal energy

FIG. 3-1. Typical office building. *Photo by Balthazar Korab.*

consumption of the building. These calculations are necessary to evaluate the savings that result from modification of existing buildings.

THE TYPICAL BUILDING

Consider an office building located in Detroit, Michigan where temperatures are expected to reach 0°F in winter. The interior of the building is maintained at 70°F.

This is a 6-story building, approximately 80 ft high, with an overall area of 138×138 ft. The east wall has no glass, the other three walls have varying amounts of glass, Figure 3-1.

HEAT LOSS THROUGH GLASS

The glass areas are as follows:

North facing	9,165 ft^2
East facing	0
South facing	787
West facing	9,189
Total glass	19,141 ft^2

Heat transmission (Q) through any surface, in Btu's per hour (Btuh), is the product of area (A) times the rate of heat transmission (U), in Btu's per hour per square foot per °F (Btuh·ft^2·°F), times the difference between the outside temperature (T_o) and the inside temperature (T_i). Thus the simple formula:

$$Q = AU(T_o - T_i) \qquad (3-1)$$

The sum of the glass areas is 19,141 ft^2. The temperature difference $(T_o - T_i)$ is 0°F − 70°F = −70. The rate of heat transfer $U = 1.06$ Btuh·ft^2·°F. This U-value is furnished by the glass manufacturer. Lacking such specific

TABLE 3-1. U-Values of Windows[3-1]

Description	Winter	Summer
single glass	1.10	1.04
insulating glass—double		
0.25-in. air space	0.58	0.61
0.5-in. air space	0.49	0.56
insulating glass—triple		
0.25-in. air spaces	0.39	0.44
0.5-in. air spaces	0.31	0.39
storm windows		
1-in. to 4-in. air space	0.50	0.50

information, refer to Table 3-1, which is extracted from the *ASHRAE Handbook, 1977 Fundamentals*, where the winter transmission rate of single-pane glass is given as 1.10.[3-1]

Applying the formula:

$$Q = 19,141 \times 1.06 \times -70 = -1.42 \times 10^6 \text{ Btuh}$$

Note that $(T_o - T_i)$ is always a negative quantity during winter operation and a positive quantity during summer operation. Therefore, a positive sign indicates cooling load, a negative sign indicates heating load.

REDUCING HEAT LOSS THROUGH GLASS

Table 3-1 shows that heat transmission through glass windows can be reduced by replacing single panes with double or triple panes or by protecting single glass with storm windows. If the building under discussion had double pane windows, with a ¼ in. space between the

two sheets of glass, the U-value becomes 0.58 rather than 1.06 and the heat loss is reduced by 45%.

Recomputing the heat loss through our glass area, using the U-value of double-pane glass

$$Q = 19,141 \times .58 \times -70 = -.777 \times 10^6 \text{ Btuh}$$

or a savings of $.64 \times 10^6$ Btuh.

HEAT LOSS THROUGH WALLS

The building under discussion is on a site that slopes toward the Detroit river. While it is 80 feet high from the basement floor to the roof, the north and east walls are 71 feet high, the south wall is 78 feet high and the west wall, which slopes at the ground line, averages 74½ feet high.

To calculate the exposed wall area, consider the building as two units, a 71 foot tall building and a partially exposed 9 foot high basement.

Seven feet of the south basement wall and none of the north and east basement walls are exposed. The west wall exposure is 0 feet at the north end and 7 feet at the south end, an average of 3½ feet. Since each wall is 138 feet long, total exposed basement wall is

$$138 \times (7 + 3\tfrac{1}{2}) = 1449 \text{ ft}^2$$

The walls of the building proper are 138 feet wide and 71 feet tall.

$$71 \times 138 \times 4 = 39,192 \text{ ft}^2$$

Subtracting the glass area, net exposed wall area

$$A_w = 1449 + 39,192 - 19,141 = 21,500 \text{ ft}^2$$

The wall is composed of 4-in. face brick and 8-in.

TABLE 3-2. Coefficients of Transmission (U) of Masonry Walls[3-1]

Coefficients are expressed in Btu per (hour) (square foot) (degree Fahrenheit difference in temperature between the air on the two sides), and are based on an outside wind velocity of 15 mph

Replace Cinder Aggregate Block with 6-in. Light-weight Aggregate Block with Cores Filled (New Item 4)

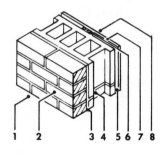

Construction	1 Between Furring	1 At Furring	2 Between Furring	2 At Furring
1. Outside surface (15 mph wind)	0.17	0.17	0.17	0.17
2. Face brick, 4 in.	0.44	0.44	0.44	0.44
3. Cement mortar, 0.5 in.	0.10	0.10	0.10	0.10
4. Concrete block, cinder aggregate, 8 in.	1.72	1.72	2.99	2.99
5. Reflective air space, 0.75 in. (50 F mean; 30 deg F temperature difference)	2.77	—	2.77	—
6. Nominal 1-in. × 3-in. vertical furring	—	0.94	—	0.94
7. Gypsum wallboard, 0.5 in., foil backed	0.45	0.45	0.45	0.45
8. Inside surface (still air)	0.68	0.68	0.68	0.68
Total Thermal Resistance (R)	R_i= 6.33	R_s= 4.50	R_i= 7.60	R_s= 5.77

Construction No. 1: U_i= 1/6.33 = 0.158; U_s = 1/4.50 = 0.222. With 20% framing (typical of 1-in. × 3-in. vertical furring on masonry @ 16-in. o.c.), U_{av} = 0.8 (0.158) + 0.2 (0.222) = 0.171
Construction No. 2: U_i = 1/7.60 = 0.132, U_s = 1/5.77 = 0.173. With framing unchanged, U_{av} = 0.8(0.132) + 0.2(0.173) = 1.40

cinder aggregate concrete block with ½ in. of mortar in between. This is a very rudimentary wall, devoid of any insulating elements.

Before trying to calculate the U-value of this wall, look at a more complex wall, Table 3-2, and the effect of modifying the cinder block. Each of the construction elements, 1 through 8, has been assigned a *thermal resistance* or R-value that has been derived from actual laboratory observation. The total thermal resistance is the sum of individual resistances of each element.

This summation of individual resistances is based on an analogy with Ohm's law of direct current where:

$$I = \frac{E}{R}$$

Or, the current flow *(I)* is equal to the voltage *(E)* divided by the resistance *(R)*.

A corrolary of Ohm's law is that the total resistance of a number of resistances in series is equal to the sum of the resistances. Restated

$$I = \frac{E}{R_1 + R_2}$$

Compare this with a simplified version of Equation 3-1.

$$Q = \frac{T_o - T_i}{R}$$

where Q (heat flow) is equivalent to I (current flow) and $T_o - T_i$ is equal to the impressed voltage. Obviously, current flow is inversely related to resistance in both cases. By inspection, $1/R$ is equal to U in Equation 3-1.

Equation 3-1 can be expressed in any of three forms:

$$Q = AU(T_o - T_i)$$

$$Q = A \cdot \frac{1}{R} \cdot (T_o - T_i)$$

$$Q = A \cdot \frac{1}{R_1 + R_2 + R_n} \cdot (T_o - T_i) \qquad (3\text{-}2)$$

To calculate the U-value, take the reciprocal of the sum of all the R values.

Returning to Table 3-2, the sum of the seven resistances in Column 1 is 6.33 and the U-value is 1/6.33 or 0.158.

Column 2 is the result of a more sophisticated method of calculating total resistance in a typical wall. Note that item 6 replaces item 5. This is due to the change in R-value that occurs when furring or studding is included in a wall. There is an air space between the studs, item 5. Where studs are used, they displace the air space and add a new value for that area, item 6. In this example, we have subtracted the value of the air space (2.77) and substituted a value of the furring (0.94). As a result, the R-value has been reduced from 6.33 to 4.50.

The usual practice is to apply 1×3 inch vertical furring at intervals of 16 inches between centers. As a result, about 20% of the wall has an R-value of 4.50. The composite value of the wall is therefore $20\% \times 4.50$ and $80\% \times 6.33$. The composite U-value is therefore 0.171.

In column 3 we have altered the composition of line 4 by replacing the 8-inch concrete block with a six-inch lightweight aggregate block and filling the cores in the block with mortar. As a result, the R-value is increased to 7.6 and the U-value becomes 0.132. Column 4 merely reflects the R-value in the 20% of the wall containing furring.

Notice that an R-value is assigned to the air film on the outside of the wall (item 1) and the inside of the wall (item 8). This again has been determined by laboratory measurements.

We can now extract the values for the wall used in our building. It is the sum of the following items:

1. Outside surface (15 mph wind)	.17
2. Face brick, 4″	.44
3. Cement mortar, .5 inch	.10
4. Concrete block, cinder aggregate, 8 in.	1.72
8. Inside surface (still air)	.68
Total thermal resistance *(R)*	3.11

$$U = 1/R = 1/3.11 = .322 \text{ Btuh·ft}^2 \cdot {}^\circ\text{F}$$

Returning to the original building, the wall area is

TABLE 3-3. Thermal Properties of Typical Building and Insulating Materials[3-1]

Description	Density (lb/ft³)	Conductivity (k)	Conductance (C)	Resistance[b] (R) Per inch thickness (1/k)	Resistance[b] (R) For thickness listed (1/C)	Specific Heat, Btu/(lb) (deg F)	SI Unit Resistance[b] (R) (m·K)/W	SI Unit Resistance[b] (R) (m²·K)/W
BUILDING BOARD								
Boards, Panels, Subflooring, Sheathing								
Woodboard Panel Products								
Asbestos-cement board	120	4.0	—	0.25	—	0.24	1.73	
Asbestos-cement board ....0.125 in.	120	—	33.00	—	0.03			0.005
Asbestos-cement board ....0.25 in.	120	—	16.50	—	0.06			0.01
Gypsum or plaster board ....0.375 in.	50	—	3.10	—	0.32	0.26		0.06
Gypsum or plaster board ....0.5 in.	50	—	2.22	—	0.45			0.08
Gypsum or plaster board ....0.625 in.	50	—	1.78	—	0.56			0.10

21,500 ft^2 and has a U-value of .322. Therefore

$$Q = 21,500 \times .322 \times -70 = -.485 \times 10^6 \text{ Btuh}$$

DETERMINING R-VALUES

In Table 3-2, the thermal resistances were expressed as R-values. However, R-values are not always available and must sometimes be derived. Notice, in Table 3-3, that the values under the column labeled *Resistance (R)* are re-expressed as 1/k or 1/C and that the values for k and C are given in the columns to the left. Note, too, that 1/k is the value per inch of thickness while 1/C is for the thickness listed. For example: the first listing, asbestos-cement board of unspecified thickness, is assigned a 1/k of 0.25 per inch. The same board, with a specified thickness of 0.125 in., can no longer have a 1/k value and instead is listed as a 1/C value of .03. This is derived by multiplying the 1/k value by the thickness or .125 × .25 = .03 = 1/C = R.

Summarizing:

$$R = 1/k = 1/C = 1/f$$

These values are defined as follows:

k—Thermal conductivity: The amount of heat (Btu) transmitted in one hour through one square foot of a homogeneous material one inch thick for a difference of temperature of one degree Fahrenheit between the two surfaces of the material.

C—Conductance: A variant of the k-value for a given thickness. Most often used with standard dimensional materials such as ¾-inch sheathing where, for a given thickness, $(x) \cdot \frac{1}{k} = \frac{1}{C}$.

f—Film coefficient: The combined surface loss due to radiation and convection. Theoretically, this value may be stated as *f* but is most often given as an *R*-value. The most frequently used film coefficients are R = 0.17 for an outside air film, assuming a 15 mph wind, and R = 0.68 for inside still air.

By deriving R-values from values for *k* or *C* and for each of the building elements, the total resistance can be found as the sum of all individual resistances.

HEAT LOSS THROUGH THE ROOF

The roof of our building is constructed of 0.375 inches of built-up roofing on a 5-inch concrete slab with an area of 138 × 138 or 19,044 ft^2. Table 3-4 shows a comparable roof with only two inches of concrete.

Note that the resistance of the inside air film (item 1) of the roof, a horizontal surface, is less than for walls, which are vertical. Item 6 has been adjusted for 5 rather than 2 in. of concrete.

The resistance of the roof is calculated the same as the walls, by summing the resistances of its elements:

1. Inside surface (still air)	0.61
6. Concrete slab, lightweight aggregate, 5 in.	5.55
8. Built-up roofing, 0.375 in.	0.33
9. Outside surface (15 mph wind)	0.17
Total Thermal Resistance (R)	6.66

$$U = 1/R = 1/6.66 = .15 \text{ Btuh} \cdot \text{ft}^2$$

The heat loss (Q) through the roof, using Equation 3-1, is

$$Q = 19,044 \times .15 \times -70 = -.2 \times 10^6 \text{ Btuh}$$

If we add a 2½ in. layer of rigid insulation, with an R-value of 8.0, to the original thermal resistance of 6.66, the new R-value becomes 14.66 and

$$U = 1/R = 1/14.66 = .068 \text{ Btuh} \cdot \text{ft}^2$$

$$Q = 19,044 \times .068 \times -70 = -.091 \times 10^6 \text{ Btuh}$$

TABLE 3-4. Coefficients of Transmission (U) of Flat Masonry Roofs with Builtup Roofing[3-1]

These Coefficients are expressed in Btu per (hour) (square foot) (degree Fahrenheit difference in temperature between the air on the two sides), and are based upon an outside wind velocity of 15 mph

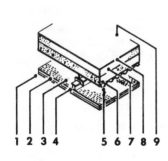

Add Rigid Roof Deck Insulation, C = 0.24 (R = 1/G) (New Item 7) Construction (Heat Flow Up)	1	2
1. Inside surface (still air)	0.61	0.61
1. Metal lath and lightweight aggregate plaster, 0.75 in.	0.47	0.47
3. Nonreflective air space, greater than 3.5 in. (50 F mean; 10 deg F temperature difference)	0.93*	0.93*
4. Metal ceiling suspension system with metal hanger rods	0**	0**
5. Corrugated metal deck	0	0
6. Concrete slab, lightweight aggregate, 2 in.	2.22	2.22
7. Rigid roof deck insulation (none)	—	4.17
8. Built-up roofing, 0.375 in.	0.33	0.33
9. Outside surface (15 mph wind)	0.17	0.17
Total Thermal Resistance (R)	4.73	8.90

By adding the insulation, we have reduced the Btuh by $.109 \times 10^6$, a savings of 55%.

CONDENSATION DUE TO INSULATION

Adding insulation at the ceiling level lowers the temperature in the plenum space between the ceiling and the roof and may lead to condensation problems where there are high latent heat loads. Moisture laden air migrates from the heated space, through ceiling tile joints and insulation blanket joints, into the colder plenum space. If the temperature on the cold side of the insulation is below the inside air dew point, moisture will condense. This condition is very common in unventilated plenums when the outside temperature is below 30°F. Installing insulation at the roof surface, with a vapor barrier facing the heated side, will avoid condensation.

HEAT LOSSES BELOW GRADE

Traditionally, heat losses below grade have been estimated using the rule of thumb of 0.1 Btuh/ft^2 for floors and 0.2 Btuh/ft^2 for walls.

ASHRAE's latest method assumes that heat from a basement wall or floor follows a radial isotherm and the heat loss is therefore dependent on the length of the arc the heat follows in reaching the surface, Figure 3-2.

At present, the new method is focused on estimating the heat loss of a small basement, not exceeding 32 feet in width or 7 feet in depth. Therefore, we will have to employ prudent approximations in applying this new procedure to the outsize basement under our major building. The calculation is shown in Table 3-5.

As you can see, there is a factor that reflects the longer path that the heat travels at each succeeding foot of depth. At the first and second foot level, the heat loss is greater than the traditional 0.2 Btuh/ft^2. Thereafter, the factor is far less than traditionally assumed. At the 8th

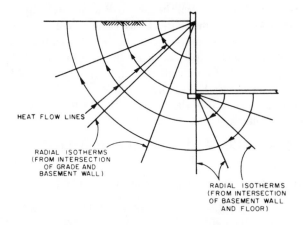

FIG. 3-2. Isothermal heat flow from a basement.[3-1]

HEAT FLOW LINES

RADIAL ISOTHERMS
(FROM INTERSECTION
OF GRADE AND
BASEMENT WALL)

RADIAL ISOTHERMS
(FROM INTERSECTION
OF BASEMENT WALL
AND FLOOR)

TABLE 3-5. Heat Loss Through Basement Walls

| Depth, ft | Wall Length | | | | Factor | Btu/°F |
	N&E	W	S	Total		
1	138	138	138	552	.410	226
2	138	138	138	552	.222	123
3	138	118	0	394	.155	61
4	138	99	0	375	.119	47
5	138	79	0	355	.096	34
6	138	59	0	335	.079	26
7	138	39	0	315	.069	22
8	138	20	0	296	.069	20
9	138	0	0	276	.069	19
					Total	578

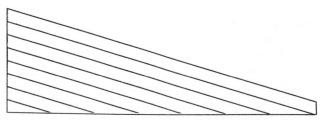

FIG. 3-3. Profile of the west basement wall.

and 9th foot level we continue to use the factor for the 7th foot level, 0.069, thus erring on the safer side. As discussed earlier, maximum depth of the south basement wall is 2 feet, of the north wall, 9 feet, Figure 3-3. Since each factor reflects the heat path to the surface, the length of the one-foot sections of the west wall must be calculated as a function of the tangent, starting with the ratio, tangent $\theta = 7/138$. The square footage of each one-foot level is multiplied by the given factor to compute the heat loss per °F. The total AU is 578 Btuh·°F and the heat loss is

$$Q = 578 \times (0 - 70) = -.041 \times 10^6 \text{ Btuh}$$

The rule-of-thumb method estimates the heat loss from the basement walls as $.049 \times 10^6$ or 20% greater.

The average depth of the basement floor is 6'4" and the width of the floor is 138 feet. To be on the safe side, we will use the ASHRAE factor for 5 feet of depth and the maximum 32 feet of width. This factor is 0.023 Btuh·ft^2·°TD. The temperature difference (TD) is the difference between the space temperature and the ground temperature, which is 48°F. For the floor

$$Q = 138^2 \times .023 \times (48 - 70) = -.010 \times 10^6 \text{ Btuh}$$

or approximately ¼ of that assumed by the rule of thumb method. Adding the two losses together, the basement heat loss is

$$Q_b = .041 + .010 = -.051 \times 10^6 \text{ Btuh}$$

HEAT LOSS DUE TO AIR INFILTRATION

Air infiltration is defined as the leakage of outside air into a building through windows, doors and interstices due to pressure differences existing between the inside and outside air. The additional heating load due to air infiltration is estimated as follows:

$$Q = AC \times Vol \times Density \times Specific\ Heat \times (T_o - T_i) \quad (3-3)$$

You perhaps recognize this as a variation of our basic equation where volume $\times$ air changes (AC) is substituted for area and density $\times$ specific heat is substituted for the U-value.

The density of air is .075 lb/ft^3 and the specific heat of air is .2445 Btu/lb. These two values are sometimes combined for convenience (.075 $\times$ .2445) to give a value of .018 Btu/ft^3.

It is difficult to estimate the amount of infiltration. Several methods for estimating air leakage through doors, windows, and walls are discussed in the *ASHRAE Handbook, 1977 Fundamentals*. The simplest one uses the number of air changes per hour.

Since the air change method is at best an approximation, we can use approximate dimensions in calculating our volume. On this basis, we estimate that our building is 138 feet square and is 80 feet high for a total of 1.524×10^6 ft^3. Applying this volume to Equation 3-3 and assuming one air change per hour,

$$\begin{aligned} Q &= 0.018\ V\ (T_o - T_i) \\ &= 0.018 \times 1.524 \times 10^6 \times -70 \\ &= -1.92 \times 10^6 \text{ Btuh} \end{aligned}$$

REDUCTION OF AIR INFILTRATION

The air infiltration in existing buildings can be reduced by simply fixing broken windows, closing holes in walls, freeing large louvered ventilation exhausts that were frozen in an open position, etc. In addition, sealing of cracks around doors and windows, around openings for electrical conduits, piping, etc. helps to reduce the rate of air infiltration. This has resulted in the recommendation of tight construction for new and existing structures and suggestions for reducing ventilation rates.

CAUTION FOR TIGHT STRUCTURES

A too tight structure is not recommended as it may have adverse health effects, such as a buildup of indoor pollutant levels and higher levels of radiation from construction materials. A tight structure may also adversely affect the combustion efficiency of a fossil fuel furnace when it does not have a separate inlet for combustion air.

These factors should be given serious consideration before recommending lower levels of ventilation in tight structures. The side effects of such an energy conservation measure should be studied in more detail and the public should be made aware of the results.

DESIGN HEAT LOSS RATE

The design heat loss rate is the sum of the transmission heat loss, the heat loss below grade, and the heat loss due to air infiltration. Here is a summary of these heat losses:

Transmission Heat Loss

Surface	A	U	$(T_o - T_i)$	Q
Walls	21,500	.322	-70	$-.48 \times 10^6$
Glass	19,141	1.06	-70	-1.42×10^6
Roof	19,044	.15	-70	$-.2 \times 10^6$
Heat Loss Below Grade				$-.05 \times 10^6$
Infiltration Heat Loss				-1.92×10^6
			TOTAL	-4.07×10^6 Btuh

DESIGN HEATING LOAD

The design heating load is equal to the design heat loss rate if the internal heat release due to lights, occupants, and other sources is nonexistent or negligible. For larger buildings, due consideration should be given to the internal rate of heat release. In the case of the building under consideration, we must calculate the heat gain from an average 425-person occupancy and from a lighting load equal to 3.25 W/ft^2 for floors 1 through 5 and a lighting level in the basement of 2.5 W/ft^2.

Heat Gains Due to Occupants The average person exudes 640 Btuh while active. Therefore the heat gain due to occupancy is

$$425 \times 640 = 0.272 \times 10^6 \text{ Btu}$$

Heat Gain Due to Lights The lighting loads are expressed in W/ft^2. To translate W to Btu, multiply by the factor 3.41 and further correct this value where the lighting is the fluorescent type. Since we are dealing with fluorescents, add a 20% factor for the heat thrown off by the ballast in the fluorescent system. Therefore, one W/ft^2 is equal to 3.41 $\times$ 1.2 or 4.092 Btu/W.

The area of each of the floors is 19,044 ft^2. Therefore, the heat due to lights on five floors is

$$\begin{aligned} Q &= 5 \times 19,044 \text{ ft}^2 \times 3.25 \text{ W/ft}^2 \times 3.41 \text{ Btu/W} \times 1.2 \\ &= 1.266 \times 10^6 \text{ Btu} \end{aligned}$$

and the heat due to lights on the remaining floor is

$$\begin{aligned} Q &= 1 \times 19,044 \text{ ft}^2 \times 2.5 \text{ W/ft}^2 \times 3.41 \text{ Btu/W} \times 1.2 \\ &= .195 \times 10^6 \text{ Btu} \end{aligned}$$

The total heat of lights is therefore 1.461×10^6 Btu.

Design Heat Load   Since the design heat load is equal to the design heat loss plus all internal gains, add the heat gains from occupancy and lights to the previously calculated design heat loss.

$$\text{Design heat load} = (-4.07 + .27 + 1.46) \times 10^6$$
$$= 2.34 \times 10^6 \text{ Btuh}$$

USE FACTOR IN INTERNAL HEAT GAINS

The foregoing calculations of heat gains due to lights and occupants implicitly assume that the building will be occupied and lighted 24 hours a day. Actually, the building management reports that full lighting is in use for 11 hours during each work day and that only 10% of the lights are in use during other work day hours and on weekends. From this information we can calculate a diversity or use factor. There are five days of 11 hours each when lights are being used 100% of the time or 55 hours. There are five days of 13 hours each and 2 weekend days of 24 hours each or 113 hours when 10% of the lights are in use. During these nonworking periods, the lights are, in effect, in use only 11.3 hours. Totaling: the lights are being used a total of 66.3 hours in 168-hour period. The diversity factor therefore is .395. Applying this diversity factor to the base light heat gain will give the average per hour heat gain. This allows us to integrate this factor with the overall heat load which is expressed in Btuh.

There is a similar situation with occupancy. The building management estimates full occupancy for 10 hours and 2% occupancy thereafter and on weekends. From this information, we can calculate another diversity factor for occupants. There are 5 days of 10 hours each when the building is 100% occupied or 50 hours. There are 5 days of 14 hours each and 2 weekend days of 24 hours each or 118 hours when the building is 2% occupied. During these nonworking periods, the building is, in effect, fully occupied only 2.36 hours. Totaling: the building is being occupied only 52.36 hours in a 168-hour period. Our diversity factor is therefore .312. Applying this diversity factor to the base occupancy heat gain will give the average per hour heat gain.

NET DESIGN HEAT LOAD

The actual design heat load can be calculated after the diversity factors are used to reduce the assumed heat gains from occupants and lights.

Occupant gain	$.272 \times .312 =$	.084
Lights gain	$1.461 \times .395 =$	.577
Internal heat gain		$.66 \times 10^6$
Transmission and ventilation load		-4.07×10^6
Net design heat load		-3.41×10^6 Btuh

ESTIMATING HEATING SEASON ENERGY REQUIREMENTS

The total energy requirements for a heating season may be estimated in many different ways, one of which is the Degree Days Method.

The degree day system is an estimating procedure that has been used by the heating industry for over 30 years. It is based on the premise that heating is not required when the outside temperature is 65°F or higher.

Degree days are determined by subtracting the average daily outside temperature from the base of 65°. Thus, a day having an average outside temperature of 45° is counted as 20 degree days $(65° - 45°)$ and a day when the average temperature is 65° or higher is counted as zero degree days.

The relationship between degree days and fuel consumption is linear since doubling the degree days usually doubles the fuel consumption. On this basis, one can compare total degree days for two locations and derive a rough estimate of comparable fuel consumption in either. For instance: Detroit has an average of 6232 degree days per year while Atlanta has 2961. Therefore, a building in Detroit will consume 110% more energy than the same building in Atlanta. Seasonal variations between the two locations are not a factor since the fuel consumed is the same whether the degree days are accumulated over two days or a week.

To estimate seasonal requirements, first divide the total heat loss by the design temperature difference. If a diversity factor has been used in adjusting internal heat gains, include the heat gains in the overall loss; since the diversity factor, in effect, distributes the heat gain over a 24-hour period. The resultant is the net heat loss per degree of temperature difference for the entire building. Implicitly, we have averaged all the U-values for each heat transfer element into an overall heat transfer rate and multiplied this new U by the total building surface. Thus

$$\frac{Q}{T_o - T_i} = UA$$

Since the first design heat loss was for one hour only, multiply this load by 24 hours and by the number of degree days, thus incorporating a new temperature difference. Detroit has 6232 degree days. Then adjust the result by a temperature correction factor (C_p), based on the outside design temperature, given in Table 3-6. Since our design temperature is 0°F, the C_p is 0.71. Lastly, correct for the efficiency of the heating system, which is usually given as 80%. The energy requirement for the heating system is

$$E = \frac{\text{Heat Load}}{T_o - T_i} \times \frac{24 \times \text{Degree Days} \times CF}{\text{Heating System Efficiency}} \quad (3\text{-}4)$$

$$E = \frac{-3.41 \times 10^6}{0 - 70} \times \frac{24 \times 6232 \times .71}{.8} = 6466.4 \times 10^6 \text{ Btu}$$

TABLE 3-6. Heat Loss vs Degree Days Interim Factor C_p[3-2]

Outdoor Design Temp, F	-20	-10	0	+10	+20
Factor C_D	0.57	0.64	0.71	0.79	0.89

TEMPERATURE SETBACK

With an average building occupancy of 10 hours a day, it appears to be advisable to maintain a 70° temperature during the occupied hours and a 60° temperature during the unoccupied hours. To be on the safe side, add another hour to the occupied time to bring the temperature from 60° to 70° due to the normal lag factor. Therefore, assume that the building will operate at 70° for 11 hours and at 60° for 13 hours. Dividing the resulting degree hours (60° × 13 + 70° × 11) by 24 hours gives an average operating temperature of 64.5°.

Knowing that the average outside temperature for Detroit is 37.2°, we can calculate the temperature setback factor as follows:

$$\frac{T_1 - T_2}{T_3 - T_2}$$

where T_1 = average indoor temperature with setback
  = 64.5°

  T_2 = average outdoor temperature, winter = 37.2°

  T_3 = indoor design temperature = 70°

$$\text{Setback factor} = \frac{64.5 - 37.2}{70 - 37.2} = .832$$

Seasonal energy consumption is estimated to be 16.8% less with night setback.

BIN METHOD

The bin method of estimating yearly energy requirements has been made feasible by the availability of more precise weather information compiled by the U.S. Air Force.[3-3] To compare, the degree day method takes the high and low temperature for every 24 hours and averages the two to determine the degree days. For example, on a day when the low is 0°F and the high is 18°F, the average of the two is 9°F. This, subtracted from 65°, gives a value of 56 degree days. On the other hand, the bin method would divide this range into 4 bins of 5° each: 0–4°, 5–9°, 10–14°, 15–19° and give the number of hours during which these temperatures prevailed. One could find, for instance, that an average winter's day had one hour in the zero bin, 4 hours in the 5° bin, 8 hours in the 10° bin, and 11 hours in the 15° bin. Seasonal requirements are determined by adding the total number of hours in each of the 5° bins.

Once again, we want to restate our design load in terms of per degree rather than per design temperature

since we are replacing T_o with the average bin temperature. Therefore:

$$Q_{bin} = \frac{Q_o \times (T_{bin} - T_i)}{T_o - T_i} \times N_h \qquad (3-5)$$

where T_{bin} is the average temperature of each bin and N_h is the number of hours in each bin or the number of hours during which this temperature range prevails.

This process is repeated for each of the 5° bins. Finally, the total of all the bins is adjusted by the energy efficiency factor, usually 80%.

An example of this method is shown in Table 3-7. Column 1 is the mean temperature for each 5° bin and Column 2 is the number of hours in each year in which the temperature is within the bin range.

TABLE 3-7. Bin Method for Calculating Seasonal Heating Requirements

1 Mean D.B. Temp	2 No. Hours	3 Heat Loss Rate × 10⁶	4 Internal Gain × 10⁶	5 Setback Factor	6 Bin Heat Loss × 10⁶
62	695	.465	.66	.313	—
57	633	.756	.66	.577	—
52	592	1.047	.66	.694	39.44
47	566	1.337	.66	.761	202.32
42	595	1.628	.66	.804	386.10
37	808	1.919	.66	.833	758.33
32	884	2.209	.66	.855	1086.17
27	618	2.500	.66	.872	939.36
22	377	2.791	.66	.885	682.38
17	248	3.082	.66	.896	521.17
12	131	3.372	.66	.905	313.31
7	61	3.663	.66	.913	163.74
2	17	3.954	.66	.919	50.55
-3	4	4.244	.66	.925	13.06
-8	1	4.535	.66	.929	3.55

TOTAL 5,159.48

Assuming 80% efficiency, input energy = $6,450 \times 10^6$ Btu

Column 3 is the design heat loss occurring in each bin. As previously described, Column 3 is the result of substituting the bin temperature for the design temperature and multiplying by the number of hours during which the bin temperature prevails. Since the previously calculated heat loss is 4.07×10^6 Btuh at $T_o - T_i = 70°$, the heat loss per degree is $.058142 \times 10^6$ and the heat loss per bin = $(T_{bin} - 70) \times .058142$. Multiplying the bin loss rate by the number of hours shown in Column 2 gives the total heat loss per bin.

Column 4 is the internal heat loss, adjusted for diversity, which reduces the heat loss shown in Column 3.

Column 5 is the night setback factor. It is equal to

$$\frac{64.5° - T_{bin}}{70° - T_{bin}}$$

Column 6 is the net energy required for each bin. Net heat loss per bin = (Column 3 × Column 5 − Column 4) × Column 2. The total of all bins is shown at the bottom of Column 6. This total, in turn, is multiplied by 1/.8 to adjust for the 80% efficiency of the heating system. The result is the total seasonal energy requirement.

HOUR-BY-HOUR CALCULATIONS

Still another, and more exact method, uses basically the same equations already employed to calculate the heat loss in our building. However, these equations are modified by weighting factors that reflect the thermal storage capacities of various construction materials and the characteristics of the energy distribution system. A computer program is used to perform these calculations, which employ hour-by-hour weather data. Some of these computer programs are cited in Chapter 11.

INTERNAL HEAT GAINS

The internal gains from occupants, lights, equipment and motors must be taken into account when determining heat gains and corresponding cooling loads. This is a rather straightforward procedure if the air conditioning system operates only during occupied hours. On the other hand, when air conditioning is operated continuously, the radiant component of the sensible heat load must be recognized.

Radiant energy only affects air temperature after it has been absorbed by the walls, floors and furniture and has warmed them to a temperature higher than the air temperature. Thereafter, it heats the air by convection and conduction. This energy stored by the structure and its contents contributes to the cooling load after a time lag and is still present after the lights have been turned off and the personnel have left. The consequence of the time lag is that the actual load may be lower than the instantaneous heat gain, but significantly affect the peak load.

The lag effect from the radiant component is recognized in the form of a Cooling Load Factor (CLF). ASHRAE lists these factors starting on page 25.16 of the *ASHRAE Handbook, 1977 Fundamentals*. Table 15 lists the CLF for lights, Table 17 lists the CLF for people, and Table 20 lists the CLF for appliances.

While the radiant component may be 70% of the sensible heat load due to people, the impact of this load is negated when people are closely seated in theaters and auditoriums. In this case, ASHRAE recommends that the CLF be taken as 1.0.

To give an example: Determine the cooling load due to recessed fluorescent lights at 1200, 1400 and 1600 hours. The lights are turned on at 0800 hours and turned off at 1800 hours. Total input to the lights is 1000W. The A/C system runs on a 24-hour schedule.

The room has ordinary office furniture, tile flooring over 3-in. concrete and a medium ventilation rate.

The CLF schedule depends on the number of hours the lights are on in the 24-hour period, the mass of the furniture and the carpeting, the velocity and mode of air delivery, whether the light fixtures are recessed or vented, and the mass of the floor.

Since the lights are on for 10 hours, Table 15B, cited above, applies and is shown here as Table 3-8. Ordinary office furniture, recessed fixtures and medium ventilation is signified by a coefficient of 0.55. A 3 in. concrete floor and medium ventilation are in Classification B. Entering Table 3-8 at the section denoted by coefficient .55 and line denoted by the letter B, the values for 4th, 6th and 8th hour of operation are 0.73, 0.78 and 0.82 respectively.

The heat gain input is 1000W × 3.41 = 3410 Btuh. The actual heat load is

TABLE 3-8. Cooling Load Factors When Lights Are on for 10 Hours[3-1]

"a" Coefficients	"b" Classification	0	1	2	3	4	5	6	7	8	9	10	11	12	13	14	15	16	17	18	19	20	21	22	23
	A	0.03	0.47	0.58	0.66	0.73	0.78	0.82	0.86	0.88	0.91	0.93	0.49	0.39	0.32	0.26	0.21	0.17	0.13	0.11	0.09	0.07	0.06	0.05	0.04
	B	0.10	0.54	0.59	0.63	0.66	0.70	0.73	0.76	0.78	0.80	0.82	0.39	0.35	0.32	0.28	0.26	0.23	0.21	0.19	0.17	0.15	0.14	0.12	0.11
0.45	C	0.15	0.59	0.61	0.64	0.66	0.68	0.70	0.72	0.73	0.75	0.76	0.33	0.31	0.29	0.27	0.26	0.24	0.23	0.21	0.20	0.19	0.18	0.17	0.16
	D	0.18	0.62	0.63	0.64	0.66	0.67	0.68	0.69	0.69	0.70	0.71	0.27	0.26	0.26	0.25	0.24	0.23	0.23	0.22	0.21	0.21	0.20	0.19	0.19
	A	0.02	0.57	0.65	0.72	0.78	0.82	0.85	0.88	0.91	0.92	0.94	0.40	0.32	0.26	0.21	0.17	0.14	0.11	0.09	0.07	0.06	0.05	0.04	0.03
	B	0.08	0.62	0.66	0.69	0.73	0.75	0.78	0.80	0.82	0.84	0.85	0.32	0.29	0.26	0.23	0.21	0.19	0.17	0.15	0.14	0.12	0.11	0.10	0.09
0.55	C	0.12	0.66	0.68	0.70	0.72	0.74	0.75	0.77	0.78	0.79	0.81	0.27	0.25	0.24	0.22	0.21	0.20	0.19	0.17	0.16	0.15	0.14	0.14	0.13
	D	0.15	0.69	0.70	0.71	0.72	0.73	0.73	0.74	0.75	0.76	0.76	0.22	0.22	0.21	0.20	0.20	0.19	0.18	0.18	0.17	0.17	0.16	0.16	0.15
	A	0.02	0.66	0.73	0.78	0.83	0.86	0.89	0.91	0.93	0.94	0.95	0.31	0.25	0.20	0.16	0.13	0.11	0.08	0.07	0.05	0.04	0.04	0.03	0.02
	B	0.06	0.71	0.74	0.76	0.79	0.81	0.83	0.84	0.86	0.87	0.89	0.25	0.22	0.20	0.18	0.16	0.15	0.13	0.12	0.11	0.10	0.09	0.08	0.07
0.65	C	0.09	0.74	0.75	0.77	0.78	0.80	0.81	0.82	0.83	0.84	0.85	0.21	0.20	0.18	0.17	0.16	0.15	0.14	0.14	0.13	0.12	0.11	0.11	0.10
	D	0.11	0.76	0.77	0.77	0.78	0.79	0.79	0.80	0.81	0.81	0.82	0.17	0.17	0.16	0.16	0.15	0.15	0.14	0.14	0.14	0.13	0.13	0.12	0.12
	A	0.01	0.76	0.81	0.84	0.88	0.90	0.92	0.93	0.95	0.96	0.97	0.22	0.18	0.14	0.12	0.09	0.08	0.06	0.05	0.04	0.03	0.03	0.02	0.02
	B	0.04	0.79	0.81	0.83	0.85	0.86	0.88	0.89	0.90	0.91	0.92	0.18	0.16	0.14	0.13	0.12	0.10	0.09	0.08	0.08	0.07	0.06	0.06	0.05
0.75	C	0.07	0.81	0.82	0.83	0.84	0.85	0.86	0.87	0.88	0.89	0.89	0.15	0.14	0.13	0.12	0.12	0.11	0.10	0.10	0.09	0.09	0.08	0.08	0.07
	D	0.08	0.83	0.83	0.84	0.84	0.85	0.85	0.86	0.86	0.87	0.87	0.12	0.12	0.12	0.11	0.11	0.11	0.10	0.10	0.10	0.09	0.09	0.09	0.09

Number of hours after lights are turned on

TABLE 3-9. Solar Intensity and Solar Heat Gain Factors for 32°N Latitude[3-1]

| Date | Solar Time am | Direct Normal Btuh/ft² | Solar Heat Gain Factors, Btuh/ft² | | | | | | | | | | | | | | | | | | Solar Time pm |
			N	NNE	NE	ENE	E	ESE	SE	SSE	S	SSW	SW	WSW	W	WNW	NW	NNW	HOR	
Oct 21	7	99	4	7	43	74	92	96	85	60	24	5	4	4	4	4	4	4	10	5
	8	229	13	15	63	143	195	217	206	162	90	17	13	13	13	13	13	13	63	4
	9	273	20	20	33	120	193	234	239	208	144	54	21	20	20	20	20	20	125	3
	10	293	24	24	26	62	147	207	234	225	183	109	32	24	24	24	24	24	173	2
	11	302	27	27	27	29	76	152	203	221	207	160	85	29	27	27	27	27	203	1
	12	304	28	28	28	28	30	78	151	199	215	199	151	78	30	28	28	28	213	12
	HALF DAY TOTALS		103	106	200	433	708	941	1038	972	753	441	226	125	104	103	103	103	679	

TABLE 3-10. Shading Coefficients for Single Glass with Indoor Shading[3-1]

			Type of Shading				
			Venetian Blinds		Roller Shade		
	Nominal Thickness[a]	Solar Trans.[b]			Opaque		Translucent
			Medium	Light	Dark	White	Light
Clear	3/32 to 1/4	0.87 to 0.80					
Clear	1/4 to 1/2	0.80 to 0.71					
Clear Pattern	1/8 to 1/2	0.87 to 0.79	0.64	0.55	0.59	0.25	0.39
Heat-Absorbing Pattern	1/8	—					
Tinted	3/16, 7/32	0.74, 0.71					
Heat-Absorbing[d]	3/16, 1/4	0.46					
Heat-Absorbing Pattern	3/16, 1/4	—	0.57	0.53	0.45	0.30	0.36
Tinted	1/8, 7/32	0.59, 0.45					
Heat-Absorbing or pattern	—	0.44 to 0.30	0.54	0.52	0.40	0.28	0.32
Heat-Absorbing[d]	3/8	0.34					
Heat-Absorbing or Pattern	—	0.29 to 0.15 / 0.24	0.42	0.40	0.36	0.28	0.31
Reflective Coated Glass							
S.C.[c] = 0.30			0.25	0.23			
0.40			0.33	0.29			
0.50			0.42	0.38			
0.60			0.50	0.44			

TABLE 3-11. Shading Coefficients for Insulating Glass with Indoor Shading[3-1]

				Type of Shading				
		Solar Trans.[b]		Venetian Blinds[c]		Roller Shade		
	Nominal Thickness in., each light					Opaque		Translucent
Type of Glass		Outer Pane	Inner Pane	Medium	Light	Dark	White	Light
Clear Out Clear In	3/32, 1/8	0.87	0.87	0.57	0.51	0.60	0.25	0.37
Clear Out Clear In	1/4	0.80	0.80					
Heat-Absorbing[d] Out Clear In	1/4	0.46	0.80	0.39	0.36	0.40	0.22	0.30
Reflective Coated Glass								
SC[c] = 0.20				0.19	0.18			
0.30				0.27	0.26			
0.40				0.34	0.33			

Time	CLF	Input × CLF
1200	0.73	2490 Btuh
1400	0.78	2660 Btuh
1600	0.82	2800 Btuh

COOLING LOAD CALCULATIONS

In cooling load calculations, one must consider, in addition to transmission and infiltration loads, solar heat gains and internal gains due to occupancy, equipment and processes. Cooling heat gains are further divided into sensible heat gains due to temperature changes and latent heat gains due to moisture.

Before discussing heat gain calculations, it is important to differentiate between *heat gain* and *cooling load*.

Heat gain is defined as the rate at which heat enters or is generated within a space.

Cooling load is the rate at which energy is removed from the space in order to maintain the design temperature. Cooling load generally differs from the heat gain due to the effect of the thermal mass of the building.

HEAT GAIN THROUGH GLASS

Glass permits solar energy to enter a building. This solar radiation provides heat in winter, for the perimeter zones, and becomes a source of additional cooling load in summer. The use of heat-absorbing glass, reflective coating, shades, awnings, drapes, etc. helps reduce solar heat gains.

The heat gain through a glass surface can be estimated from the following equation:

$$Q_g = A_g \times SHG \times SC + U_g A_g \times \Delta T \qquad (3\text{-}6)$$

where

Q_g = Heat gain, Btuh
A_g = Glass area
SHG = Solar heat gain factor
SC = Shading coefficient
U_g = Overall transmission coefficient of glass
ΔT = Temperature difference between inside and outside at design conditions

The year-round effect of heat gain and loss should be considered when specifying glass and its associated shading or reflecting devices.

Table 3-9 gives the SHG for various orientations. Tables 3-10 and 3-11 give the shading coefficients for various types of glass, with or without shades.

The cooling load due to heat gain through the fenestration area is estimated as follows:

$$Q_g = U_g A_g \times CLTD + A_g \times SC \times SHG \times CLF \qquad (3\text{-}7)$$

where
  CLTD = Cooling Load Temperature Difference
        (Table 9, p. 25.11, Ref. 3-1)
  CLF = Cooling Load Factor (Tables 11 and 12,
        p. 25.13, Ref. 3-1.)

CLF and *CLTD* were developed from transient heat flow equations for various types of fenestration surfaces.

Example 3-1

An office building in Atlanta, Georgia, has the following glass areas;
        8,000 ft^2 facing north and south
        6,000 ft^2 facing east and west
The summer outdoor design conditions for this location are 95°F D.B. and 78°F W.B. The inside of the building is maintained at 75°F. Calculate the heat gain at 2:00 p.m. on October 21 for each of the following choices in fenestration and compare the annual operating costs of the various types of glass.

Energy consumption is 1 hp per ton of air conditioning and the cost of electricity is $.06/kWh. The first cost of an air conditioning system having an assumed life of 25 years is $1,100 per ton and the interest rate is 12%. The estimated annual hours of cooling is 750 hours.

	U-value	SC	Cost/ft^2
1. Single glass	1.04	1.00	$2.50
2. Single glass, dark drapes	0.81	0.63	$3.50
3. Single glass, reflective coating	0.85	0.35	$4.00
4. Double-pane (insulating) glass with ¼" air space	0.61	0.88	$4.50
5. Double-pane glass with reflective coating	0.50	0.26	$6.00

Solution 3-1

The first step is to calculate the solar gain at 2:00 p.m. on October 21. Atlanta is located at 33°N latitude, so from Table 3-9, take the SHG values given for 32°N latitude.

Orientation	Area	SHG	SC	$A_g \times SHG \times SC$	Total Btuh
North	8000	24	1.00	192,000	
South	8000	183	1.00	1,464,000	
West	6000	147	1.00	882,000	
East	6000	24	1.00	144,000	2,634,000

2,634,000 Btuh is the solar heat gain with clear glass having a SC of 1.00. The gains with the other options are as follows:
  2. Single glass with drapes (SC=0.63)
        2,634,000 × 0.63 = 1,659,420 Btuh
  3. Single glass with reflective coating (SC=0.35)

2,634,000 × 0.35 = 921,900 Btuh
4. Double pane glass (SC = 0.88)
   2,634,000 × 0.88 = 2,317,920 Btuh
5. Double pane glass, reflective coating (SC = 0.26)
   2,634,000 × 0.26 = 684,840 Btuh

Total heat gains through the five glass options are:
1. Clear glass.

$$Q_1 = U_gA_g(T_o - T_i) + Q_{solar}$$
$$= 1.04 \times 28000 \times 20 + 2,634,000$$
$$= 1.04 \times 560,000 + 2,634,000$$
$$= 3,216,400 \text{ Btuh}$$

2. Single glass with drapes.

$$Q_2 = 0.81 \times 560,000 + 1,659,420$$
$$= 2,113,020 \text{ Btuh}$$

3. Single glass with reflective coating.

$$Q_3 = 0.85 \times 560,000 + 921,900$$
$$= 1,397,900 \text{ Btuh}$$

4. Double pane glass

$$Q_4 = 0.61 \times 560,000 + 2,317,920$$
$$= 2,659,520 \text{ Btuh}$$

5. Double glass with reflective coating

$$Q_5 = 0.50 \times 560,000 + 684,840$$
$$= 964,840 \text{ Btuh}$$

Now compute the first cost, operating cost, and annual operating cost where

Tons of A/C = Btuh/12,000 Btuh
Yearly KWh = $.7456/\eta \times$ tons $\times 750$ hrs (75 to 125 tons, $\eta = .90$. Over 150 tons, $\eta = .91$.)
First cost = 28,000 ft$^2 \times$ glass cost + $1,100 \times$ tons
Annual cost = first cost × capital investment factor + energy cost. At 12% interest and 25 years of life, the capital investment factor $= \frac{.12(1.12)^{25}}{(1.12)^{25} - 1} = .1275$

Option	Tons A/C	Yearly kWh	Energy @.06	First Cost	Annual Cost
1	268	164,687	$9,881	$364,000	$56,291
2	176	108,153	6,489	291,000	43,591
3	116	72,074	4,324	239,600	34,873
4	222	136,420	8,185	370,200	55,385
5	80	49,707	2,982	256,000	35,622

It would appear that we have two choices, Option 3, with the lowest annual cost, and Option 5, with the lowest energy cost. However, the picture becomes clearer when the heating season cost of the two Options is calculated, using $U = 1.1$ with Option 3 and $U = .58$ with Option 5. Atlanta has 2961 degree days and a winter design temperature of 22°F. Cost of gas is $3.00 per million Btu.

	Heating energy	Cooling energy	Annual Cost
Option 3	$8,208	$4,324	$43,081
Option 5	4,328	2,982	39,950

On a year-round basis, Option 5 is both energy efficient and cost effective and should be selected.

HEAT GAIN FROM OCCUPANTS

The heat gain from occupants is a function of their activity level, age and sex and the environmental influence. Table 3-12 lists the sensible and latent heat gains during various activities.

The heat gain from people is calculated from the following equations:

$$Q_{ps} = XQ_s \times CLF_p \times N_p \times D_p$$
$$Q_{pl} = XQ_l \times N_p \times D_p$$
$$Q_p = Q_{ps} + Q_{pl} \tag{3-8}$$

where

Q_{ps} = Sensible heat gain from people
XQ_s = Sensible heat gain per person, Table 3-12
CLF_p = Cooling load factor

TABLE 3-12. Rates of Heat Gain from Occupants of Conditioned Spaces[3-1]

Degree of Activity	Typical Application	Total Heat Adults, Male			Total Heat Adjusted[b]			Sensible Heat			Latent Heat		
		Watts	Btuh	kcal/hr	Watts	Btuh	kcal/hr	Watts	Btuh	kcal/hr	Watts	Btuh	kcal/hr
Seated at rest	Theater, movie	115	400	100	100	350	90	60	210	55	40	140	30
Seated, very light work writing	Offices, hotels, apts	140	480	120	120	420	105	65	230	55	55	190	50
Seated, eating	Restaurant[c]	150	520	130	170	580[c]	145	75	255	60	95	325	80
Seated, light work, typing	Offices, hotels, apts	185	640	160	150	510	130	75	255	60	75	255	65
Standing, light work or walking slowly	Retail Store, bank	235	800	200	185	640	160	90	315	80	95	325	80
Light bench work	Factory	255	880	220	230	780	195	100	345	90	130	435	110
Walking, 3 mph, light machine work	Factory	305	1040	260	305	1040	260	100	345	90	205	695	170
Bowling[d]	Bowling alley	350	1200	300	280	960	240	100	345	90	180	615	150

N_p = Number of people
D_p = Diversity factor for people
Q_{pl} = Latent heat gain from people
XQ_l = Latent heat gain per person, Table 3-12
Q_p = Total heat load from people

If air conditioning is operated only during the hours of occupancy, $CLF_p = 1.0$. If the air conditioning is operated over a 24 hour period, CLF_p is taken from Table 17, p. 25.17, Ref. 3-1.

HEAT GAIN FROM LIGHTS

The heat gain from lights is calculated from

$$Q_l \text{ (Btuh)} = 3.41W \times F_u \times IP \tag{3-9}$$

where

W = Total installed light wattage
F_u = Use factor = $\frac{\text{wattage in use}}{\text{installed wattage}}$
IP = $\frac{\text{input watts}}{\text{rated output watts}}$

The factor IP is a function of the type of light fixture and is discussed in Chapter 6.

The cooling load due to the heat gain from lights is

$$Q_{cl} = Q_l \times CLF_l$$

As mentioned earlier, CLF_l is shown in Table 15, p. 25.16 ff., Ref. 3-1.

HEAT GAIN FROM APPLIANCES

The heat gain from hooded appliances that employ electricity or steam is calculated from

$$Q_a = Q_r \times F_r \times F_u = 0.16\,Q_r \text{ in most cases} \tag{3-10}$$

where

Q_a = Heat gain from appliances, Btuh
Q_r = Rated input energy
F_r = Fraction of input energy converted to radiation, 0.32 in most cases. Convection and conduction heat are negligible with an effective hood.
F_u = Utilization factor = 0.5 if no other information is available.

When hooded appliances are direct fired

$$Q_a = Q_r \times F_r/F_{fl} \times F_u = 0.1\,Q_r \text{ in most cases}$$

where

F_{fl} = Flue loss factor, 1.6 in most cases.

When appliances are unhooded, the F_r factor no longer applies in the above equation and the total heat gain due latent and sensible heat must be considered. When food preparation is involved, 34% of the heat gain is assumed to be latent and 66% to be sensible.

If the space is conditioned at all times, the CLF is

applied to the sensible portion of the heat gain. The CLF for hooded and unhooded appliances are found in Tables 20 and 21, p. 25.21, Ref. 3-1.

HEAT GAIN FROM ELECTRIC MOTORS

When both the motor and the driven equipment are within the conditioned space, the heat gain is

$$Q_m = \frac{hp}{\eta} \times F_L \times 2545 \tag{3-11}$$

where

Q_m = Heat gain from the motor
hp = Rated horsepower, motor
η = Motor efficiency, decimal
F_L = Load factor = $\frac{\text{actual load}}{\text{rated load}}$
2545 = horsepower to Btuh conversion factor

When the motor is outside the conditioned space and the equipment is within

$$Q_m = hp \times F_L \times 2545$$

Conversely, when the motor is inside the conditioned space and the equipment is outside

$$Q_m = hp \times F_L \times 2545 \times \frac{1-\eta}{\eta}$$

The cooling load from motors is the same as that of an unhooded appliance and the same CLF applies.

HEAT GAIN FROM VENTILATION

The heat gain due to air infiltration and outside air used for ventilation is estimated as follows:

Sensible heat gain

$$Q_{vs} = (.075)(0.24 + 0.45W)(n_a V)(T_o - T_i) \tag{3-12}$$

where

.075 = lbs of dry air per ft^3
0.24 = specific heat of dry air, Btu/lb/°F
0.45 = specific heat of water vapor, Btu/lb/°F
W = humidity ratio, usually 0.01 to 0.012 lb of moisture per lb of dry air
n_a = number of air changes per hour
V = volume of conditioned space
$T_o - T_i$ = Temperature difference between outside and inside air

When $W = 0.012$

C_p = the specific heat of humid air
= $0.24 + 0.45(.012) = .245$ Btu/lb/°F

Therefore,

$$Q_{vs} = 0.0184 \times n_a V(T_o - T_i)$$

Latent Heat Gain

$$Q_{vl} = (.075)(1076) \times n_a V(W_o - W_i) \tag{3-13}$$

where

1076 = latent heat from condensing 75°F vapor to

TABLE 3-13. Cooling Load TD for Calculating Cooling Load from Sunlit Walls[3-1]

North Latitude Wall Facing	1	2	3	4	5	6	7	8	9	10	11	12	13	14	15	16	17	18	19	20	21	22	23	24	Hr of Maximum CLTD	Minimum CLTD	Maximum CLTD	Difference CLTD
Group A Walls																												
N	14	14	14	13	13	13	12	12	11	11	10	10	10	10	10	10	11	11	12	12	13	13	14	14	2	10	14	4
NE	19	19	19	18	17	17	16	15	15	15	15	15	16	16	17	18	18	18	19	19	20	20	20	20	22	15	20	5
E	24	24	23	23	22	21	20	19	19	18	19	19	20	21	22	23	24	24	25	25	25	25	25	22	18	25	7
SE	24	23	23	22	21	20	20	19	18	18	18	18	18	19	20	21	22	23	23	24	24	24	24	22	18	24	6
S	20	20	19	19	18	18	17	16	16	15	14	14	14	14	14	15	16	17	18	19	19	20	20	20	23	14	20	6
SW	25	25	25	24	24	23	22	21	20	19	19	18	17	17	17	17	18	19	20	22	23	24	25	25	1	17	25	8
W	27	27	26	26	25	24	24	23	22	21	20	19	19	18	18	18	18	19	20	22	23	25	26	26	1	18	27	9
NW	21	21	21	20	20	19	19	18	17	16	16	15	15	14	14	14	15	15	16	17	18	19	20	21	1	14	21	7
Group B Walls																												
N	15	14	14	13	12	11	11	10	9	9	9	8	9	9	9	10	11	12	13	14	14	15	15	15	24	8	15	7
NE	19	18	17	16	15	14	13	12	12	13	14	15	16	17	18	19	19	20	20	21	21	21	20	20	21	12	21	9
E	23	22	21	20	18	17	16	15	15	15	17	19	21	22	24	25	26	26	27	27	26	26	25	24	20	15	27	12
SE	23	22	21	20	18	17	16	15	14	14	15	16	18	20	21	23	24	25	26	26	26	25	24	24	21	14	26	12
S	21	20	19	18	17	15	14	13	12	11	11	11	11	12	14	15	17	19	20	21	22	22	21	23	11	22	11
SW	27	26	25	24	22	21	19	18	16	15	14	14	13	13	14	15	17	20	22	25	27	28	28	24	23	13	28	15
W	29	28	27	26	24	23	21	19	18	17	16	15	14	14	14	15	17	19	22	25	27	29	29	30	24	14	30	16
NW	23	22	21	20	19	18	17	15	14	13	12	12	12	11	12	12	13	15	17	19	21	22	23	23	24	11	23	9
Group C Walls																												
N	15	14	13	12	11	10	9	8	8	7	7	8	8	9	10	12	13	14	15	16	17	17	17	16	22	7	17	10
NE	19	17	16	14	13	11	10	10	11	13	15	17	19	20	21	22	22	23	23	23	23	22	21	20	20	10	23	13
E	22	21	19	17	15	14	12	12	14	16	19	22	25	27	29	29	30	30	30	29	28	27	26	24	18	12	30	18
SE	22	21	19	17	15	14	12	12	13	13	16	19	22	24	26	28	29	29	29	28	27	26	24	19	12	29	17
S	21	19	18	16	15	13	12	10	9	9	9	10	11	14	17	20	22	24	25	26	25	25	24	22	20	9	26	17
SW	29	27	25	22	20	18	16	15	13	12	11	11	11	13	15	18	22	26	29	32	33	33	32	31	22	11	33	22
W	31	29	27	25	22	20	18	16	14	13	12	12	12	13	14	16	20	24	29	32	35	35	35	33	22	12	35	23
NW	25	23	21	20	18	16	14	13	11	10	10	10	10	11	12	13	15	18	22	25	27	27	27	26	22	10	27	17
Group D Walls																												
N	15	13	12	10	9	7	6	6	6	6	6	7	8	10	12	13	15	17	18	19	19	19	18	16	21	6	19	13
NE	17	15	13	11	10	8	7	8	10	14	17	20	22	23	24	24	25	25	24	23	22	20	18	19	7	25	18
E	19	17	15	13	11	9	8	9	12	17	22	27	30	32	33	33	32	32	31	30	28	26	24	22	16	8	33	25
SE	20	17	15	13	11	10	8	8	10	13	17	22	26	29	31	32	32	32	31	30	28	26	24	22	17	8	32	24
S	19	17	15	13	11	9	8	7	6	7	9	12	16	20	24	27	29	29	29	28	26	24	22	19	19	6	29	23
SW	28	25	22	19	16	14	12	10	9	8	8	8	10	12	16	21	27	32	36	38	38	37	34	31	21	8	38	30
W	31	27	24	21	18	15	13	11	10	9	9	9	10	11	14	18	24	30	36	40	41	40	38	34	21	9	41	32
NW	25	22	19	17	14	12	10	9	8	7	7	8	9	10	12	14	18	22	27	31	32	32	30	27	22	7	32	25
Group E Walls																												
N	12	10	8	7	5	4	3	4	5	6	7	9	11	13	15	17	19	20	21	23	20	18	16	14	20	3	22	19
NE	13	11	9	7	6	4	5	9	15	20	24	25	25	26	26	26	26	26	25	24	22	19	17	15	16	4	26	22
E	14	12	10	8	6	5	6	11	18	26	33	36	38	37	36	34	33	32	30	28	25	22	20	17	13	5	38	33
SE	15	12	10	8	7	5	5	8	12	19	25	31	35	37	37	36	34	33	31	28	26	23	20	17	15	5	37	32
S	15	12	10	8	7	5	4	3	4	5	9	13	19	24	29	32	34	33	31	29	26	23	20	17	17	3	34	31
SW	22	18	15	12	10	8	6	5	5	6	7	9	12	18	24	32	38	43	45	44	40	35	30	26	19	5	45	40
W	25	21	17	14	11	9	7	6	6	6	7	9	11	14	20	27	36	43	49	49	45	40	34	29	20	6	49	43
NW	20	17	14	11	9	7	6	5	5	5	6	8	10	13	16	20	26	32	37	38	36	32	28	24	20	5	38	33

TABLE 3-14. Cooling Load TD for Calculating Cooling Load from Flat Roofs[3-1]

Roof No	Description of Construction	Weight lb/ft²	U-value Btu/(h·ft²·°F)	1	2	3	4	5	6	7	8	9	10	11	12	13	14	15	16	17	18	19	20	21	22	23	24	Hour of Maximum CLTD	Minimum CLTD	Maximum CLTD	Difference CLTD
	Without Suspended Ceiling																													
1	Steel sheet with 1-in. (or 2-in.) insulation	7 (8)	0.213 (0.124)	1	−2	−3	−3	−5	−3	6	19	34	49	61	71	78	79	77	70	59	45	30	18	12	8	5	3	14	−5	79	84
2	1-in. wood with 1-in. insulation	8	0.170	6	3	0	−1	−3	−3	−2	4	14	27	39	52	62	70	74	74	70	62	51	38	28	20	14	9	16	−3	74	77
3	4-in. l.w. concrete	18	0.213	9	5	2	0	−2	−3	−3	1	9	20	32	44	55	64	70	73	71	66	57	45	34	25	18	13	16	−3	73	76
4	2-in. h.w. concrete with 1-in. (or 2-in.) insulation	29	0.206 (0.122)	12	8	5	3	0	−1	−1	3	11	20	30	41	51	59	65	66	66	62	54	45	36	29	22	17	16	−1	67	68
5	1-in. wood with 2-in. insulation	19	0.109	3	0	−3	−4	−5	−7	−6	−3	5	16	27	39	49	57	63	64	62	57	48	37	26	18	11	7	16	−7	64	71
6	6-in. l.w. concrete	24	0.158	22	17	13	9	6	3	1	1	3	7	15	23	33	43	51	58	62	64	62	57	50	42	35	28	18	1	54	63
7	2.5-in. wood with 1-insulation	13	0.130	29	24	20	16	13	10	7	6	6	9	13	20	27	34	42	48	53	55	56	54	49	44	39	34	19	6	56	50
8	8-in. l.w. concrete	31	0.126	35	30	26	22	18	14	11	9	7	7	9	13	19	25	33	39	46	50	53	54	53	49	45	40	20	7	54	47
9	4-in. h.w. concrete with 1-in. (or 2-in.) insulation	52 (52)	0.200 (0.120)	25	22	18	15	12	9	8	8	10	14	20	26	33	40	46	50	53	53	52	48	43	38	34	30	18	8	53	45
10	2.5-in. wood with 2-in. insulation	13	0.093	30	26	23	19	16	13	10	9	8	9	13	17	23	29	36	41	46	49	51	50	47	43	39	35	19	8	51	43
11	Roof terrace system	75	0.106	34	31	28	25	22	19	16	14	13	13	15	18	22	26	31	36	40	44	45	46	45	43	40	37	20	13	46	33
12	6-in. h.w. concrete with 1-in. (or 2-in.) insulation	(75) 75	0.192 (0.117)	31	28	25	22	20	17	15	14	14	16	18	22	26	31	36	40	43	45	45	44	42	40	37	34	19	14	45	31
13	4-in. wood with 1-in. (or 2-in) insulation	17 (18)	0.106 (0.078)	38	36	33	30	28	25	22	20	18	17	16	17	18	21	24	28	32	36	39	41	43	43	42	40	22	16	43	27

50°F liquid

$W_o - W_i$ = change in humidity ratio between outside and inside air

Restated,

$$Q_{vl} = 80.7 \times n_a V(W_o - W_i)$$

If W_o is less than W_i, Q_{vl} equals zero.

The cooling load from infiltration and ventilation is equal to the heat gain from these sources. Total cooling load

$$Q_v = Q_{vs} + Q_{vl}$$

INTERNAL PARTITIONS

When two spaces in a building, or a space and a corridor, are at different temperatures, the heat flow from space 1 to space 2, through the common partition, is

$$Q = UA(T_1 - T_2)$$

It is assumed that T_1 and T_2 remain constant over a long period of time and that we have a steady state condition.

ADJUSTMENT FACTOR

The total cooling load is the sum of the heat gains through the walls and roof, glass, internal partitions, internal heat sources, such as people, lights, appliances and motors, and infiltration and ventilation.

A fraction of the sensible heat gain from these sources, except infiltration and ventilation, is lost to the surroundings and reduces the overall cooling load. The adjustment factor, Fc, is calculated as follows:

$$F_c = 1 - 0.02K_t \qquad (3\text{-}14)$$

The K_t value is derived from the sum of the heat transfers through the building envelope, partitions and corridors, which is then divided by the perimeter of each story of the building, denoted by L_F. Thus

$$K_t = \frac{1}{L_F}(U_rA_r + U_wA_w + U_gA_g + U_dA_d + U_pA_p)$$

with the subscripts denoting roof, walls, glass, doors and partitions, respectively.

The adjusted sensible load is equal to

$$Q_c = Q_s \times F_c$$

Example 3-2

Calculate the design cooling load for the building shown in Figure 3-1. Base the calculations on the fact that the windows are equipped with light colored shear drapes that give the windows a shade coefficient of 0.55.

Base the calculation on August 21 at 2:00 p.m. and outside design temperatures of 91° D.B. and 73° W.B. The building is maintained at 75° during the occupied

hours of 7:30 a.m. and 5:30 p.m. The chillers are shut off at all other times.

Solution 3-2

To summarize the previous discussion, cooling load is produced by the heat gain

- through exterior opaque surfaces
- through exterior glass surfaces
- from air infiltration and ventilation
- from internal lights and equipment
- from occupants
- through internal partitions

Opaque Surfaces

The heat gain through opaque walls and the roof is equal to the area, in ft^2, times the U-factor of the walls and roof times the Cooling Load Temperature Difference (CLTD).

The CLTD values were computed on the basis of an outdoor maximum temperature of 95° and a mean of 85°. The assumed indoor temperature is 78°. When the design temperatures deviate from these norms, as in our case, the CLTD must be adjusted.

The mean daily range for Detroit is 20° and the daily mean is therefore $91° - (.5 \times 20) = 81°$.

Since both the indoor and outdoor temperatures vary from the norm, correct for CLTD in the following fashion. Subtract the indoor temperature from 78°, subtract 85° from the daily mean.

$$
\begin{array}{ll}
78° - 75° = & +3 \\
81° - 85° = & \underline{-4} \\
\text{Correction} & -1
\end{array}
$$

The U-factor must also be adjusted to include an increase in R-value between a 15 mph winter wind velocity and a 7.5 mph summer velocity. The difference is R = +.08. Therefore the U-value for walls becomes $1/3.19 = .313$ and that for the roof becomes $1/6.74 = .148$.

A wall composed of face brick and 8″ concrete block is classified as a Group D wall and the CLTD is given in Table 3-13. Reading down the 14 hour column for each orientation, and subtracting one from each value, we can construct the following tabulation.

Surface	Area	U	CLTD	Heat Gain
North wall	633	.313	9	1,783
West wall	1092	.313	10	3,418
South wall	9977	.313	15	46,842
East wall	9798	.313	31	95,070
Roof	19044	.148	63	177,566
			Total	$324,679 = Q_{tr}$

With the roof, the U value was once again adjusted for

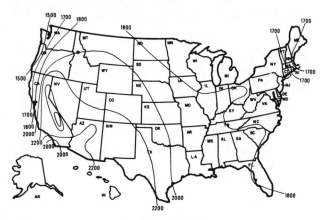

FIG. 3-4. Values of Y'.[3-4]

$$Q_{tr} = (UA_w + UA_r + UA_g)(T_{bin} - T_i) \qquad (3-15)$$
$$= (6729.5 + 2818.5 + 19141)(T_{bin} - T_i)$$
$$= 28,689(T_{bin} - T_i)$$

The roof solar gain,

$$Q_{sr} = UA_r(0.18 \times SHF - 7) \times Y \qquad (3-16)$$

The SHF for a horizontal surface at 40°N latitude is given, in Table 3-15, as 87 Btuh/ft^2,

$$Q_{sr} = 2818.5(0.18 \times 87 - 7) \times .75$$
$$= 18,306.2 \text{ Btuh}$$

The wall solar gain depends on the orientation of the wall and the SHF values are listed in Table 3-15.

$$Q_{sw} = \Sigma U_w A_w \times 0.18 \times SHF \times Y$$
$$= \Sigma .313 \times A_w \times 0.18 \times SHF \times .75$$
$$= \Sigma .042255 \times A_w \times SHF$$

Wall	U-factor	Area	SHF	Q_s
North	.042255	633	17	454.7
West	.042255	1092	48	2,214.8
East	.042255	9798	48	19,872.7
South	.042255	9977	31	13,068.9
			$Q_{sw} =$	35,611 Btuh

The glass solar gain again depends on the orientation of the glass and the shading coefficient, SC = .55.

$$Q_{sg} = \Sigma SC \times A_g \times SHF \times Y$$
$$= \Sigma .55 \times A_g \times SHF \times .75$$
$$= \Sigma .4125 \times A_g \times SHF$$

Glass	SC Factor	Area	SHF	Q_g
North	.4125	9165	17	64,269.6
West	.4125	9189	48	181,942.2
South	.4125	787	31	10,063.8
			$Q_{sg} =$	256,276 Btuh

The total solar gain

TABLE 3-15. Summer Solar Heat Factor (SHF) for Vertical and Horizontal Surfaces,[3-3] Btuh·ft^2.

	NORTH LATITUDE				
ORIENTATION	24°	32°	40°	48°	56°
N	20	18	17	17	18
NE	37	35	33	32	32
E	46	47	48	50	51
SE	33	39	45	51	56
S	17	23	31	41	51
SW	33	39	45	51	56
W	46	47	48	50	51
NW	37	35	33	32	32
HORIZ	89	89	87	84	78

$$Q_s = Q_{sr} + Q_{sw} + Q_{sg}$$
$$= 18,306 + 35,611 + 256,276$$
$$= .310 \times 10^6 \text{ Btuh}$$

Infiltration load is the same as in Example 3-2.

$$Q_{vs} = 28,041(T_{bin} - T_i)$$
$$Q_{vl} = 122.99 \times 10^6 \times (W_{bin} - .012) \text{ Btuh}$$

Total sensible heat gain

$$Q_s = (Q_{tr} + Q_{vs})(T_{bin} - T_i)$$
$$= (28,689 + 28,041)(T_{bin} - T_i)$$
$$= 56,730(T_{bin} - T_i) \text{ Btuh}$$

Internal heat loads, including diversity factor, from Example 3-2, is

$$Q_i = Q_p + Q_1$$
$$= .084 \times 10^6 + .577 \times 10^6$$
$$= .661 \times 10^6 \text{ Btuh}$$

The dry bulb temperature, the wet bulb temperature and the number of hours in each 5° bin are taken from Air Force data collected at Selfridge AFB near Detroit and listed in Table 3-16.

Column 1 lists the mean dry bulb temperature for each 5° bin above 60°. Column 2 lists the number of hours when the given temperature range occurs. Column 3 lists the coincident wet bulb temperature for each bin.

Column 4 is the sensible heat gain from transmission and infiltration which is equal to $T_{bin} - T_i$ (Col. 1−75°) times bin hours (Col. 2) times Q_s (56,730 Btuh).

Column 5 totals the solar gain which is equal to Q_s (.310 × 10^6) times the total number of hours in the cooling season (3015). Column 6 totals the internal gains which are equal to .661 × 10^6 times the total number of hours in the cooling season.

Column 7 lists the humidity ratio derived from dry bulb and wet bulb temperatures listed in Columns 1 and the change in wind R and the correction of -1 to the CLTD. Table 3-14 lists two roofs comparable to the one on our building. Since they use $4''$ of 40 lb concrete while we are using $5''$ of 30 lb concrete, the masses are comparable and the value chosen is for *Roof No. 3* or $64-1 = 63$.

Glass

The glass in our building is equipped with light colored sheer drapes that create a shade coefficient (SC) of 0.55. The summer U-value for the glass is 1.00 and the solar gain is shown in Table 3-9. Cooling Load Factors are from p. 25.13, Ref. 3-1.

The heat gain through glass is the sum of the conduction heat transfer and the direct solar heat gain, Equation 3-7.

$$Q_g = U_g A_g \times CLTD_g + A_g \times SC \times SHG \times CLF_g$$
$$U_g = 1.00$$
$$CLTD_g = 13$$
$$SC = 0.55$$
$$Q_g = 1.00 \times 13 \times A_g + .55 \times SHG \times CLF \times A_g$$

Orientation	Area	SHG	CLF	Q_g
South	787	149	.58	47,638
North	9165	35	.75	251,465
West	9189	216	.29	436,036
			Total	735,139

Lights

The cooling load due to lights is equal to the heat gain due to lights multiplied by the CLF. Since the system operates only during the load period, the CLF, in this case, is 1.0 and the lighting load is equal to the heat gain calculated for the design heating load. Therefore

$$Q_1 = 1.461 \times 10^6$$

People

When calculating the cooling load due to occupants, the sensible and latent loads are calculated separately. Using the values from Table 3-12,

Sensible heat gain $= Q_{ps} = 315 \times 425 = .134 \times 10^6$ Btuh
Latent heat gain $= Q_{pl} = 325 \times 425 = .138 \times 10^6$ Btuh

Infiltration

Cooling load due to infiltration has both a sensible and a latent component. Infiltration is one air change per hour. Assume that

$$W_o = 0.0134 \text{ lb of water vapor/lb of air}$$
$$W_i = 0.012 \text{ lb of water vapor/lb of air}$$

It follows that

$$Q_{vs} = 0.0184 \times 1.524 \times 10^6 (91 - 75)$$
$$= 0.449 \times 10^6 \text{ Btuh}$$

$$Q_{vl} = 0.075 \times 1.524 \times 10^6 \times 1076(.0134 - .012)$$
$$= 0.172 \times 10^6 \text{ Btuh}$$

The total sensible load

$$Q_s = Q_{tr} + Q_g + Q_1 + Q_{ps} + Q_{vl}$$
$$= (.325 + .735 + 1.463 + .134 + .449) \times 10^6$$
$$= 3.106 \times 10^6 \text{ Btuh}$$

Total latent load

$$Q_1 = Q_{pl} + Q_{vl} = (.138 + .172) \times 10^6$$
$$= .310 \times 10^6 \text{ Btuh}$$

Adjustment Factor

Since doors and partitions are inconsequential, the adjustment factor for the sensible cooling load is based on

$$K_t = 1/L_f (U_r A_r + U_w A_w + U_g A_g)$$

With an unexposed basement, $L_f = 138 \times 4$ walls $\times 5$ stories. However the exposed basement of our building adds the equivalent of another 1½ walls. Therefore

$$L_f = 138(5 \times 4 + 1.5) = 2967$$
$$1/L_f = 1/2967$$
$$K_t = 1/2967(.148 \times 19,044 + .313 \times 21,500$$
$$+ 1.0 \times 19,141)$$
$$= 1/2967(2819 + 6730 + 19,141)$$
$$= 9.67$$
$$F_c = 1 - 0.02K_t$$
$$= 1 - .02 \times 9.67 = .807$$

The adjusted sensible cooling load

$$Q_{sa} = Q_s F_c = 3.106 \times 10^6 \times .807 = 2.507 \times 10^6$$

Total cooling load

$$Q_c = Q_{sa} + Q_1 = (2.507 + .310) \times 10^6$$
$$= 2.817 \times 10^6 \text{ Btuh}$$

Example 3-3

Using the Bin Method, calculate the seasonal cooling requirements for the building described in Example 3-2.

With the Bin Method, the Solar Heat Factor (SHF), Table 3-15, and the Y' factor, Figure 3-4, are used in varying combinations, to estimate the solar heat gain through walls, roof and glass.

From Figure 3-4, the Y' value for the Detroit area is 1800. Y is given as Y'/2400, therefore, Y = 1800/2400 = .75

To calculate heat gain by the Bin Method, begin with the transmission gain Q_{tr}. From Example 3-2,

	Summer U	Area	UA
Wall	.313	21,500	6729.5
Roof	.148	19,044	2818.5
Glass	1.00	19,141	19141.0

TABLE 3-16. Bin Method of Calculating Cooling Energy Requirements

1 D.B. Temp	2 No. Hours	3 W.B. Temp	4 Sensible Gain $\times 10^6$	5 Solar Gain $\times 10^6$	6 Internal Gain $\times 10^6$	7 W_o	8 Latent Gain $\times 10^6$	9 Load $\times 10^6$
92	33	75	31.8			.0148	11.36	
87	115	70	78.3			*		
82	261	68	103.7			*		
77	456	66	51.7			*		
72	669	62	− 113.9			*		
67	791	59	− 359.0			*		
62	690	56	− 508.9			*		
TOTAL	3015		− 716.3	934.7	1992.9		11.36	2,222.7

Seasonal energy requirements at .7 COP $= 3.175 \times 10^6$ Btu $= 3.175 \times 10^6$ lb of steam

3. There is no latent heat gain when W_o is less than .012. Therefore, a W_o of less than .012 is indicated by *. Column 8 lists the latent heat gain which is $122.99 \times 10^6 \times (W_{bin} - .012)$ times Column 2.

Column 9 is the sum of Columns 4, 5, 6 and 8. The seasonal energy requirement is determined by multiplying the total heat gain by 1/.7 to adjust for a COP of .7.

SUMMARY

From the foregoing examples, we have seen that heat transfer through the building envelope can be minimized by

- reducing the overall heat transmission coefficient, or

U factor, of walls, roof and floor by increasing insulation and avoiding excessive glass.
- reducing the U factor for glass by using double- and triple-pane insulating glass.
- reducing solar heat gain through the glass by using drapes, blinds and reflective glass.
- reducing infiltration air by sealing leaks around windows, doors, and wall and roof penetrations.
- controlling internal heat gains by using more efficient illumination as discussed in Chapter 6.
- orienting the building, where possible, so that less glass faces west and more glass faces south in the northern hemisphere.

Chapter 4

HEATING, VENTILATION AND AIR CONDITIONING SYSTEMS

The primary purpose of a heating, ventilating and air conditioning (HVAC) system is to maintain a comfortable environment for the occupants or optimum conditions for the processes carried out in that space.

The procedures for calculating heating and cooling loads were reviewed in Chapter 3. However, those energy calculations did not take into account the characteristics of individual HVAC systems and an overall efficiency of the system was assumed. Now we will review the characteristics and energy requirements of the following types of HVAC systems.

- Air systems
- Air/water systems
- Water systems

AIR SYSTEMS

Some systems use air to control the conditions in a space for comfort or process work. Air systems are divided into two basic categories:

- Single duct systems
- Dual duct systems

Single duct systems contain both the main heating and cooling coils in one duct and provide air at an average temperature. The conditioned air is then distributed to the various zones or spaces through terminal units.

Dual duct systems use separate ducts to contain either the main heating or cooling coils. The two air streams (one hot and one cold) are mixed at either the terminal unit of each zone or at the main fan which has separate supply ducts.

Air systems are simple and easy to design, install and operate. They offer a wide choice of temperature and humidity control under all operating conditions, including simultaneous heating and cooling. These systems can be equipped to operate on an economizer cycle during cooling, using outside air when its temperature is less than the desired space-supply temperature. During marginal weather, this cycle conserves refrigeration energy.

These systems are used for air conditioning all types of buildings, including offices, stores, schools, universities, hospitals, laboratories, computer rooms and factories.

A schematic layout for a single duct, constant volume system for a single zone is shown in Figure 4-1. This system is used to control one set of space conditions, such as a computer room, small individual shop or classroom. A set of dampers controls the fractions of outside air and return air to be mixed, after which the blended air is heated to the desired minimum air temperature by the preheating coil during winter operation. The reheat coil helps control the final supply temperature after the humidifier and can be located before the supply fan. The cooling temperature is controlled by the cooling coil.

Figure 4-2 shows the layout for a single duct, constant volume multizone system. The reheat coils are located at the terminal units in each zone to control the final supply temperature.

Single Duct Energy Calculations

The following procedure is used to estimate the energy requirements of a single duct, constant volume system. Steps 1 through 4 are applicable to all systems.

1. LOAD CALCULATION   Calculate the hourly and peak heating and cooling loads using the methods out-

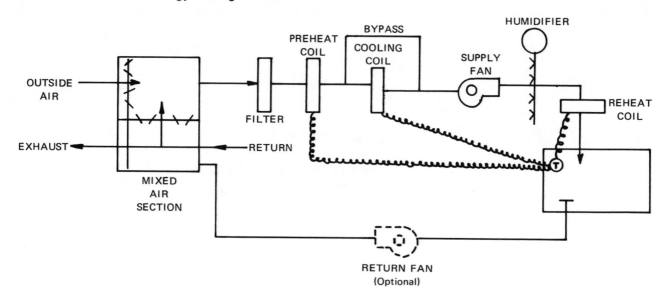

FIG. 4-1. Schematic of a single duct, constant volume, single zone system.

lined in Chapter 3. The main components are:
- Heat transmission through exterior walls and roofs
- Infiltration heat transfer
- Direct solar heat gain through windows
- Indirect solar heat gain through exterior walls and roofs
- Interior heat gains due to occupancy, lights, equipment operation and other processes within the space.

2. FAN SIZE   The supply fan size is determined by the peak heating or cooling load or the peak ventilation requirement. Let

T_s = Supply air temperature

T_{sh} = Upper limit of heating supply air, usually 110° to 130°

T_{sc} = Lower limit of cooling supply air, usually 50° to 55°

T_{sp} = Desired space temperature or thermostat setting

T_r = Return air temperature

Q_{ph} = Peak hourly heating load

Q_{pc} = Peak hourly sensible cooling load

n_a = Number of air changes per hour due to ventilation

V = Space volume

W = Humidity ratio (lb of water vapor/lb of dry air)

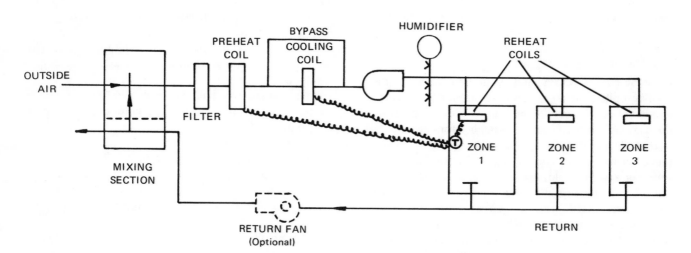

FIG. 4-2. Schematic of a single duct, constant volume, multizone system.

D = Density of air

= (0.075 lb/ft^3 at 70°F, 1 atmosphere)

cfm_s = Supply fan air flow rate in cubic feet per minute

C_p = Specific heat of moist air which is equal to the specific heat of dry air plus the specific heat of water vapor times the humidity ratio.

= .24 + .45W = .24 + .45(.010)

= .2445 Btu/lb·°F

Let us make an energy balance on a space.

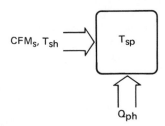

At design heating conditions:

$$\text{Heat rate of supply air} = \text{Heat load (peak)}$$
$$Mass \times specific\ heat\ (T_{sp} - T_{sh}) = Q_{ph}$$
$$.075 \times cfm_s \times 60 \times 0.2445\ (T_{sp} - T_{sh}) = Q_{ph}$$
$$1.1\ cfm_s\ (T_{sp} - T_{sh}) = Q_{ph}$$

$$cfm_s\ heating = \frac{Q_{ph}}{1.1\ (T_{sp} - T_{sh})} \qquad (4\text{-}1)$$

Similarly, for satisfying the peak cooling load, the air flow rate of the supply fan is given by:

$$cfm_s\ cooling = \frac{Q_{pc}}{1.1\ (T_{sp} - T_{sc})} \qquad (4\text{-}2)$$

The air flow rate for satisfying the minimum ventilation air requirements is given by:

$$cfm_s\ ventilation = \frac{n_a \times V}{60} \qquad (4\text{-}3)$$

The size of the supply fan is based on the highest of the three numbers given by Equations 4-1, 4-2 and 4-3.

3. HEATING AND COOLING CAPACITY

The size of the heating system (boiler or furnace) and the cooling system (chiller or air conditioning unit) is selected to satisfy the design heating and cooling loads. The sizes of auxiliary equipment, such as chiller and boiler pumps, are proportional to the size of the boiler or chiller. The required input energy (gas, oil or electricity) is established from the manufacturer's data as a function of rated output. The equipment with higher ratio of output to input energy, at the average load, will have a higher efficiency and a lower energy consumption. Refer to the *ASHRAE Handbook, 1981 Fundamentals* and *1980 Systems*, for further details.

4. HORSEPOWER FOR PUMPS AND FANS

The motor horsepower required for various pumps and fans is estimated as follows. For pump horsepower, let

hp = horsepower (33,000 lb-ft/min)

H_d = Pump head, in feet of water. This is difference in pressure between the pump outlet and inlet

= $p_o - p_i$

η_p = Pump efficiency

η_m = Electric motor efficiency

gpm = water flow in gallons per minute

8.33 = lb per gallon of water

$$pump\ hp = \frac{mass\ flow\ rate\ \times pressure\ change}{hp \times \eta_p}$$
$$= \frac{gpm \times 8.33 \times H_d}{33000 \times \eta_p}$$
$$= \frac{gpm \times H_d}{3962 \times \eta_p} \qquad (4\text{-}4)$$

$$pump\ motor\ hp = \frac{gpm \times H_d}{3962 \times \eta_p \times \eta_m} \qquad (4\text{-}5)$$

For fan horsepower, let

cfm = Air flow, in cubic feet per minute

D_p = Fan pressure, in inches of water. This is the difference in pressure between the fan outlet and inlet

$$fan\ hp = \frac{mass\ flow\ rate \times pressure\ change}{hp \times \eta_m}$$
$$= \frac{cfm \times D_p \times \dfrac{inches\ of\ water}{407\ inches/atmosphere} \times \dfrac{14.7\ lb/in^2}{atmosphere} \times \dfrac{144\ in^2}{ft^2}}{33000\ \dfrac{ft\text{-}lb}{minute/hp} \times \eta_m}$$
$$= \frac{cfm \times D_p}{6346 \times \eta_p} \qquad (4\text{-}6)$$

The electric power required for running the fan or pump at a given hour is given by

$$E_e = hp \times 0.746\ kW \qquad (4\text{-}7)$$

5. SUPPLY AIR TEMPERATURE

After establishing the size of the primary equipment (the size of existing equipment is given by the manufacturer) the next step is to establish the desired air supply temperature.

Let Q_s = sensible thermal load (Btuh) at a given hour

Using the energy balance procedure shown in Step 2,

the supply air temperature is calculated as follows.

$$(1.1) \ (cfm_s) \ (T_{sp} - T_s) = Q_s$$

$$T_s = T_{sp} - \frac{Q_s}{(1.1) \ (cfm_s)} \qquad (4\text{-}8)$$

6. RETURN AIR TEMPERATURE

The return air temperature is equal to the space air temperature in the absence of a plenum above the space. For buildings where the plenum is treated as a separate zone during heating and cooling load calculations, the temperature changes in return air due to plenum load, Q_p, from lights, heat transmission, and air infiltration, are calculated as follows.

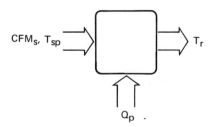

$$(1.1) \ (cfm_s) \ (T_r - T_{sp}) = Q_p$$

$$T_r = T_{sp} + \frac{Q_p}{1.1 \ (cfm_s)} \qquad (4\text{-}9)$$

If the plenum is not treated as a separate zone,

$$T_r = T_{sp} \qquad (4\text{-}10)$$

The change of air temperature due to flow across the return fan is generally small and negligible in most cases.

7. ECONOMIZER AIR TEMPERATURE AT FAN ENTRANCE

For maximum economy, the desired entering fan temperature, T_{ef}, Figure 4-3, is equal to the supply temperature, T_s, minus DT_s, the temperature rise across the fan. DT_s is due to energy released by the fan motor, Q_m.

$$Q_m = 0.746 \ \frac{kW}{hp} \times 3414 \ \frac{Btuh}{kW} \times hp$$

Substituting in Equation 4-8 . . .

$$(1.1) \ (cfm_s) \ (DT_s) = 2547 \times hp$$

Again substituting in Equation 4-6 . . .

$$(1.1) \ (cfm_s) \ (DT_s) = 2547 \times \frac{cfm_s(D_{ps})}{6346 \ \eta_m}$$

$$DT_s = \frac{0.401 D_{ps}}{1.1 \eta_m} \qquad (4\text{-}11)$$

Hence,

$$(T_{ef})_{economical} = T_s - DT_s \qquad (4\text{-}12)$$

If the heat gain caused by the motor has been accounted for in the hourly thermal load calculations, then

$$(T_{ef})_{economical} = T_s \qquad (4\text{-}13)$$

8. MIXED AIR CONDITIONS

The mixed air section can be operated in the following three modes:
- fixed fractions
- temperature-type economizer
- enthalpy-type economizer

Fixed Outside-Air and Return-Air Fractions: The mixed air temperature and humidity ratios are calculated

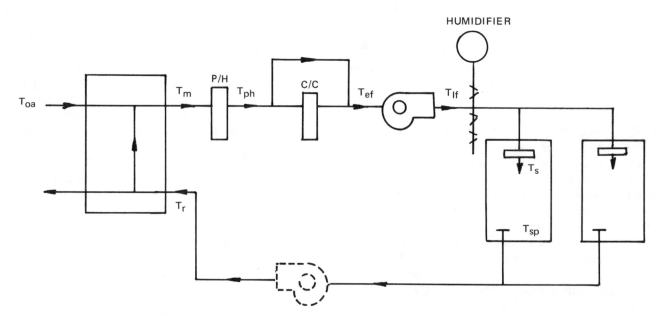

FIG. 4-3. Schematic of a system designed for minimum energy consumption.

by making an energy balance and a mass balance on the 3 streams as shown below.

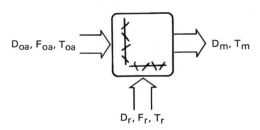

Let
F = Fraction of air
T = Temperature of air
D = Density of air
W = Humidity ratio of air
C_p = Specific heat of air (assumed to be constant for outside, return and mixed air)
subscript oa = outside air
r = return air
m = mixed air

Energy Balance

$$C_p D_{oa} F_{oa} T_{oa} + C_p D_r F_r T_r = C_p D_m T_m$$
$$= C_p (D_{oa} F_{oa} + D_r F_r) T_m$$

The middle equality reflects the fact that

$$F_{oa} + F_r = 1$$

Hence

$$T_m = \frac{D_{oa} F_{oa} T_{oa} + D_r F_r T_r}{D_{oa} F_{oa} + D_r F_r} \qquad (4\text{-}14)$$

Similarly

$$W_m = \frac{D_{oa} F_{oa} W_{oa} + D_r F_r W_r}{D_{oa} F_{oa} + D_r F_r} \qquad (4\text{-}15)$$

The density of air is calculated by using the ideal gas equation of state, which is applicable for most of the conditions encountered in the heating/air conditioning temperature range.

$$D = \frac{P}{RT} = \frac{P \times 144}{53.3 \times T} \qquad (4\text{-}16)$$

where

P = Pressure in lb/in² absolute
T = Temperature °R
R = Gas constant for air
= 53.3 ft-lb/(lb-°R)

Temperature-type Economizer Cycle   This mode increases the fraction of outside air above the minimum limit, by changing the position of the dampers, to approach T_{ef} (the economizer supply fan temperature) as closely as possible. The fractions of outside air are set, for various conditions, as follows.

Condition 1:   Return air temperature, T_r, is greater than outside temperature, T_{oa}.

(a) If the required temperature at the fan inlet, T_{ef}, is equal to or less than the temperature of the outside air, T_{oa}, use 100% outside air.

(b) If the required temperature, T_{ef}, is greater than the temperature of the outside air, use a mixture of outside air and return air. The proportions are calculated from the energy balance, Equation 4-14.

$$F_{oa} = \frac{D_r(T_r - T_{ef})}{D_r T_r - D_{oa} T_{oa} + T_{ef}(D_{oa} - D_r)} \qquad (4\text{-}17)$$

where

D_r = Return air density (lb/ft³)
D_{oa} = Outside air density (lb/ft³)
T_{ef} = Economizer air temperature for the supply fan entrance

(c) If the required temperature, T_{ef}, is equal to or less than the return air temperature, T_r, use a maximum amount of return air and the legally required amount of outside air.

Condition 2:   If the return air temperature, T_r, equals the outside air temperature, T_{oa}, use 100% outside air.

Condition 3:   Return air temperature, T_r, is less than the outside air temperature, T_{oa}.

(a) If the required temperature at the fan inlet, T_{ef}, is equal to or less than the return air, use the legally required minimum of outside air.

(b) If the required temperature, T_{ef}, is greater than the temperature of the return air and less than the temperature of the outside air, use mixed air. The proportions are calculated using Equation 4-17.

(c) If the required temperature, T_{ef}, is equal to or greater than the outside air, use 100% outside air.

The calculated F_{oa} is set at 1.0 at the upper limit and to $(F_{oa})_{minimum}$ at the lower limit. After establishing F_{oa} and F_r, the mixed air conditions are calculated like mixed air fractions.

Enthalpy-Type Economizer Cycle   This cycle controls the fraction of outside air, and the return air enthalpy of the mixed air approaches that of the supply air. If H_{oa} (the enthalpy of the outside air) is less than H_r (the enthalpy of the return air) then F_{oa} is calculated in the same manner as for temperature-type economizers. If H_{oa} is greater than H_r, then F_{oa} is set at the minimum ventilation limit.

9. PREHEAT ENERGY REQUIRED   The minimum desired air temperature after the preheat coil is generally set at $T_{min} = 50$ to $55°$. If the mixed air temperature, T_m, is less than T_{min}, energy, Q_{ph}, will be supplied by the preheat coil. The amount of energy added is calculated from the following equation which is adapted from Equation 4-1.

$$Q_{ph} = 1.1 \, cfm_s(T_m - T_{min}) \, Btuh \qquad (4\text{-}18)$$

If T_m is equal to or greater than T_{min}, then $Q_{ph} = 0$.

10. COOLING COIL ENERGY REQUIRED Compare T_m, the temperature of air leaving the preheat coil, with the desired temperature at the fan inlet, T_{ef}. If T_{ef} is greater than T_m, then cooling coil energy $Q_{cc} = 0$. If T_{ef} is less than T_m, then cooling energy is required and is calculated from the following equation.

$$Q_{cc} = 1.1 cfm_s(T_m - T_{ef}) \, Btuh \qquad (4\text{-}19)$$

This is the sensible energy required for cooling.

11. HEATING COIL ENERGY REQUIRED Again, compare the temperature of the air leaving the preheat coil, T_m, with the desired air temperature at the fan entrance, T_{ef}. If T_{ef} is greater than T_m, heating energy, Q_h, is required and is calculated by substituting Q_h for Q_{cc} in Equation 4-19. If T_{ef} is less than T_m, then $Q_h = 0$.

12. ENERGY REQUIRED FOR HUMIDITY CONTROL Compare the humidity ratio of the mixed air, W_m, with the humidity ratio desired in the space, W_s. If W_m is greater than W_s, the excess water vapor must be condensed. The cooling energy required, Q_{ce}, for condensation is calculated by taking a mass and energy balance around the humidification controller.

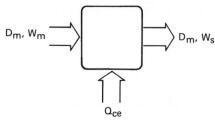

$$\text{Enthalpy in} + Q_{ce} = \text{Enthalpy out}$$
$$\dot{m}_a W_m \times 1076 + Q_{ce} = \dot{m}_a W_s \times 1076$$

Hence

$$Q_{ce} = \dot{m}_a \times (W_s - W_m) \times 1076$$
$$= (.075 \times cfm_s \times 60) \times (W_s - W_m) \times 1076 \qquad (4\text{-}20)$$
$$W_s = 0.01 \text{ to } 0.012 \text{ lb of water vapor per lb of air}$$
$$Q_{ce} = 4842 \times cfm_s (.012 - W_m) \qquad (4\text{-}21)$$

13. RESOURCE REQUIREMENT The energy required for heating or cooling is translated into a resource requirement by factoring in the efficiency of the primary equipment and the coefficient of performance, COP, of the cooling equipment. For example, if gas is used for heating,

$$\text{Gas (therms)} = \frac{\text{Heating energy } (Q_{ph} + Q_h) \text{ for all zones}}{\text{Furnace efficiency} \times 100,000}$$

$$\text{Electricity (kWh)} = \frac{\text{Cooling energy } (Q_{cc} + Q_{ce}) \text{ for all zones}}{3414 \times \text{COP}}$$

$$\text{Auxiliary (kWh)} = \frac{\text{hp (fan + pump + other)} \times \text{operating hours}}{0.746}$$

The furnace or boiler efficiency and the COP of the chillers is a function of the ratio of the actual load to full rated capacity at a given hour. Part-load efficiency is best established from manufacturer's data.

Example 4-1. Single Duct System
The second story of a building has a design cooling load of 0.65×10^6 and a design heating load of -0.45×10^6.

Each floor of the building has an independent air handling unit with a fan capacity of 31,000 cfm and a total fan pressure of 5 in. The supply fan is rated at 40 hp, the return fan at 10 hp.

The fractions of outside air and return air are regulated by a temperature-type economizer cycle control. The minimum setting of the outside air damper is 15%.

Calculate the energy consumed by this single duct system, Figure 4-1.

Solution 4-1
The load and the fan size are specified, so begin by calculating the supply temperature, T_s, in the cooling mode. Assuming a space temperature, T_{sp}, of $75°$ and applying Equation 4-8,

$$T_s = T_{sp} - \frac{Q_c}{(1.1)(cfm_s)}$$
$$= 75 - \frac{0.65 \times 10^6}{(1.1)(31,000)}$$
$$= 75 - 19.1 = 55.9°F$$

T_s is the temperature leaving the fan. The temperature entering the fan, T_{ef}, must be reduced by the heat rise through the fan which is calculated using Equations 4-11 and 4-12.

$$DT_s = \frac{0.401 \times 5}{1.1 \times .8} = 2.28$$

$$T_{ef} = T_s - DT_s = 55.9 - 2.28 = 53.62°$$

The return temperature is set equal to the space temperature since all space loads are included in Q_c.

$$T_r = T_{sp} = 75°$$

With outside air at $91°$ and return air at $75°$, the air mixer will be operating at 15% outside air. To calculate the temperature of the mixed air, T_m, first calculate the density of the two air sources using Equation 4-16.

$$D_{oa} = \frac{P}{RT_{oa}} = \frac{14.7 \times 144}{53.3 \times (460 + 91)} = 0.072 \ lb/ft^3$$

$$D_r = \frac{P}{RT_r} = \frac{14.7 \times 144}{53.3 \times (460 + 75)} = 0.074 \ lb/ft^3$$

Substituting these values in Equation 4-14,

$$T_m = \frac{0.072 \times 0.15 \times 91 + 0.074 \times 0.85 \times 75}{0.072 \times .15 + 0.074 \times 0.85}$$

$$= \frac{0.9828 + 4.7175}{0.0108 + 0.0629} = \frac{5.7}{0.0737}$$

$$= 77.35$$

The sensible heat load can now be calculated from Equation 4-19.

$$Q_c = 1.1 \ cfm \ (T_m - T_{ef})$$
$$= 1.1 \ (31,000) \ (77.35 - 53.62)$$
$$= .809 \times 10^6 \ Btuh$$

The humidity ratio of the outside air, at 91° DB and 73° WB, is .0134. The humidity ratio of the return air is established at its saturated temperature over the coil. At 53.6°, W_r = .0088. The humidity ratio of the mixed air W_m is calculated using Equation 4-15.

$$W_m = \frac{.072 \times 0.15 \times 0.0134 + 0.074 \times .85 \times .0088}{0.0737}$$

$$= .0094 \ lb \ water \ vapor/lb \ dry \ air$$

The latent heat is calculated using a variation of Equation 4-21.

$$Q_{ce} = 4842 \times cfm_s \ (W_m - W_s)$$
$$= 4842 \times 31,000 \times (.0094 - .0088)$$
$$= .090 \times 10^6$$

Adding the sensible and latent load, the total cooling load

$$Q_c = .809 \times 10^6 + .090 \times 10^6 = .899 \times 10^6$$

The total energy input required by an absorption chiller having a COP of 0.7 is

$$E_c = \frac{.899 \times 10^6}{.7} = 1284 \times 10^6 = 1284 \ lb \ of \ steam/hr.$$

The power required to run the fans is

$$kWh = \frac{Fan \ Horsepower \times 0.746}{\eta_m}$$

$$= \frac{(40 + 10) \times 0.746}{0.85}$$

$$= 43.88$$

Heating Energy

Outside air temperature	=	0°F
Set space temperature	=	70°F
Design heating load	=	$-.46 \times 10^6$ Btu
Q_b, 150 ft of baseboard radiation @ 1000 Btu/ft	=	$-.15 \times 10^6$ Btu
Net heating load	=	$-.31 \times 10^6$ Btu

Desired supply temperature $= T_{sp} - \dfrac{Q_h}{1.1 \ cfm_s}$

$$= 70 - \frac{-310,000}{(1.1)(31,000)}$$
$$= 70 + 9.1$$
$$= 79.1°F$$

Since the outside air temperature is less than the desired supply temperature of 79.1°F, the damper will be set for 15% outside air. The return air is at the set space temperature or 70°. The mixed air temperature is given by Equation 4-14.

$$D_{oa} = \frac{14.7 \times 144}{53.3 \times (460 + 0)} = 0.086 \ lb/ft^3$$

$$D_r = \frac{14.7 \times 144}{53.3 \times (460 + 70)} = 0.075 \ lb/ft^3$$

$$T_m = \frac{0.086 \times 0.15 \times (460 + 0) + 0.075 \times 0.85 \times (460 + 70)}{0.086 \times 0.15 + 0.075 \times 0.85}$$

$$= 518.22°R = 58.22°F$$

In calculating the cooling load, the temperature rise across the fan was found to be 2.28°F. Therefore, the temperature of the air leaving the fan is equal to $58.22 + 2.28 = 60.5°F$.

The energy required at the heating coil

$$Q_{hc} = 1.1 \ cfm_s(T_{lf} - T_s)$$
$$= 1.1(31,000)(60.5 - 79.1)$$
$$= -.634 \times 10^6 \ Btuh$$

Total heating energy required $= Q_{hc} + Q_b$
$$= -.634 \times 10^6 - .15 \times 10^6$$
$$= -.784 \times 10^6$$

At 80% overall efficiency, the input energy required

$$= \frac{-.784 \times 10^6}{0.8}$$
$$= -.98 \times 10^6 \ Btuh$$
$$= 980 \ lb \ of \ steam/hr$$

Total Energy Input

Steam for chiller	=	1284 lb/hr
Steam for heating	=	980 lb/hr
Total steam	=	2264 lb/hr
Fan power	=	43.88 kWh

Dual-Duct System

A typical dual-duct, multizone system is shown in Figure 4-4. In comparing it with a single duct system, Figure 4-2, you will notice that the return, mixing and supply system up to the fan are essentially the same in both designs. However, the cooling coil is housed in a separate duct, called the cold deck, while the reheat coils for each space have been consolidated into one heating coil housed in a parallel hot deck.

If the mixed air temperature is less than the desired cold deck temperature, the preheating coil is activated and the mixed air is heated to the desired temperature before entering the fan. After leaving the fan, the air is divided between the hot deck and the cold deck. The hot

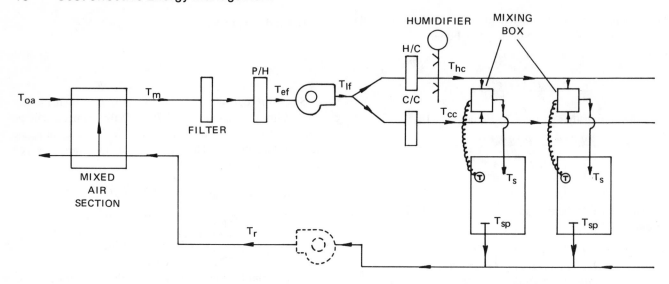

FIG. 4-4. Schematic of a dual-duct, constant volume, multizone system.

deck temperature and humidity are controlled by the hot deck heating coil and humidifier: the temperature of the cold deck by the cooling coil. Space temperature in each of the zones is controlled by mixing boxes that blend air from the hot and cold decks to meet the individual space requirements.

The temperature of the air supplied by the hot deck and the cold deck may be controlled in the following three ways:

(1) The temperature of the hot deck (110° to 130°F) and the cold deck (50 to 55°F) are uniform year-round.

(2) Temperatures of the hot deck and the cold deck are controlled by the space requirements. The hot deck temperature is set equal to the warmest temperature required by any of the several zones. The cold deck temperature is set equal to the coolest temperature required by any zone.

(3) The cold deck temperature remains constant while the hot deck temperature varies with the temperature of the outside air.

In general, system 2 is the most economical.

Example 4-2. Constant Temperature, Dual-duct System.

Using the data generated in Example 4-1, replace the single duct system with a dual duct system having (a) a fixed hot deck temperature of 120°F and a fixed cold deck temperature of 50°F (b) a hot deck temperature equal to the highest temperature required by any of several zones or the temperature leaving the supply fan, whichever is higher, and a cold deck temperature equal to the lowest temperature required by any of several

zones or the temperature leaving the supply fan, whichever is lower.

For the sake of simplicity, assume that all the zones have a proportionate heat loss and heat gain and that the average return air temperature is the same as that of a single duct system.

If one or more zones have a disproportionate heat loss and gain, calculate an energy balance, Equation 4-8, for all zones. Assuming that T_{sp}, the space temperature, is equal to T_r, the return temperature, any deviations from the average will be reflected in the overall value of T_r.

The average return air temperature from all zones is the sum of the mass of air from each zone times the temperature of the zone divided by the mass of air from all zones or

$$T_r = \frac{\Sigma \dot{m}_{ri} T_{ri}}{\Sigma \dot{m}_{ri}}$$

where
    i signifies one of several zones

$\dot{m}_{ri}$ = mass of return air from any zone

T_{ri} = temperature of the return air from any of several zones

Σ = sum of all zones

Solution–Option (a)

From Example 4-1, in the cooling mode

$$Q_c = .65 \times 10^6 \text{ Btuh}$$
$$T_{sp} = T_r = 75°$$
$$T_m = 77.35$$
$$T_{lf} = 77.35 + 2.28 = 79.63$$
$$T_s = 55.9°$$

With a dual duct system, the hot deck air, $T_{hd} = 120°$, is mixed with the cold deck air, $T_{cd} = 50°$, to create 55.9° supply air.

The fraction of cold air, F_c, can be calculated for each zone or, for simplicity's sake, for the entire system.

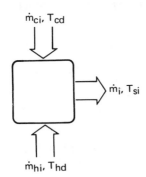

Let

T_{si} = The supply air temperature for one of several zones

$\dot{m}_i$ = Total air supply for one of several zones which is determined by the size of the mixing box

$\dot{m}_{ci}$ = Cold air supplied to one of several zones

$\dot{m}_{hi}$ = Hot air supplied to one of several zones

F_{ci} = Fraction of cold air supplied to one of several zones

$\quad = \dfrac{\dot{m}_{ci}}{\dot{m}_i}$

F_c is derived from the energy balance at the mixing box, under steady state conditions.

$$\dot{m}_{ci} \times C_p \times T_{cd} + \dot{m}_{hi} \times C_p \times T_{hd} = \dot{m}_i \times C_p \times T_{si}$$

Since $\dot{m}_{hi} = \dot{m}_i - \dot{m}_{ci}$

$$\dot{m}_{ci} \times T_{cd} + (\dot{m}_i - \dot{m}_{ci})T_{hd} = \dot{m}_i \times T_{si}$$

$$\dot{m}_{ci} \times (T_{cd} - T_{hd}) = \dot{m}_i(T_{si} - T_{hd})$$

$$F_{ci} = \frac{\dot{m}_{ci}}{\dot{m}_i} = \frac{T_{si} - T_{hd}}{T_{cd} - T_{hd}} \qquad (4\text{-}22)$$

$$\dot{m}_{ci} = F_{ci} \times \dot{m}_i$$

If F_{ci} is less than zero, it is taken as zero. This indicates that the heating load is not being met at the given hour because of an inadequate hot deck temperature or insufficient air supply. As a result, the space temperature will be below the desired space temperature at that hour. In this case,

$$T_{si} = T_{hd} \qquad F_{ci} = 0 \qquad \dot{m}_{hi} = \dot{m}_i$$

If F_{ci} is greater than one, it is taken as one. This indicates that the cooling load is not being met and that the space temperature is greater than the desired space temperature at that hour. In this case,

$$T_{si} = T_{cd} \qquad F_{ci} = 1.0 \qquad \dot{m}_{ci} = \dot{m}_i$$

The total air flow in the cold deck is the sum of the cold air supplied to the various mixing boxes.

$$\dot{m}_c = \Sigma \dot{m}_{ci} \qquad\qquad F_c = \dot{m}_c / \dot{m}_s$$

The total air flow in the hot deck is the sum of the hot air supplied to the various mixing boxes.

$$\dot{m}_h = \Sigma \dot{m}_{hi} = \dot{m}_s - \dot{m}_c$$

Returning to the solution of Example 4-2(a), the fraction of cold air for the entire dual-duct system can be calculated from Equation 4-22.

$$F_c = \frac{T_s - T_{hd}}{T_c - T_{hd}} = \frac{55.9 - 120}{50 - 120} = 0.916$$

This fraction is used to calculate the heating coil energy, using the following equation:

$$\begin{aligned} Q_{hc} &= 1.1 \, cfm_s \times (1 - F_c)(T_{lf} - T_{hd}) \qquad (4\text{-}23) \\ &= 1.1 \times 31,000 \times (1 - .916)(79.6 - 120) \\ &= -115,722 \, Btuh \end{aligned}$$

Similarly, the sensible component of the cooling coil energy is

$$\begin{aligned} Q_{cc} &= 1.1 \times 31000 \times .916 \times (79.6 - 50) \\ Q_{cc} &= 1.1 \, cfm_s \times F_c \times (T_{lf} - T_c) \\ &= 924,574 \, Btuh \end{aligned}$$

The humidity ratio of the air leaving the cold deck is established by the saturation temperature over the coil. At 50°, $W_r = .0076$. From Example 4-1, we know that the portion of outside air is 15% and its $W = .0134$. Therefore

$$\begin{aligned} W_m &= \frac{0.72 \times .15 \times 0.0134 + .074 \times .85 \times .0076}{0.0737} \\ &= .0085 \end{aligned}$$

The latent cooling in a dual-duct system is

$$\begin{aligned} Q_{ce} &= 4842 \times cfm_s \times F_c \times (W_m - W_s) \\ &= 4842 \times 31,000 \times .916 \times (.0085 - .0076) \\ &= 123,744 \, Btuh \end{aligned}$$

The total cooling energy requirement is the sum of the sensible and latent load or

$$Q_{cc} = 924,574 + 123,744 = 1,048,318$$

From Example 4-1, in the heating mode,

$$\begin{aligned} Q_h &= -310,000 \, Btuh \\ T_{sp} = T_r &= 70° \\ T_m &= 58.22° \\ T_{lf} &= 58.22 + 2.28 = 60.5 \\ T_s &= 79.1° \end{aligned}$$

The cold deck fraction, from Equation 4-22,

$$F_c = \frac{79.1 - 120}{50 - 120} = .584$$

Since the mixed air temperature exceeds the cold deck

temperature of 50°, the preheat coil is inactive. The heating coil energy requirement is

$$Q_{hc} = 1.1 \; cfm_s (1 - .584)(60.5 - 120)$$
$$= -844,043 \; Btuh$$

The sensible cooling coil load is

$$Q_c = 1.1 \times 31,000 \times .584 \times (60.5 - 50)$$
$$= 209,101 \; Btuh$$

Under steady state conditions, there is no latent cooling load with the outside air at 0° and its $W = .0000$. After a number of passes over the cooling coil, the cold deck W and the space W will equalize at $W = .0076$. At 70°, this produces a relative humidity of 38% which is well within the current ASHRAE 90-75 recommendation of 20 to 65% RH.

Resource Requirements

Assuming an 80% efficiency for the heating system, the energy required by the heating coil in the cooling mode is

$$Q_{hc} = \frac{115,722}{.8} = 144,653 = 145 \; lb \; steam/hr$$

With a COP of .7, the energy required by the cooling coil in the cooling mode is

$$Q_{cc} = \frac{1,048,318}{.7} = 1,497,597 = 1,498 \; lb \; steam/hr$$

Total energy required in the cooling mode is

$$145 + 1,498 = 1,643 \; lb \; steam/hr$$

The energy required by the heating coil in the heating mode is

$$Q_{hc} = \frac{844,043}{.8} = 1,055,054 = 1,055 \; lb \; steam/hr$$

The energy required by the cooling coil in the heating mode is

$$Q_{cc} = \frac{209,101}{.7} = 298,715 = 299 \; lb \; steam/hr$$

The total energy required in the heating mode, including baseboard radiation, is

$$150/.8 + 1,055 + 299 = 1,542 \; lb \; steam/hr$$

Solution—Option (b)

With this option, the hot deck temperature, during cooling, is equal to the temperature of the air leaving the fan and the cold deck temperature is equal to the cold air supply temperature. From Option (a), during the cooling mode,

$$T_{hd} = T_{lf} = 79.63°$$
$$T_{cd} = T_s = 55.9°$$

Therefore

$$F_c = \frac{T_s - T_{hd}}{T_{cd} - T_{hd}} = \frac{55.9 - 79.63}{55.9 - 79.63} = 1$$

$$= 100\% \; cold \; air$$

Cooling coil energy for the sensible heat load is

$$Q_{cc} = 1.1 \times 31,000 \times 1 \times (79.63 - 55.9)$$
$$= 809,193 \; Btuh$$

With a 55.9° coil, the $W_s = .0096$. Therefore

$$W_m = \frac{.072 \times .15 \times .0134 + .074 \times .85 \times .0096}{.0737}$$
$$= .0102$$
$$Q_{ce} = 4842 \times 31000 \times 1 \times (.0102 - .0096)$$
$$= 90,061$$

The total sensible and latent heat load during the cooling mode is

$$809,193 + 90,061 = 899,254$$

From Option (a), during the heating mode,

$$T_{hd} = T_s = 79.1°$$
$$T_{cd} = T_{lf} = 60.5°$$
$$F_c = \frac{T_s - T_{hd}}{T_{cd} - T_{hd}} = \frac{79.1 - 79.1}{60.5 - 79.1}$$
$$= \frac{0}{-18.6} = 0 \; cold \; air$$

Heating coil energy is

$$Q_{hc} = 1.1 \times 31,000 \times (60.5 - 79.1)$$
$$= -634,260 \; Btuh$$

Resource Requirements

Apply the heating efficiency factor and the cooling COP

$$Q_{cc} = \frac{899,254}{.7} = 1285 \; lb \; steam/hr$$
$$Q_h = Q_{hc} + Q_b$$
$$= \frac{634,260 + 150,000}{.8} = 980 \; lb \; steam/hr$$

A comparison of the steam requirements for the two dual-duct options and a single-duct system is shown in Table 4-1. The comparison shows that the energy required for a constant temperature dual-duct system is far

TABLE 4-1. Steam Required, lb/hr

	Dual duct Fixed Temperature	Dual duct Variable Temperature	Constant Volume Single Duct
Heating Mode— Design Point	1542	980	980
Cooling Mode— Design Point	1643	1285	1284

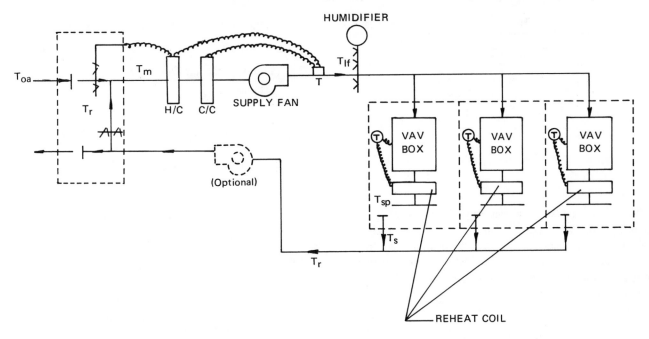

FIG. 4-5. Schematic of a variable air volume system with optional reheat.

in excess of the energy required by the other two systems. However, by resetting the maximum and minimum temperatures, a dual-duct system can exhibit the same economy as a single-duct system while operating at the maximum and minimum design points. At other temperatures, when hot and cold air are blended, the dual-duct system is the most uneconomic of systems.

The comparison can be refined still further by using a computer and hour-by-hour load and weather data.

VARIABLE AIR VOLUME SYSTEM

In single and dual-duct air systems, the space temperature is controlled by varying the supply air temperature as the load changes, as opposed to varying the volume of supply air and using a constant temperature. The variable air volume system has gained popularity in recent years because it has the potential of saving energy when properly selected.

A typical layout for a variable volume fan system with optional reheat is shown in Figure 4-5. It consists of a central fan supplying air (at a temperature determined by the user) to the variable air volume (VAV) boxes of each zone. The VAV boxes vary the amount of air given to each space (in response to their respective room thermostats) in order to achieve the desired temperature control. When the space requires peak cooling, the VAV boxes allow maximum air flow. As the space cooling load decreases, the air flow is reduced proportionally until minimum air flow (set by the user and fan stability) is

reached. If less cooling is required than that given at minimum air flow rate, either the supply air temperature is increased (which is more desirable from the point of energy conservation) or the reheat coil is activated. The other components, such as the mixed air section, humidifier, and heating and cooling coils, operate in a fashion similar to those of single and dual-duct fan systems. Some of the zones, in particular the perimeter zones, may have baseboard radiation to supplement the VAV system.

VAV Energy Requirements

With a VAV system, the temperature leaving the supply fan, T_{lf}, is controlled in one of two ways:

1. Constant T_{lf}, whereby T_{lf} is set equal to the temperature leaving the chiller plus DT_s.
2. Variable T_{lf}, whereby T_{lf} is adjusted to equal the temperature of the coldest air required by any zone.

The air flow rate is determined by first identifying the minimum supply air temperature, T_{sm}, for each zone, which is set equal to T_{lf}. For instance, here is the energy balance for the ith zone:

$$1.1 \; cfm_i \, (T_{spi} - T_{si}) = Q_{si}$$

If no reheat is required, $T_{si} = T_{lf}$ and

$$1.1 \; cfm_i \, (T_{spi} - T_{lf}) = Q_{si}$$

The quantity of air then, is

$$cfm_i = \frac{Q_{si}}{1.1 \, (T_{spi} - T_{lf})} \qquad (4\text{-}24)$$

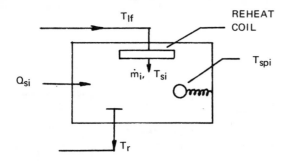

This calculated air flow rate must be compared with the minimum and maximum air flow rates for a given zone. If cfm_i is greater or less than the design cfm_d, then reset the cfm_i equal to cfm_d and adjust the desired supply air temperature to meet the load. The modified supply air temperature becomes

$$T_{si} = T_{spi} - \frac{Q_{si}}{1.1 \times cfm_i}$$

The total air flow rate, cfm_t, is calculated by summing the air flow rates of all zones. The fraction of the air flow rate, compared to the design or rated air flow rate, is

$$F_r = \frac{cfm_t}{cfm_d} \qquad (4\text{-}25)$$

Reheat energy is required for zones where the desired air temperature, T_{si}, (at the minimum air flow setting) is greater than T_{lf}. The reheat energy required equals

$$Q_{rh} = 1.1 \times \Sigma cfm_i \times (T_{lf} - T_{si})$$

Example 4-3. VAV System

If the dual duct system in Example 4-2 is replaced by a VAV System, calculate the input energy (steam and electric power) needed at the design heating and cooling points. The minimum air flow setting for the fan is 30% of the rated flow rate. The minimum air temperature at the cooling coil is 50°F. The temperature of the air leaving the supply fan is set equal to the coldest air temperature required by any zone.

Solution 4-3
From earlier examples

$$Q_{si} = .65 \times 10^6 \text{ Btuh}$$
$$max\ cfm = 31,000$$
$$@31,000, DT_s = 2.28°$$

For the VAV System

$$T_{cc} = 50°$$
$$W_r = .0076$$

min outside air = $31,000 \times .15 = 4650$ cfm

When $DT_s = 2.28°$, $T_{lf} = 52.58°$, and

$$cfm = \frac{.65 \times 10^6}{1.1 \times (75 - 52.28)} = 26,008 \text{ cfm}$$

The fraction of the air flow rate

$$F_f = \frac{26,008}{31,000} = .839$$

When $F_f = .839$, DT_s becomes $2.28 \times .839 = 1.91°$. Through the process of iteration, the final value of $F_f = .824$ and DT_s becomes $2.28 \times .824 = 1.88°$. Thus,

$$cfm = \frac{.65 \times 10^6}{1.1 \times (75 - 51.88)} = 25,558 \text{ cfm}$$

$$T_m = \frac{.072 \times 4650 \times 91 + .074 \times 75 \times (25558 - 4650)}{.072 \times 4650 + .074 \times (25558 - 4650)}$$
$$= 77.846°$$

$$W_m = \frac{.072 \times .0134 \times 4650 + .074 \times .0076 \times (25558 - 4650)}{.072 \times 4650 + .074 \times (25558 - 4650)}$$
$$= .00863$$

$$Q_{cc} = 1.1 \times 25558 \times (77.846 - 50) = 782,857 \text{ Btuh}$$
$$Q_{ce} = 4842 \times 25558 \times (.00863 - .0076)$$
$$= 127,465 \text{ Btuh}$$

Power required to run the fans, with an F_f of .824, is

$$kWh = \frac{.824 \times 50 \times 0.746}{0.85} = 36.16$$

Total energy consumed

$$Q_{cc} + Q_{ce} = 782,857 + 127,464 = 910,321 \text{ Btuh}$$

Steam required, at .7 COP = 1,300 lbs.

Steam required for heating is the same as a single duct system.

Table 4-2 shows that the VAV system has the same steam consumption in the heating mode and slightly more in the cooling mode than the most economical constant volume system. However, the Table does not reflect the savings in fan energy due to VAV operation. This reduction in fan energy is the primary benefit from a VAV system and permits a trade-off of low cost coil energy for high cost fan energy.

TABLE 4-2. Steam required, lb/hr

	Constant Volume	Dual Duct Fixed Temp.	Dual Duct Variable Temp.	Variable Volume
Heating Design Point	980	1542	980	980
Cooling Design Point	1284	1643	1285	1300

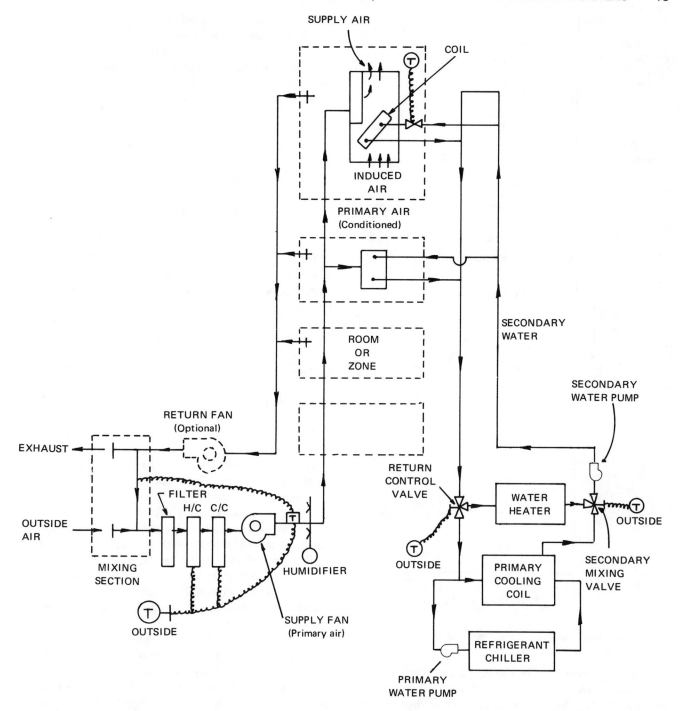

FIG. 4-6. Schematic of a two-pipe induction unit system.

AIR/WATER SYSTEMS

Air/water systems use both air and water to heat and cool a space. These systems are primarily used in the perimeter rooms of multistory, multiroom buildings such as hotels, office buildings, hospitals and apartments. Some of the advantages of air/water systems are:

Compactness   Because the density and specific heat of water are greater than those of air, distribution pipes are considerably smaller. The use of water also reduces the air quantity supplied to each space when compared to that supplied by all-air systems. Thus, less space is required for air/water systems.

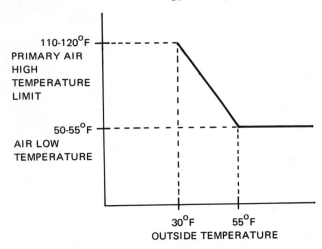

FIG. 4-7. Primary air temperature range.

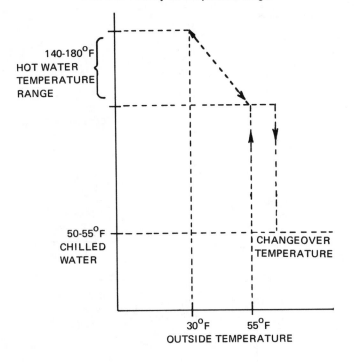

FIG. 4-8. Induction coil water temperature range.

Less horsepower   The pumping horsepower needed to circulate water throughout a building is significantly less than the fan horsepower required for air distribution with all-air systems.

Versatility   Each room is treated as a zone with air/water systems. This enables the system to simultaneously heat or cool adjacent rooms by taking advantage of the dual distribution system.

The dual air/water system retains the major performance capabilities of more versatile air systems, including positive ventilation, central dehumidification, winter humidification, and good temperature control over widely fluctuating conditions.

Quiet operation   The central location of the filtration, air handling, and humidification units of an induction-type system leads to quiet operation and minimal odor and corrosion problems.

System Layout

Air/water systems are classified according to the number of pipes used for water distribution. There are

- Two-pipe systems
- Three-pipe systems
- Four-pipe systems

These systems are subdivided into induction and non-induction systems. Non-induction systems, using fan coil units, are associated with all-water systems and will be discussed later.

Two-Pipe Induction Unit System

Figure 4-6 shows a typical layout for a two-pipe induction unit system with its three main subsystems:

1. Primary air supply system
2. Water supply system
3. Induction unit box

The primary air supply system is similar to all-air systems. It consists of a section for mixing outside air and return air, a supply fan or air handler with heating and cooling coils, a humidifier, and an optional return fan. The supply fan delivers the primary, conditioned air to the induction unit located in each zone. This primary air is discharged through nozzles and induces room (secondary) air across a coil which is supplied with hot or chilled (secondary) water from the water supply system. The temperature of the primary air being supplied to the induction unit is generally controlled as a function of outside air temperature. A typical temperature setting is shown in Figure 4-7.

The temperature of the secondary water being supplied to the induction coils is also controlled as a function of outside air. A typical temperature setting for secondary water is shown in Figure 4-8. This water temperature is maintained by using a refrigeration chiller, primary water cooler, primary water pump, water heater, secondary water mixing valve, and secondary water pump.

The primary chilled water is mixed with hot water at the secondary water mixing valve to maintain its temperature at the desired point. The return control valve returns a quantity of water to the primary system equal to that taken in through the mixing valve. The quantity of

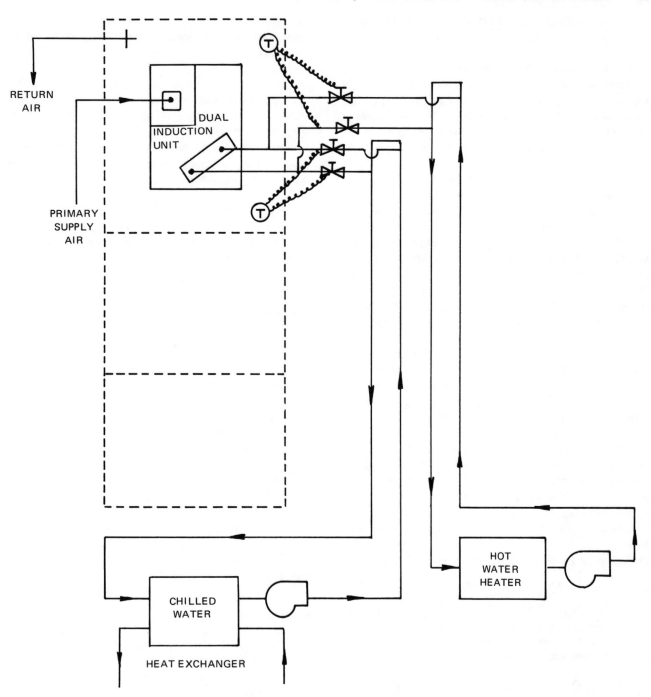

FIG. 4-9. Schematic of a four-pipe, induction unit system.

secondary water is constant or variable depending upon the control system. The *ASHRAE Handbook, 1980 Systems* should be referred to for details of design and control operation.

Four-pipe Induction Unit System

The primary air supply for a four-pipe induction system is similar to that for the two-pipe induction system. On the water side, there are two pipes: two for the supply and two for the return of hot and cold water. The induction units have dual coils, making the supply of hot and cold water available as needed by different zones.

The maximum and minimum allowable temperature leaving the induction coils are independent of the outside temperature. A typical layout for the water side is shown in Figure 4-9.

Three-pipe Induction Unit System

These systems use three pipes: one to supply hot water, one to supply cold water, and a common pipe for the return of water from all the coils. The cold water is at a constant temperature, while the temperature of the hot water is reset as a function of outdoor temperature.

Induction System Energy Calculations

The procedures used in establishing the energy rquirements of an induction system are similar to those used for other air systems.

1. LOAD CALCULATION   Calculate both the hourly and design heating and cooling loads using the procedures outlined in Chapter 3.

2. EQUIPMENT SIZES   Establish the design air and water flow rates for fans and water pumps, the chiller and heater sizes, the ratio of induced air to primary air, etc., to satisfy the heating and cooling loads.

3. TEMPERATURE SCHEDULES   Identify the primary air temperature and secondary water temperature schedules, similar to the ones shown in Figures 4-7 and 4-8. This helps determine the temperature of the primary air as it leaves the supply fan.

4. PRIMARY AIR TEMPERATURES   Calculate the return air temperature and the mixed air temperature, using one of the three mixing options discussed under single duct energy requirements. The economizer control will tend to adjust the fraction of outside air above the minimum outside air requirement to approach T_{ef}.

5. ENERGY REQUIREMENTS   If T_{ef} is greater or less than T_m, energy will be supplied, by either the heating or cooling coil, to satisfy the primary load, Q_p.

$$Q_p = 1.1 \ cfm_s \ (T_m - T_{ef})$$

6. INDUCTION COIL LOAD   The induction coil sensible load in a given space or zone is obtained by subtracting the energy contribution of the primary air from the zone sensible load or

$$Q_{iu} = Q_{si} - Q_{pi} = Q_{si} - 1.1 \ cfm_{si}(T_{spi} - T_{lf})$$

where T_{spi} = set point temperature of the ith zone
The induced air flow rate is given by

$$\dot{m}_i = \dot{m}_p \times RI$$

where $\dot{m}_p$ = primary air flow rate
     RI = ratio of induced air flow rate to the primary air flow rate

7. INDUCTION COIL ENERGY   The energy to be supplied by the induction coil is calculated by taking an

energy balance for the zone, as with all air systems. First, establish the desired supply temperature of the secondary air

$$T_{si} = T_{spi} - \frac{Q_{iu}}{1.1 \times cfm}$$

If T_{si} is greater than the maximum secondary coil temperature, Figure 4-8, T_{si} is set equal to this maximum temperature. Similarly, if T_{si} is less than the minimum air temperature supplied by the secondary coil, T_{si} is set equal to this minimum temperature. The energy to be supplied by the induction coil is

$$Q_i = 1.1 \times cfm \ (T_{spi} - T_{si})$$

A positive sign will indicate a cooling energy requirement and a negative sign will indicate a heating energy requirement.

Example 4-4. Induction Coil System

Assume that the building described in Example 4-1 is equipped with a two-pipe induction coil system, Figure 4-6. Calculate the energy required at the heating and cooling design points, based on the following assumptions.

Primary air supply fan	= 10,000 cfm, 15 hp
Supply fan pressure	= 4 inch water column
Primary air supply temperature	= 53°F
Induced air maximum temperature	= 120°F
Induced air minimum temperature	= 60°F
Ratio of primary air to induced air flow	= 0.4
Minimum outside air flow rate	= 4650 cfm
Maximum secondary water temperature	= 130°F
Minimum secondary water temperature	= 50°F

Solution 4-4

From Example 4-1,
$$Q_c = 0.65 \times 10^6 \ \text{Btuh}$$
$$T_r = 75°$$
$$T_{oa} = 91°$$
$$D_{oa} = 0.072$$
$$D_r = 0.074$$

First, calculate the temperature rise across the fan which is

$$= \frac{0.401 \times 4}{1.1 \times 0.8} = 1.82°$$

Since the primary air supply temperature $= T_{lf} = 53°$

$$T_{ef} = 53° - 1.82 = 51.18°$$

With a minimum air flow rate of 4650 cfm,

$$F_{oa} = \frac{4650}{10,000} = .465 \text{ and } F_r = .535$$

Therefore

$$T_m = \frac{.072 \times .465 \times 91 + .074 \times .535 \times 75}{.072 \times .465 + .074 \times .535} = 82.33°$$

The energy supplied by the primary coil

$$\begin{aligned}
Q_{cc} &= 1.1 \text{ cfm } (T_m - T_{ef}) \\
&= 1.1 \times 10,000 \ (82.33 - 51.18) \\
&= 342,650 \text{ Btuh}
\end{aligned}$$

The induced air load

$$\begin{aligned}
Q_{iu} &= Q_{si} - 1.1 \text{ cfm}_s \ (T_{sp} - T_{lf}) \\
&= 0.65 \times 10^6 - 1.1 \times 10,000 \times (75 - 53) \\
&= .408 \times 10^6 \text{ Btuh}
\end{aligned}$$

The induced air flow $= \dfrac{10,000}{.4} = 25,000$ cfm

The supply temperature of the induced air

$$\begin{aligned}
T_{si} &= T_{sp} - \frac{Q_{iu}}{1.1 \text{ cfm}_i} \\
&= 75 - \frac{.408 \times 10^6}{1.1 \times 25,000} \\
&= 75 - 14.84 \\
&= 60.16°
\end{aligned}$$

The humidity ratio of the air leaving the primary coil is established by the saturation temperature of the coil. At 51.2°, $W_r = .0081$. From Example 4-1, we know that $W_{oa} = .0134$. Therefore,

$$\begin{aligned}
W_m &= \frac{.072 \times .465 \times .0134 + .074 \times .535 \times .0081}{.072 \times .465 + .074 \times .535} \\
&= .0105
\end{aligned}$$

Since the temperature of the induction coil, at 60.2°, is higher than that of the primary coil, at 51.2°, the entire latent heat load is at the primary coil.

$$\begin{aligned}
Q_{ce} &= 4842 \times \text{cfm}_s \times (W_m - W_s) \\
&= 4842 \times 10,000 \times (.0105 - .0081) \\
&= 116,208
\end{aligned}$$

Adding the sensible load at both the primary and induction coils and the latent load, the steam requirement at a .7 COP

$$= \frac{(.343 + .408 + .116) \times 10^6}{.7 \times 1,000}$$
$$= 1239 \text{ lb steam/hr}$$

The fan horsepower is considerably less than the previous systems.

$$kWh = \frac{15 \times 0.746}{0.85} = 13.16$$

From Example 4-1, the net heating load $= .31 \times 10^6$ Btuh

$$\begin{aligned}
T_{oa} &= 0° \\
T_{sp} &= 70° \\
D_{oa} &= .086 \\
D_r &= .075
\end{aligned}$$

Therefore

$$\begin{aligned}
T_m &= \frac{.086 \times .465 \times (460 + 0) + .075 \times .535 \times (460 + 70)}{.086 \times .465 + .075 \times .535} \\
&= 495.06 - 460 = 35.06°
\end{aligned}$$

The energy supplied by the primary heating coil

$$\begin{aligned}
Q_{hp} &= 1.1 \times 10,000 \ (T_m - T_{ef}) \\
&= 1.1 \times 10,000 \times (35.06 - 51.18) \\
&= -177,320 \text{ Btuh}
\end{aligned}$$

The induced air load

$$\begin{aligned}
&= -.31 \times 10^6 - 1.1 \times 10,000 \ (70 - 53) \\
&= -.497 \times 10^6
\end{aligned}$$

The supply temperature of the induced air

$$\begin{aligned}
T_{si} &= 70 - \frac{-.497 \times 10^6}{1.1 \times 25,000} \\
&= 70 + 18.07 = 88.07°
\end{aligned}$$

Including the baseboard load, the steam required at 80% efficiency,

$$= \frac{(.15 + .177 + .497) \times 10^6}{.8 \times 1,000}$$
$$= 1,030 \text{ lb steam/hr}$$

The input steam requirements for a two-pipe induction system are summarized in Table 4-3.

ALL-WATER SYSTEMS

All-water systems use fan-coil or ventilator-type room terminal units. Heating is provided by supplying hot water through the finned coil and cooling by circulating chilled water through the coil. Two, three or four-pipe water distribution systems, comparable to those employed with induction systems, may be used.

The all-water systems are particularly suitable to multiroom buildings, where space for ductwork is limited, or where relatively low temperature hot water is supplied by a central heat pump system using trhe interior zone loads as a heat source. The all-water systems are not re-

TABLE 4-3. Steam Required, lb/hr

	Constant Volume	Dual Duct Fixed Temp.	Dual Duct Variable Temp.	Variable Volume	Two-pipe Induction
Heating Design Point	980	1542	980	980	1030
Cooling Design Point	1284	1643	1285	1300	1239

commended where there are high latent heat loads.

Typical layouts for two and four-pipe fan coil systems are shown in Figures 4-10 and 4-11. A two-pipe fan-coil system requires a changeover valve whose temperature is set as a function of the outside air temperature. The four-pipe fan-coil system simultaneously supplies hot or cold water as required by various zones at a given time. The energy requirement calculations are similar to the secondary air energy requirement calculations employed with two and four-pipe induction systems.

UNITARY SYSTEMS

Window air conditioners, through-the-wall conditioners, rooftop systems, water-loop heat pumps, and absorption systems are classified as unitary systems. They are used for conditioning zones having a cooling load 0.5 to 25 tons.

Multiple unit systems, employing one unit for each zone, have the following advantages over central system alternatives:

1. Individual room control is provided simply and in-

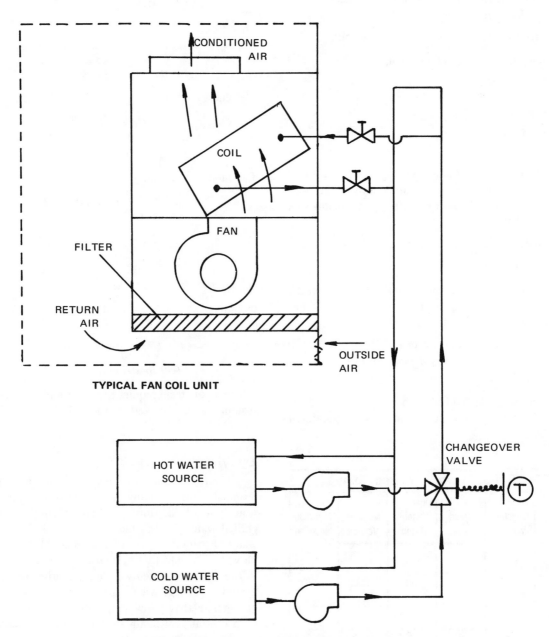

FIG. 4-10. Schematic of a two-pipe, fan coil system.

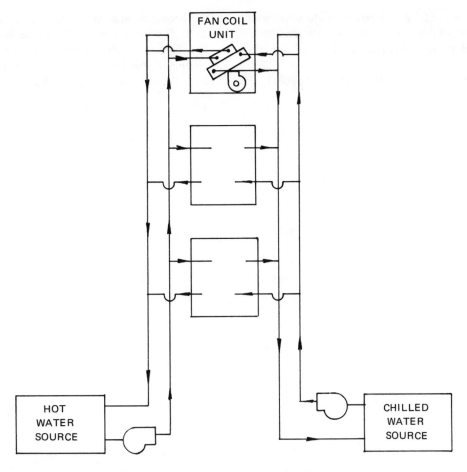

FIG. 4-11. Schematic of a four-pipe, fan coil system.

expensively.

2. Individual air distribution systems are provided for each room.
3. Heating and cooling capabilities are available at all times.
4. Manufacturer-matched components assure consistent performance, improved reliability and quicker installation.

Unitary systems do have some limitations.

1. Relatively few options are available with respect to coil and condenser sizes, blower and control choices.
2. Unitary sustems are inadequate where there are exceptioanlly high or low latent heat loads.

Unitary System Energy Calculations

The procedure for calculating the energy requirements of unitary systems is simpler than for other systems.

1. LOADS   Establish the sensible thermal load, including infiltration and ventilation loads, for each zone,

Q_{si}, at a given hour. The fraction of time (F_r) that the system will be on at a given hour is

$$F_r = \frac{Q_{si}}{Q_{cap}}$$

where Q_{cap} is the cooling or heating capacity of the system. If F_r is greater than 1, the load is not met and the room temperature will be more or less than the set point temperature.

2. RESOURCE REQUIREMENT   The input enegy for heating or cooling is estimated from the manufacturer's data.

$$Input\ energy = \frac{Rated\ full\ load\ input \times F_r}{Part\ load\ efficiency}$$

3. AUXILIARY ENERGY   The electrical input for the auxiliary equipment is

$$Electrical\ input\ (kW) = 0.759 \times rated\ hp \times F_r$$

CONCLUSION

The selection of a particular type of system should be based on the annual energy consumption. To calculate this precisely requires a computer solution in which hour-by-hour loads, weather data, system energy requirements and primary equipment efficiency are taken into account.

The energy requirement analysis is then coupled with the economic analysis to determine the annual operating cost. This helps one to choose the most energy-efficient and cost-effective system. The equations required for computer simulations of various types of systems have been formulated in this chapter.

Chapter 5

ENERGY EFFICIENT LIGHTING SYSTEMS

Lighting is the most visible form of energy use and requires about 5% of the total national energy output. However, it consumes roughly 23% of the electricity generated in the U.S.

It is simplistic to conclude that energy can be saved by turning off lights. This arbitrary practice confounds the lighting designer's efforts to economically produce an esthetic environment that aids in visual comfort, safety and security so that people can perform their jobs efficiently and accurately. Energy used by lighting can be better saved by energy efficient lighting designs and a good maintenance program.

The procedure for designing efficient lighting systems can be reduced to three basic phases.[5-1]

Phase I.   Investigation, analysis and selection of the design criteria for a luminous environment.

Phase II. Selection of alternative systems that satisfy the basic design criteria.

Phase III. Comparative evaluation of alternative systems based on life-cycle cost analysis.

I. SELECTION OF DESIGN CRITERIA

Design criteria are based on a high level of performance by persons undertaking a task. They include specific quantities of light and the psychological reaction to a lighting system. Design criteria are dependent upon the following variables.

a. Function of a space. A hospital operating room functions differently than a hospital waiting room.

b. User of a space. An older eye requires more luminance than the younger eye.

c. Task definition. A task is the combination of both visual and nonvisual components. For example, reading can be considered a 100% visual task, while waiting for an elevator might be considered a nonvisual task.

d. How much light or visibility.

Probably the most difficult step in the selection of design criteria is to quantify *how much light* is necessary to perform a task. Theoretically, for a given task and user, there is an optimum light level, at which the user can perform the task with maximum speed and accuracy. Until recent years, the footcandle (fc) has been defined as the metric for visibility with performance criteria being based on a quantity of footcandles. For example: let's say that, in order to read a No. 2 pencil on white paper, there must be 70 footcandles on the paper. The 70 footcandles, then, defines the criterion in terms of how much light is needed to perform the task.

There are several ways to predetermine the footcandle levels of a particular design. The one most commonly used is the Zonal Cavity method. However, it gives only a one-number average for the space, which limits the designer's information about visibility in that space. A more accurate method is a point-by-point calculation to determine the footcandle level at various points within the space. The more information the designer has about the available visibility, the easier it is to optimize the lighting design in terms of performance criteria.

Recent experiments have shown that, for many tasks, the footcandle is not a good metric for visibility. Therefore, criteria based only on footcandles will not give an accurate indication of a user's ability to perform a task. Experiments performed by Dr. R. H. Blackwell and others have shown that the visibility of a task depends on:

a. The physical characteristics of the task itself, including the task contrast and how the task reflects light.

b. How the light reaches the task; that is, from what angles and direction.

The metric for visibility that evolved from Blackwell's experiments is Equivalent Sphere Illumination (ESI). The definition of ESI is: the footcandle level, in a photometric sphere, that renders a task equally visible in the sphere as in the real lighting environment. ESI, then, is a metric for visibility. However, we are still left with the problem *how much ESI* to perform a task. It can be concluded that a quantity of 50 ESI on a task produces more visibility than quantity of 30 ESI on the same task. This relationship says only that, on a relative basis, one system produces more visibility than another. However, the fundamental question still exists and that is, "How much performance?" Lighting designers are not really concerned with the actual numeric quantity of light, but rather, how well users can perform their tasks. Further experiments are beginning to provide the answers to *how much performance* for a given quantity of ESI. Dr. Blackwell has developed performance curves like those shown in Figure 5-1.

Figure 5-1 shows three curves, each representing a different task or family of tasks. Curve 1 is a family of easy tasks, Curve 2 is a family of medium tasks, and Curve 3 is a family of difficult tasks. The curves relate Relative Visual Performance to Visibility Level (VL). ESI can easily be calculated from [5-2] Visibility Level (VL), so that VL actually represents a quantity of light. The Relative Visual Performance axis rates, from 0 to 100 percent, the visual accuracy in performing a task.

For example, if the criterion for a space is a user

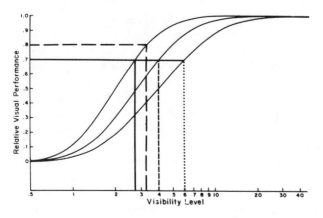

FIG. 5-1. Visual performance curves.[5-2]

performance of 70 percent, Figure 5-1 shows that a **VL of 2.7 is needed for Task 1, a VL of 4.0 is needed for Task 2, and a VL of 6.0 is needed for Task 3. If task 1 performance is to be 80% instead of 70%, then VL must be 3.1 rather than 2.7. Since VL corresponds to a quantity of light, it can be seen that, to increase performance, more light is needed. If the task becomes more difficult, while maintaining the same performance level, again, more light is needed.**

c. **Psychological Aspects of the Space**

The psychological aspects of lighting are the least understood at this time. Visual Comfort Probability (VCP) is an attempt to establish some index of comfort or noncomfort produced by a lighting system. Since VCP is the percentage of people who will not be uncomfortable in a particular luminous environment, VCP is used as a psychological design criteria.

II. SELECTION OF ALTERNATE SYSTEMS

Phase II consists of three basic steps:

1. Listing the various lighting concepts.
2. Selecting the most feasible lighting concepts.
3. Designing and developing the selected systems.

The first step, and **perhaps the most important, is the** listing of various radically different lighting concepts without regard to their feasibility.

The second step is an initial review of the alternate lighting concepts and the elimination of those that obviously cannot work. This evaluation should eliminate only those systems that are in obvious conflict with known physical or cost constraints, there being no possibility of circumventing these constraints.

Now, the designer can begin to develop actual lighting systems from the surviving concepts. The primary objective, at this point, is to develop alternate systems that provide basically the same lighting performance in order to simplify the cost/benefit analysis in Phase III of the design process.

To properly evaluate the benefits of a particular lighting design, the lighting performance must be predetermined to a high degree of accuracy. Until recently, the only method of predetermining performance was the Zonal Cavity method. Because more accurate and extensive informaiton is needed for the design of energy-efficient lighting systems, this method is inappropriate and is not recommended for energy-conserving illumination systems. With the advent of the digital computer and the availability of sophisticated programs, it is possible to accurately simulate the performance of a particular design on a point-by-point basis. The designer can now design a *first-guess* system and, through an iterative process using the computer, eventually refine the system until it satisfies the basic lighting criteria. By repeating this process with each alternative, a variety of lighting systems can be designed . . . all having approximately equal lighting performance. This computer-aided design capability is not restricted to large firms that have their own in-house computer systems. These computer programs can also be accessed and used by smaller firms through one of several national time-sharing services.

III. EVALUATION OF ALTERNATIVES

The final phase of the design procedure is a comparative evaluation of the alternative lighting systems and the selection of the optimum system. The primary method of evaluation is a cost/benefit analysis. Since the alternate lighting systems were designed to provide approximately equal lighting performance, or equal *benefits*, the cost/benefit analysis becomes a basic cost analysis. To properly evaluate the impact of energy-related costs, the life-cycle costing procedure must be used, as described in Chapter 7. The lighting system that delivers the required performance, at the lowest life-cycle cost, becomes the recommended system.

ANALYSIS OF LIGHTING SYSTEMS

As already noted, two methods are most commonly used to analyze lighting systems:

1. Zonal Cavity Method or Lumen Method
2. Point-by-Point Method

A familiarity with light loss factors is a prerequisite to using either method.

Light Loss Factors

Light loss factors are an important consideration in the analysis of lighting systems. Efficiency of the system at some future time can be calculated if at least eight light loss factors can be accounted for. Some of the light loss factors that are not recoverable are extremes in ambient temperature, ballast factor, voltage variations and luminaire surface depreciation. The light loss factors that are recoverable are:

Room Surface Dirt Depreciation Accumulation of dirt on the room surfaces reduces the amount of luminous flux reflected and interreflected to the work plane. To take this into account, Room Surface Dirt Depreciation (RSDD) is included in the analysis of lighting systems. RSDD can be calculated by using Figure 9-5 of the *IES Lighting Handbook, Fifth Edition.*[5-3]

Lamp Lumen Depreciation Data on lumen depreciation of lamps, due to aging, can generally be obtained from the manufacturer. Lamp Lumen Depreciation

(LLD) should be minimal if the greatest possible lighting is to be maintained.

Luminaire Dirt Depreciation The accumulation of dirt on luminaires reduces the light output and hence the light available on the work plane. This loss of light is taken into account by the Luminaire Dirt Depreciation (LDD) factor. The method for determining LDD is given in the *IES Lighting Handbook.* Figures 5-2 through 5-5 show ranges for LLD and LDD factors for fluorescent, mercury vapor, metal halide and high pressure sodium lamps respectively.[5-4]

Burnouts Lamp manufacturers furnish statistical data on the mortality rates of their lamps. Since lamp burnouts contribute to the loss of light, a good lamp service program should be incorporated into the design of the system. Figures 5-2 through 5-5 show ranges for these burnout factors.

Light loss is the product of all eight factors. Of the eight, LDD and LLD are the main components of the light loss factor.

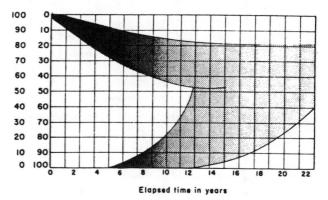

Elapsed time in years

FIG. 5-2. LLD, LDD and burnout factors (BF) for fluorescent lamps. [5-4]

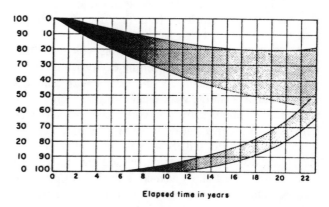

Elapsed time in years

FIG. 5-3. LLD, LDD and burnout factors (BF) for mercury vapor lamps. [5-4]

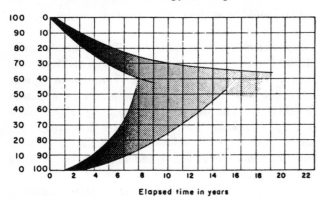

FIG. 5-4. LLD, LDD and burnout factors (BF) for metal halide lamps. [5-4]

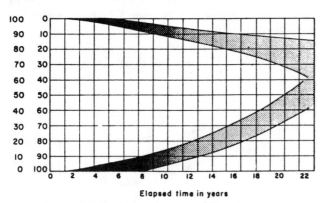

FIG. 5-5. LLD, LDD and burnout factors (BF) for high pressure sodium lamps. [5-4]

Lumen and Zonal Cavity Methods

The Lumen method is based on the definition

$$\text{illumination} = \frac{\text{luminous flux}}{\text{area}}$$

By knowing the initial lumen output of each lamp, the number of lamps installed in the area, and the square feet of the area, the lumens per square foot is derived. However, this value differs from the footcandles in the area because of light-loss factors.

The zonal cavity method divides the volume of the room into three *cavities*: the ceiling cavity which is the volume above the luminaire, the floor cavity which is the volume beneath the work plane, and the room cavity which is the volume above the work plane and below the luminaire. A Coefficient of Utilization is based on the Room Cavity Ratio and incorporates the reflectance of the ceiling cavity, the floor cavity and the walls. This refinement of the Lumen method is further corrected by incorporating both lamp lumen depreciation (LLD) and luminaire dirt depreciation (LDD) factors. Thus

$$\text{footcandles} = \frac{LL \times CU \times LLD \times LDD}{\text{Area}}$$

where
$\quad$ LL = lamp lumens = number of lamps × lumens per lamp
$\quad$ CU = coefficient of utilization derived from the zonal cavity procedure
$\quad$ LLD = lamp lumen depreciation factor
$\quad$ LDD = Luminaire dirt depreciation factor

Neither of these methods is recommended for the analysis of lighting systems as they give only an average value of all the footcandles in the space which is not comprehensive enough for proper energy conservation.

Point-by-point Method

When the candlepower distribution curve of the luminaire is available and the distance from the light source is at least five times the maximum dimension of the source, the following formulas may be used to calculate the point-by-point illumination on horizontal and vertical surfaces. [5-3]

For horizontal surfaces, Figure 5-6,

$$\text{footcandles} = \frac{\text{candle power} \times \cos \Theta}{\text{distance squared}} = \frac{cd \times \cos \Theta}{D^2}$$

Since $\cos \Theta = \frac{H}{D}$ $\qquad$ $fc = \frac{cd \times H}{D^3} = \frac{cd \times \cos^3 \Theta}{H^2}$

For vertical surfaces, Figure 5-7,

$$fc = \frac{cd \times \sin \Theta}{D^2}$$

Since $\sin \Theta = \frac{R}{D}$ $\qquad$ $fc = \frac{cd \times R}{D^3} = \frac{cd \times \cos^2 \Theta \times \sin \Theta}{H^2}$

Table 5-1 is used to facilitate the calculation of horizontal footcandle intensities. To use the table:

1. Locate the horizontal distance, R, along the upper margin of the table.
2. Locate the height, H, of the luminaire along the left margin of the table.
3. Where these two columns intersect, you will find the angle Θ and the percentage of light available at the given distance.
4. From the distribution table for the light source, find the candlepower available when the angle equals Θ.

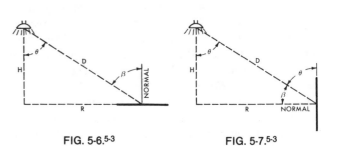

FIG. 5-6. [5-3] $\qquad\qquad$ FIG. 5-7. [5-3]

TABLE 5-1[5-5]

Horizontal Distance From Axis of Light Source—Feet																
Height	**0**	**1**	**2**	**3**	**4**	**5**	**6**	**7**	**8**	**9**	**10**	**12**	**15**	**20**	**26**	**32**
	Footcandles for Each 100 Candlepower															
4	0°0' 6.250	14° 5.707	27° 4.472	37° 3.200	45° 2.210	51° 1.524	56° 1.066	60° .764	63° .559	66° .419	68° .320	72° .198	75° .107	79° .047	81° .022	83° .012
5	0°0' 4.000	11° 3.771	22° 3.202	31° 2.522	39° 1.904	45° 1.411	50° 1.050	54° .785	58° .595	61° .458	63° .358	67° .228	72° .126	76° .057	79° .027	81° .015
6	0°0' 2.778	9° 2.673	18° 2.372	27° 1.987	34° 1.600	40° 1.260	45° .982	49° .766	53° .600	56° .474	59° .378	63° .249	68° .142	73° .066	77° .032	79° .017
7	0°0' 2.041	8° 1.980	16° 1.814	23° 1.585	30° 1.336	36° 1.100	41° .893	45° .722	49° .583	52° .473	55° .385	60° .261	65° .154	71° .074	75° .036	78° .020
8	0°0' 1.563	7° 1.527	14° 1.427	21° 1.283	27° 1.118	32° .953	37° .800	41° .666	45° .552	48° .458	51° .381	56° .267	62° .163	68° .080	73° .040	76° .022
9	0°0' 1.235	6° 1.212	13° 1.148	18° 1.054	24° .943	29° .825	34° .711	38° .607	42° .515	45° .437	48° .370	53° .267	59° .168	66° .085	71° .043	74° .025
10	0°0' 1.000	5°43' .985	11° .943	17° .879	22° .801	27° .716	31° .631	35° .550	39° .476	42° .411	45° .354	50° .263	56° .171	63° .089	69° .046	73° .027
11	0°0' .826	5°12' .816	10° .787	15° .742	20° .686	24° .623	29° .559	32° .496	36° .437	39° .383	42° .335	48° .255	54° .171	61° .092	67° .049	71° .028
12	0°0' .694	4°46' .687	9° .668	14° .634	18° .593	23° .546	27° .497	30° .448	34° .400	37° .356	40° .315	45° .246	51° .169	59° .094	65° .051	69° .030
13	0°0' .592	4°24' .587	9° .571	13° .547	17° .517	21° .481	25° .447	28° .404	32° .366	35° .329	38° .295	43° .235	49° .166	57° .096	63° .053	68° .032
14	0°0' .510	4°5' .506	8° .495	12° .477	16° .454	20° .426	23° .396	27° .365	30° .334	33° .304	36° .275	41° .223	47° .162	55° .096	62° .054	66° .033
15	0°0' .444	3°49' .442	8° .433	11° .419	15° .401	18° .380	22° .356	25° .331	28° .305	31° .280	34° .256	39° .212	45° .157	53° .096	60° .055	65° .034
16	0°0' .391	3°35' .388	7° .382	11° .371	14° .357	17° .339	21° .321	24° .300	27° .280	29° .259	32° .238	37° .200	43° .152	51° .095	58° .056	63° .035
17	0°0' .346	3°22' .344	7° .339	10° .331	13° .319	16° .306	19° .290	22° .274	25° .256	28° .239	30° .222	35° .189	41° .146	50° .094	57° .057	62° .036
18	0°0' .309	3°11' .307	6° .303	9° .297	13° .287	16° .276	18° .264	21° .250	24° .236	27° .221	29° .206	34° .178	40° .140	48° .092	55° .057	61° .036
19	0°0' .277	3°1' .276	6° .273	9° .267	12° .260	15° .251	18° .240	20° .229	23° .217	25° .205	28° .192	32° .167	38° .134	46° .090	54° .057	59° .037

Height of Light Source Above Surface—Feet (left axis)

5. Multiply the candlepower intensity by the percentage, which is the lower figure in the table and divide by 100. The result is the footcandles available at the given point.

As an example, consider a 400W luminaire having the following light distribution.

Vertical Angle	Average Candelas
135	52
125	238
115	542
105	369
95	399
85	732
75	1785
65	3987
55	7676
45	9613
35	7687
25	7687
15	7854
5	7081
0	6902

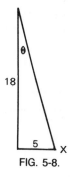

FIG. 5-8.

Let $H = 18'$ and $R = 5'$, Figure 5-8. Tan $\Theta = 5/18$ and $\Theta = 15.52°$. From Table 5-1, $\Theta = 16°$ and the percent available is .276. The distribution curve shows the candlepower at 15° is 7854. The illumination at X

$$= \frac{.276 \times 7854}{100} = 21.67 \text{ fc}$$

Example 5-1

Develop a lighting system for an unenclosed office and drafting area measuring $60' \times 140'$. Tasks within the space include office duties, with a pencil, and drafting, using plastic lead. Design constraints are a pinwheel desk arrangement, a nonstandard ceiling grid and a low suspended ceiling. Working plane height is 3.0′.

Visual Criteria

ESI Equivalent Sphere Illumination will be 50 minimum maintained ESI footcandles, 60 average maintained ESI fc.

VCP Visual Comfort Probability must be a minimum of 70, or greater.

fc Maintain 70 to 100 footcandles, though not as important as ESI.

Psychological Criteria

Maintain low luminance ratios; i.e., 3 to 1 ratio between task luminance and immediate surrounding luminance and 5 to 1 ratio between task luminance and remote surrounding luminance.

Test Area

A 9-bay test area measures 48′×63′ and has a 9.25′ ceiling. Reflectances are; walls 0.8, floor 0.2, ceiling 0.7.

The following four alternative systems are to be evaluated.

SYSTEM 1: 2′×4′ troffer, 2 lamp, 40W fluorescent luminaires in the configuration shown in Figure 5-9.

LLF (light loss factor) = LDD (luminaire dirt depreciation)+LLD (lamp lumen depreciation+RSDD (room surface dirt depreciation)

1 year LLF = $.94 \times .90 \times .99 = .84$
2 year LLF = $.89 \times .85 \times .98 = .74$
3 year LLF = $.87 \times .82 \times .97 = .69$

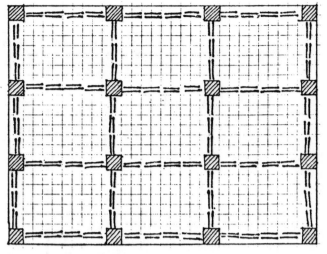

FIG. 5-10. Luminaire configuration, System 2.

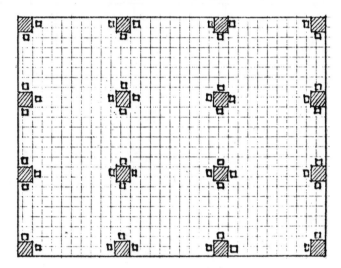

FIG. 5-11. Luminaire configuration, System 3.

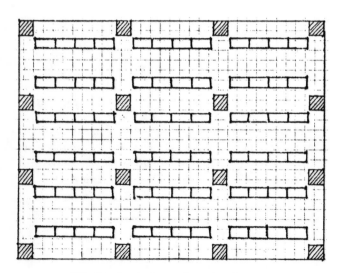

FIG. 5-9. Luminaire configuration, System 1.

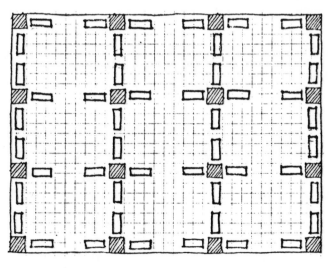

FIG. 5-12. Luminaire configuration, System 4.

SYSTEM 2: indirect, 2 lamp, 40W fluorescent luminaires in the configuration shown in Figure 5-10.

1 year LLF = $.94 \times .90 \times .93 = 0.79$
2 year LLF = $.89 \times .85 \times .90 = 0.68$
3 year LLF = $.87 \times .82 \times .87 = 0.62$

SYSTEM 3: column mounted, indirect, single, 175W high intensity discharge (HID) mercury lamp luminaires in the configuration shown in Figure 5-11.

1 year LLF = $.94 \times .90 \times .93 = .78$
2 year LLF = $.89 \times .85 \times .90 = .68$
3 year LLF = $.87 \times .82 \times .87 = .62$

SYSTEM 4: $2' \times 4'$ troffer with parabolic louver, 3 lamp, 40W fluorescent luminaries in the configuration shown in Figure 5-12.

1 year LLF = $.97 \times .90 \times .99 = .86$
2 year LLF = $.95 \times .85 \times .98 = .79$
3 year LLF = $.92 \times .82 \times .97 = .73$

The four alternate systems were analyzed using the Lumen II computer program. Lumen II generates point-by-point calculations for footcandle values, equivalent sphere illumination (ESI) and visual comfort probability (VCP). Figures 5-13, 14 and 15 show a portion of the the output of the program for System 1. Results of the computer program for all four systems are summarized in Table 5-2. Table 5-3 is a summary of first cost, operating cost and maintenance cost for each system. An economic analysis of the four systems, using the method developed in Chapter 7, was calculated over a 20-year period. Escalation rates of 8 and 6 percent were taken for energy costs, operating costs and maintenance costs respectively. The results of life-cycle cost analysis are given in Table 5-4. System 2 was picked as the lighting system for the building.

EFFICIENT LIGHT SOURCES

Table 5-5 lists the various types of lamps and their initial lumens, lamp life, efficacy and lamp lumen depre-

```
                          ILLUMINATION
                          ------------


     WORKING PLANE HEIGHT:  3.00


       AVERAGE: 132.453   MINIMUM: 109.952   MAXIMUM: 153.176   MEAN DEVIATION:  10.470


       ABS. Y      ABSOLUTE X-COORDINATE(S)
        COOR.    24.0   25.0   26.0   27.0   28.0   35.0   36.0   37.0   38.0   39.0


            ****************************************************************

         29.0 * 119.9  129.3  138.8  146.9  153.2  153.2  146.9  138.7  129.4  119.9


         28.0 * 118.6  128.1  137.3  145.1  151.1  151.1  145.1  137.3  128.1  118.6


         23.0 * 110.0  118.0  126.4  133.9  139.6  139.6  133.9  126.4  118.0  110.0


         22.0 * 111.4  119.7  128.2  135.7  141.5  141.5  135.7  128.2  119.7  111.4


         21.0 * 114.2  123.1  131.8  139.4  145.1  145.2  139.4  131.8  123.1  114.2


         20.0 * 116.7  126.0  134.9  142.5  148.3  148.3  142.5  134.9  126.0  116.7


         19.0 * 118.7  128.3  137.5  145.3  151.3  151.3  145.3  137.5  128.3  118.7
```

FIG. 5-13. Printout from Lumen II program, illumination.

EQUIVALENT SPHERE ILLUMINATION

TARGET DESCRIPTION: PENCIL TARGET - CONCENTRIC RINGS @ 25 DEGREE VIEWING ANGLE
SPHERE CONTRAST: 0.1675

WORKING PLANE HEIGHT: 3.00

	NORTH	EAST	SOUTH	WEST	TOTAL
AVERAGE=	83.877	66.453	53.913	66.443	67.672
MINIMUM=	49.798	21.275	29.645	21.272	21.272
MAXIMUM=	110.929	142.644	89.312	142.580	142.644
MEAN DEVIATION=	12.627	29.315	15.993	29.310	2?.711

ABS. Y   ABSOLUTE X-COORDINATE(S)

COOR.		24.0	25.0	26.0	27.0	28.0	35.0	36.0	37.0	38.0	39.0
29.0	N	93.0	87.1	88.0	92.6	96.6	96.6	92.6	88.0	87.1	93.0
	E	38.8	44.9	51.5	56.4	60.5	66.3	70.6	79.3	89.2	90.1
	S	54.4	41.9	46.0	49.1	56.3	51.5	44.8	41.8	89.2	93.0
	W	90.1	89.2	79.3	70.6	66.3	60.5	56.4	51.5	44.9	38.8
28.0	N	87.7	82.3	82.5	85.7	88.9	88.8	85.8	82.2	82.2	87.7
	E	21.9	26.9	31.6	32.5	33.1	34.4	38.0	47.1	59.6	61.9
	S	74.9	65.0	65.2	69.1	72.6	72.5	69.2	65.2	65.0	74.9
	W	61.9	59.6	47.1	38.0	34.4	33.1	32.5	31.6	26.9	21.9
23.0	N	63.3	49.8	50.4	55.3	59.5	59.5	55.3	50.4	49.8	63.3
	E	94.4	100.3	108.9	120.3	131.4	142.6	140.3	140.0	138.5	133.2
	S	46.1	33.8	33.6	37.0	40.0	40.0	37.0	33.7	33.8	46.1
	W	133.1	138.5	140.0	140.4	142.6	131.4	120.3	108.9	100.2	94.4
22.0	N	89.9	81.2	81.2	88.4	93.2	93.2	88.5	81.2	81.2	89.9
	E	69.1	74.2	90.7	99.6	110.7	113.3	110.7	113.3	117.6	121.2
	S	40.6	30.0	29.6	32.5	35.0	35.0	29.7	30.0	40.6	40.6
	W	118.1	121.2	117.6	113.2	110.7	99.6	90.7	74.2	69.1	
21.0	N	104.2	100.5	102.1	106.7	110.9	110.9	106.7	102.1	100.5	104.2
	E	36.6	41.8	48.4	52.2	56.1	61.5	66.0	74.7	84.6	85.7
	S	48.6	38.1	37.5	40.2	42.8	42.8	40.3	37.5	38.1	48.6
	W	85.7	84.6	74.7	65.9	61.5	56.1	52.9	48.4	41.8	36.6
20.0	N	95.0	89.6	90.3	94.0	97.9	98.0	94.0	90.3	89.6	95.0
	E	21.3	26.1	30.6	31.4	31.8	33.3	36.7	45.2	58.1	60.4
	S	67.6	59.5	58.8	61.6	64.5	64.5	61.5	58.8	59.5	67.6
	W	60.4	58.1	45.2	36.7	33.1	31.4	30.6	26.1	21.3	
19.0	N	74.8	65.0	65.2	69.1	72.4	72.4	69.1	65.1	65.0	74.8

FIG. 5-14. Printout from Lumen II program, ESI.

VISUAL COMFORT PROBABILITY

VISUAL PLANE HEIGHT: 4.00

NORTH	EAST	SOUTH	WEST	TOTAL
AVERAGE= 79.189	AVERAGE= 73.982	AVERAGE= 79.569	AVERAGE= 73.982	AVERAGE= 76.680
MINIMUM= 74.540	MINIMUM= 70.540	MINIMUM= 73.890	MINIMUM= 70.540	MINIMUM= 70.540
MAXIMUM= 84.380	MAXIMUM= 77.040	MAXIMUM= 83.150	MAXIMUM= 77.040	MAXIMUM= 84.380
MEAN DEVIATION= 2.816	MEAN DEVIATION= 1.814	MEAN DEVIATION= 2.820	MEAN DEVIATION= 1.814	MEAN DEVIATION= 3.195

ABS. Y COOR.		ABSOLUTE X-COORDINATE(S) 24.0	25.0	26.0	27.0	28.0	35.0	36.0	37.0	38.0	39.0
29.0	N	84.4	84.1	83.6	83.2	82.9	82.9	83.2	83.6	84.1	84.4
	E	71.2	71.9	72.6	73.2	72.6	76.7	75.2	75.2	75.8	76.7
	S	74.5	74.5	74.5	74.2	73.9	73.9	74.2	74.5	74.5	74.5
	W	76.7	75.8	75.2	75.2	76.7	72.6	73.2	72.6	71.9	71.2
28.0	N	82.9	82.9	82.4	81.9	81.6	81.6	81.9	82.4	82.9	82.9
	E	71.6	72.2	72.9	73.2	72.6	77.0	75.2	75.5	75.8	77.0
	S	75.5	75.2	75.2	74.9	74.9	74.9	74.9	75.2	75.2	75.5
	W	77.0	75.8	75.5	75.2	77.0	72.6	73.2	72.9	72.2	71.6
23.0	N	81.6	81.6	81.3	81.3	81.1	81.1	81.3	81.3	81.6	81.6
	E	70.5	71.2	71.9	72.6	71.6	75.8	74.2	74.9	75.5	76.4
	S	83.2	83.2	82.9	82.6	82.6	82.6	82.6	82.9	83.2	83.2
	W	76.4	75.5	74.9	74.2	75.8	71.6	72.6	71.9	71.2	70.5
22.0	N	79.4	79.4	79.1	78.8	78.5	78.5	78.8	79.1	79.4	79.4
	E	70.5	71.6	71.9	72.6	71.9	76.1	74.5	74.9	75.5	76.4
	S	80.0	80.0	79.7	79.1	79.1	79.1	79.1	79.7	80.0	80.0
	W	76.4	75.5	74.9	74.5	76.1	71.9	72.6	71.9	71.6	70.5
21.0	N	77.3	77.6	77.3	77.0	76.7	76.7	77.0	77.3	77.6	77.3
	E	70.9	71.9	72.2	72.9	72.2	76.4	74.5	74.9	75.5	76.7
	S	81.3	81.3	81.1	80.8	80.5	80.5	80.8	81.1	81.3	81.3
	W	76.7	75.5	74.9	74.5	76.4	72.2	72.9	72.2	71.9	70.9
20.0	N	76.1	76.1	75.8	75.8	75.5	75.5	75.8	75.8	76.1	76.1
	E	71.6	72.2	72.6	73.6	72.6	76.7	75.2	75.2	75.8	77.0
	S	82.1	81.9	81.6	81.3	81.1	81.1	81.3	81.6	81.9	82.1
	W	77.0	75.8	75.2	75.2	76.7	72.6	73.6	72.6	72.2	71.6
19.0	N	75.2	75.2	74.9	74.5	74.5	74.5	74.5	74.9	75.2	75.2
	E	71.6	72.2	72.9	73.6	72.6	77.0	75.2	75.5	75.8	77.0
	S	82.9	82.9	82.6	82.1	81.9	81.9	82.1	82.6	82.9	82.9
	W	77.0	75.8	75.5	75.2	77.0	72.6	73.6	72.9	72.2	71.6

FIG. 5-15. Printout from Lumen II program, VCP.

TABLE 5-2. Results of Computer Program Lumen II

System	LLF	ESI				FC				VCP			w/ft²	Cleaning Schedule
		Int. Avg.	Max.	Min.	Main Avg.	Int. Avg.	Max.	Min.	Main Avg.	Int. Avg.	Max.	Min.		
1	.84	68	143	21	54	132	153	110	111	75	82	72	2.5	Less than one year*
2	.79	94	120	72	64	96	103	85	65	100	100	100	1.86	Two years**
3	.72	103	134	75	69	89	112	69	55	100	100	100	2.2	2.5 years†
4	.86	85	100	66	65	105	138	56	81	96	98	94	3.0	Three years††

* Luminaire cannot meet criteria minimum and needs and LLF of 0.93 to maintain the minimum ESI criteria. The luminaire must be cleaned more often than once a year.

** Luminaire needs LLF of 0.70 to maintain minimum ESI criteria. The luminaire must be cleaned every two years.

† Luminaire needs LLF of 0.66 to maintain minimum criteria and should be cleaned every 2.5 years.

†† Luminaire needs LLF of 0.75 to maintain minimum ESI criteria. Luminaire must be cleaned every three years.

TABLE 5-3. Estimate of First Cost, Operating and Maintenance Costs of Four Systems

	System 1	System 2	System 3	System 4
1. Luminaire type	2-40W-CW	2-40C-CW	175HID-MS	3-40W-CW
2. Number of luminaires	224	261	94	180
3. Total luminaire cost	11,200	22,860	21,150	11,700
4. Total lamp cost	672	472	940	810
5. Total installation cost	6,048	2,925	4,700	4,860
6. Total wiring cost	3,150	2,800	2,800	2,800
7. Total first cost	21,070	29,057	29,590	20,170
8. Cost per kWh	0.030	0.030	0.030	0.030
9. Burning hours/year	5,000	5,000	5,000	5,000
10. Total input watts/luminaire	96	96	195	140
11. Total operating cost/year	3,225.00	3,758.40	2,749.50	3,780.00
12. Average lamp replacement cost/year	124.44	87.50	470.00	150.00
13. Lamp replacement labor cost/year	580.74	408.33	329.00	700.00
14. Cleaning labor cost/year	280.00	27.50	47.50	112.50
15. Total maintenance cost/year	985.19	523.33	846.50	962.50

TABLE 5-4. Economic Comparison of Four Lighting Systems

Systems Compared	Rate of Return on Differential Investment	Payback Period of Differential Investment
2 over 1	21.1%	5 years, 6 months
2 over 3*	-	-
2 over 4	26.6%	4 years, 4 months

*System 3 has higher first cost and higher operating and maintenance cost than System 2.

ciation factor. Each of these lamps has different characteristics; i.e., color, life, physical size, and light output per watt input. The choice should be the most efficient source that is appropriate for the application. It is clear, from Table 5-5, that with incandescent, fluorescent and HID lamps, the higher the wattage the more efficient the lamp. For overall design, prime consideration should be given to the more efficient sources such as fluorescent and HID.[5-5]

Table 5-6 is a comparison of the power requirements of alternate light sources capable of maintaining a lighting level of 100 fc over 10,000 ft^2 of factory area.[5-6] This data will hold true for similar interior lighting applications, provided there is sufficient ceiling height for proper mounting of the fixtures.

Another example is the lighting design for a $200' \times 100'$ factory area with a $27'$ mounting height. Figure 5-16 is an economic comparison of the various lighting systems, using mercury vapor, metal halide, high pressure sodium and fluorescent lamps for a maintained 100 fc level.[5-7]

EFFICIENT LUMINAIRES

Based on their light distribution characteristics, luminaires are classified into five categories: direct, semi-direct, general diffuse, semi-indirect and indirect. An efficient luminaire produces a greater amount of light with less wattage. For example: incandescent, indirect luminaires may require 11 watts per square foot of floor area to produce a 50 footcandle level whereas direct

TABLE 5-5. Lamp Efficacies for Light Sources

Light Source	Lamp Life	(a) Initial Lumens	Lamp & Ballast Wattage	(b) Efficacy	(c) LLD
HID MERCURY DELUXE WHITE					
1000W	24,000	63,000	1,075	59	.52
400W	24,000	22,500	455	50	.70
250W	24,000	13,000	285	46	.74
175W	24,000	8,500	205	41	.75

Because of poor LLD, the use of 1000W HID mercury lamps is not recommended.

Light Source	Lamp Life	(a) Initial Lumens	Lamp & Ballast Wattage	(b) Efficacy	(c) LLD
METALLIC HALIDE CLEAR LAMPS					
1000W	10,000	125,000	1,075	116	.74
400W(V)	15,000	34,000	465	73	.67
400W(H)	15,000	40,000	465	86	.70
250W(V)	10,000	20,500	295	69	.70
175W(H)	7,500	15,000	210	71	.66
HIGH PRESSURE SODIUM					
1000W	24,000	140,000	1,150	122	.83
400W	24,000	50,000	478	105	.83
310W	24,000	32,000	370	100	.83
250W	24,000	30,000	320	94	.83
200W	24,000	22,000	240	92	.83
150W	24,000	16,000	185	86	.83
100W	20,000	9,500	125	76	.84
70W	20,000	5,600	90	64	.84
LOW PRESSURE SODIUM					
180W T-21	20,000	33,000	222	149	.92
135W T-21	20,000	22,000	178	124	.95
90W T-21	20,000	12,750	**108	**118	.91
55W T-16	20,000	7,650	82	93	.93
35W T-16	20,000	4,600	62	74	.92
FLUORESCENT					
40W/RS/CW	26,000*	3,150	46	68	.84
110W/RS/CW 800MA(HO)	18,000*	9,050	122	74	.76
215W/RS/CW 1500MA T-12	15,000*	16,000	231	69	.62
215W/RS/CW 1500MA T-17	15,000*	16,000	231	69	.58

(a) Initial lumens (after 100 hours)
(b) Lumens per watt (including ballast)
(c) 70 percent rated life
* – 12 burning hours per start   ** – Estimated

TABLE 5-6. Relative Power Required to Maintain 100 FC in a 10,000 ft^2 Factory Area

	Fluorescent Super-Hi F96T12/CW	High Pressure Sodium 400W	Metal Halide 400W	Mercury Deluxe Twins 400W	Incandescent 1000W
No. of fixtures	73	40	65	59	70
Power required (kW)	33.2	19	30	52	70
Relative power Requirement %	175	100	158	274	368

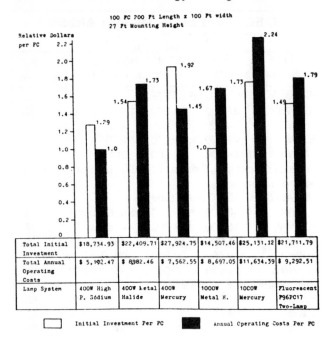

100 FC 200 Ft Length x 100 Ft width
27 Ft Mounting Height

Total Initial Investment	$18,734.93	$22,409.71	$27,924.75	$14,507.46	$25,131.12	$21,711.79
Total Annual Operating Costs	$ 5,192.47	$ 8982.46	$ 7,562.55	$ 8,697.05	$11,634.39	$ 9,292.51
Lamp System	400W High P. Sodium	400W Metal Halide	400W Mercury	1000W Metal H.	1000W Mercury	Fluorescent F96PC17 Two-Lamp

☐ Initial Investment Per FC   ■ Annual Operating Costs Per FC

FIG. 5-16.

fluorescent troffers may only require 2.5 watts per square foot.

LUMINAIRE MAINTENANCE

Easily cleaned luminaires should be selected since good lighting maintenance always assures better utilization of lighting systems. In a study of one fluorescent lighting system, in which different maintenance procedures were used, the results were as follows:

When luminaires were cleaned and relamped once every three years, the illumination dropped to 60 percent after three years.

When luminaires were cleaned every 18 months and relamped every three years, the illumination dropped to 68 percent after three years.

When luminaires were cleaned annually and one-third of the lamps were replaced every year, the illumination dropped to 78 percent after three years and no lower than 75 percent after 12 years.

DAY LIGHTING

Skylights should be used to supplement artificial lighting systems. Depending upon the area, they can produce a substantial energy savings. The method of calculating average horizontal footcandles from skylights on the work plane is called the Lumen method of top lighting and is described in the IES Recommended Practice of Daylighting (RPD) brochure. This procedure is similar to the Lumen-Zonal Cavity method of artificial lighting design in that it integrates the room and skylight into a single coefficient of utilization. The coefficient of utilization is found by using Table VI in the IES RPD brochure.[5-8]

The maintenance factor used with daylight calculations has two components: RSDD, room surface dirt depreciation, and SDD, skylight dirt depreciation. For a clean room, RSDD is 0.95. The values of SDD are determined from Table IV of the IES RPD brochure.

The illumination on the work place is equal to

$$E_s = E_H \times \frac{A_s}{A_w} \times K_u \times K_m$$

where:

E_s = Work plane illumination from skylighting
E_H = Horizontal illumination of the skylight
A_s = Area of the skylight
A_w = Area of the work plane
K_u = Coefficient of utilization
K_m = Maintenance factor

Horizontal illumination from skylights is different for an overcast sky and for a clear sky.[5-9]

An overcast sky does not have constant luminance with respect to the viewing angle. In daylighting design, a single value is assigned to the equivalent uniform sky luminance, to represent an entire overcast sky. The values of equivalent overcast sky luminance, as a function of time, date and latitude, are given in Table IX of the IES RPD. These luminance values have been plotted in Figure 5-17 as a function of solar altitude and an average curve has been drawn through the points. To use Figure 5-17, determine the solar altitude, which varies with latitude, date and time, as shown in Table 5-7. For

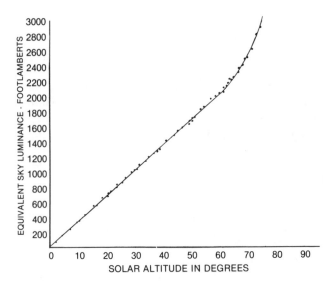

FIG. 5-17. Equivalent overcast sky luminance.

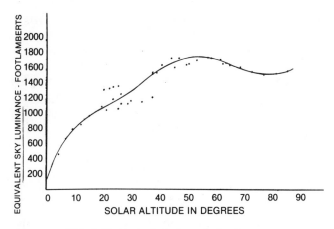

FIG. 5-18. Equivalent clear sky luminance.

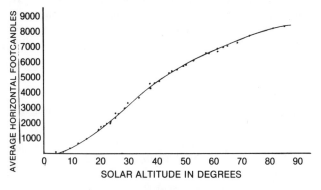

FIG. 5-19. Average horizontal illumination from solar radiation.

TABLE 5-7. Degrees of Solar Altitude vs. Latitude, Date and Time*

Latitude	Date	AM PM	6 6	7 5	8 4	9 3	10 2	11 1	noon noon
30°N	June 21		12	24	37	50	63	75	83
	Mar-Sept 21			13	26	38	49	57	60
	Dec 21				12	21	29	35	37
34°N	June 21		13	25	37	50	62	74	79
	Mar-Sept 21			12	25	36	46	53	56
	Dec 21				9	18	26	31	33
38°N	June 21		14	26	37	49	61	71	75
	Mar-Sept 21			12	23	34	43	50	52
	Dec 21				7	16	23	27	28
42°N	June 21		16	26	38	49	60	68	71
	Mar-Sept 21			11	22	32	40	46	48
	Dec 21				4	13	19	23	25
46°N	June 21		17	27	37	48	57	65	67
	Mar-Sept 21			10	20	30	37	42	44
	Dec 21				2	10	15	20	21

*Taken from Table XI, IES Recommended Practice of Daylighting.

TABLE 5-8: Illumination Due to Skylights

Time of Day AM PM	Footcandles obtained from skylights on cloudy days	Footcandles obtained from skylights on clear days
8 4	2.0	9.0
9 3	3.0	15.2
10 2	3.8	18.70
11 1	4.32	20.54
12	4.56	21.24

example, at 10:00 a.m. on December 21 at 30 degrees N latitude, the solar altitude is 29 degrees. From Figure 5-17, the equivalent uniform luminance is given as 1,000 footlamberts. The horizontal illumination in footcandles is equal to this value.

The values of clear sky luminance, as a function of time, date and latitude, are presented in Table X of the IES RPD. Four directional luminances have been averaged and plotted versus solar altitude in Figure 5-18 and an average curve drawn through the plotted points.

On a clear day, the contribution of direct sunlight to the horizontal illumination exceeds the contribution of the clear sky luminance for solar altitudes greater than 15 degrees. Horizontal illumination from solar radiation, as a function of date, time and latitude is given in Table XII of the IES RPD. These data are plotted as a function of solar altitude in Figure 5-19.

Computer programs are being developed for the point-by-point calculation of work plane illumination from skylighting.

Studies were conducted at a midwest industrial plant having skylights. Placement of the $4' \times 6'$ skylights in the $40' \times 60'$ bays is shown in Figure 5-20. The skylights are double-glazed 1/4″ plexiglass with a 1/4″ air space and have a light transmittance factor of 53%. Average footcandle levels, using the Lumen method of

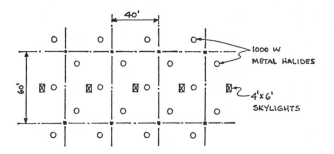

FIG. 5-20.

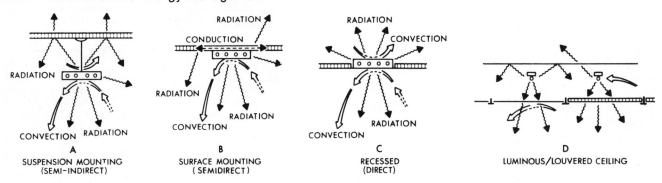

FIG. 5-21. Effect of ceiling to luminaire relationship on heat transfer.[5-10]

calculating top lighting, are shown in Table 5-8.

HEAT TRANSFER LUMINAIRES

Heat is transferred from a surface mounted, semi-indirect luminaire by radiation, conduction, and convection, Figure 5-21. Assuming good contact with the ceiling, the upper surfaces of the luminaire will absorb energy and transfer it to the ceiling by conduction. Since most

FIG. 5-22. Impact of plenum temperature on light output.

of the acoustical ceiling materials are insulators, the temperature within the luminaire will be elevated.

In the case of recessed luminaires, the temperature above the suspended ceiling increases and reduces the efficiency of the luminaire as shown in Figure 5-22.

The heat of light from a number of luminaires is[5-10]

$$Q_l = 3.413 \frac{A \times fc \times W \times tf}{lm \times cu \times mf}$$

where

Q_l = Sensible lighting heat load in conditioned spaces.
A = Total area in square feet.
fc = Average footcandles in service.
lm = Total lumens per luminaire.
cu = Coefficient of utilization.
mf = Maintenance factor.
W = Actual watts per luminaire in service.
tf = Thermal factor, ratio of energy in conditioned space to total power input.

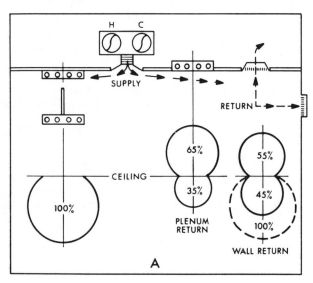

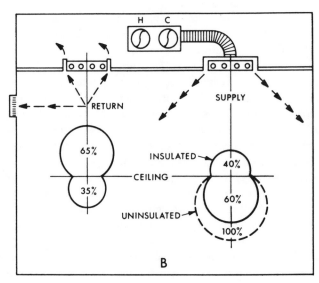

FIG. 5-23. Space conditioning systems that integrate the luminaires.[5-10]

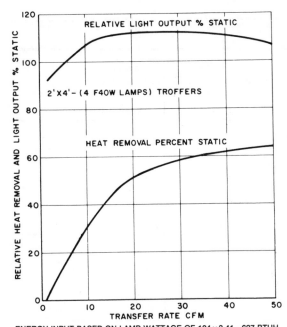

ENERGY INPUT BASED ON LAMP WATTAGE OF 184 × 3.41 = 627 BTUH
HEAT REMOVAL AT 20 CFM = APPROXIMATELY 324 BTUH (95 WATTS)

FIG. 5-24. Heat removal from a 4-lamp troffer increases light output by up to 12%.[5-11]

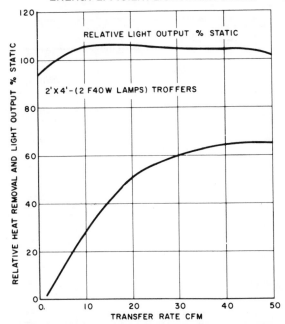

ENERGY INPUT BASED ON LAMP WATTAGE OF 92 × 3.41 = 374 BTUH
HEAT REMOVAL AT 20 CFM = APPROXIMATELY 160 BTUH (47 WATTS)

FIG. 5-25. Heat removal from a 2-lamp troffer increases light output by up to 6%.[5-11]

The ambient temperature of the luminaires can be reduced and the heat from the luminaires utilized by space conditioning systems that integrate the luminaires as shown in Figure 5-23. The recessed luminaire can be operated as a supply air diffuser or return air outlet, with the supply luminaire being connected to a downstream duct running from an air mixing unit. Room air is returned to the plenum by other luminaires and picked up by stub ducts serving the zone.

Tests have shown that about 50% to 60% of the lamp and ballast heat can be removed through the plenum air return, thus preventing the lamp heat from entering the occupied space during the cooling season. Although total building refrigeration requirements are not reduced, ventilation fan loads and electric energy use are lower when the lighting heat load is reduced in the occupied area through the use of heat transfer luminaires.

Fluorescent heat transfer systems produce more light than static lighting systems. This means that fewer luminaires are required, with resultant energy savings.

Figure 5-24 plots the heat transfer and relative light output from a 4-lamp troffer when 20 cfm of air flow over the lamps. Approximately 12% more light is provided by the four 40W fluorescent bulbs when compared to a luminaire with no air supply. Figure 5-25 shows a 6% increase in light output with a comparable 2-lamp luminaire.

Heat transfer luminaires were compared with static luminaires in a typical study.[5-11] The comparative data is shown in Figure 5-26 and costs are compared in Figure 5-27. It is evident that the 2-lamp, heat transfer luminaire uses less energy and has a higher visual comfort probability. However, the number of luminaires that are required increase the annual owning and operating costs over that of the 4-lamp, heat transfer luminaires.

REFLECTIVE FINISHES

Light finishes reflect light to the task area while dark finishes on the ceiling, walls and floors absorb light. The reflectances of offices and schools should be in the following ranges:

Ceiling finishes	80-90%
Walls	40-60%
Furniture and office equipment	25-45%
Floors	20-40%

Upgrading all surface reflectances can produce a major improvement in the lighting levels. In one installation, the ceilings, walls and floors were repainted and the furniture was refinished in a lighter color. As a result, the average illumination level increased from less than 10 footcandles to over 40 footcandles.

CONTROLLED FENESTRATION

A good lighting design can be achieved by skillful utilization of daylighting. Daylighting may be redirected with venetian blinds, diffused with drapes, or reduced

NO. OF SYSTEMS 4		ANNUAL OPERATING HOURS 3120		AVERAGE COST/KWH .0357		ANNUAL DEPRECIATION .20	
DESCRIPTION		BASE SYSTEM 4-F40 H.T	II SYSTEM 4-F40 Static	III SYSTEM 2-F40 H.T	IV SYSTEM 2-F40 Static	V SYSTEM	
1	RATED INITIAL LUMENS/LAMP	12600	12600	6300	6300		
2	RATED LAMP LIFE	24000	24000	24000	24000		
3	LAMPS/LUMINAIRE	4	4	2	2		
4	TOTAL WATTS LUMINAIRE	184	184	92	92		
5	VISUAL COMFORT PROBABILITY	.71	.71	.77	.77		
6	COEFFICIENT OF UTILIZATION	.64	.57	.68	.64		
7	LUMINAIRE DIRT DEPRECIATION (LLD)	.88	.88	.88	.88		
8	LAMP LUMEN DEPRECIATION (LLD)	.84	.84	.84	.84		
9	ROOM SURFACE DIRT DEPRECIATION (RSDD)	.98	.98	.98	.98		
10	LIST PRICE/LUMINAIRE	63.35	61.85	50.10	51.60		
11	LIST PRICE/LAMP	1.53	1.53	1.53	1.53		
12	NUMBER OF LUMINAIRIES REQUIRED	171	192	322	342		

ILLUMINATION LEVEL (MAINTAINED) 13 100 Footcandles	AREA 14 10,000	WIDTH 15 100'	LENGTH 16 100'	MOUNTING HEIGHT 17 9'	

COMMENTS
Room Reflectances - Ceiling 80% -- Walls 50% -- Floor 20%

FIG. 5-26. Comparison of heat transfer and static luminaires.[5-11]

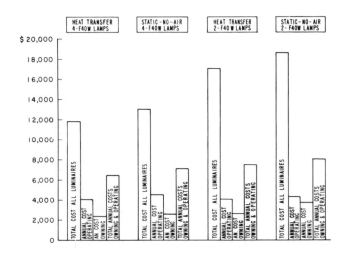

FIG. 5-27. Comparative costs of heat transfer and static luminaires.[5-11]

by shades, screens, blinds and low transmission glasses. These same devices can be used to reduce the air conditioning load.

LIGHTING CONTROLS

To be energy effective, electric lighting must be turned off when natural light is sufficient or when lighting is no longer needed. To be most energy conserving, only the minimum of electric lighting should be added, when necessary, to provide adequate illumination for the task. This requires modulation of the lighting source by either continuous control or by switching over several discrete levels.[5-12]

The most common electric light control is the on-off switch which can control selected groups of fixtures in an area or a portion of the lamps within each fixture. For example, energy can be saved by switching perimeter lighting in offices and industrial plants when natural light is available. A specific zone can be controlled from several locations by using 3-way and 4-way switches.

When lights must come on instantly, either incandescent or rapid start fluorescent lamps must be used although, for energy conservation, incandescent lamps should be avoided.

It takes 3 to 5 minutes for high intensity discharge lamps to come up to full brightness. Control of these sources can be achieved with a two-level ballast. Thus, the lamps are maintained at one-half intensity, and about one-half power for standby or nontask lighting, then switched to full brightness when needed.

Another electronic switching system is called EASE, Energy Activation Switching Equipment. The system is based on the potential disturbance of a field of inaudible ultrasonic sound waves that fill the monitored space. Motion of a person is detected by the system's sensor and a cessation of motion causes EASE to switch the lights off after an adjustable delay of from 90 to 360 seconds.

The visible parts of the system are the air couplers and sensors. The air coupler permeates the area being monitored with inaudible airborne ultrasonics. The movement or presence of a person is detected by the sensor, causing the electrical system to respond. The system is sensitive to the most minute movements: the breathing of a perfectly stationary person is sufficient to maintain room lighting in the normal *on* condition.

The EASE system is compatible with fluorescent lighting, which restrikes instantly, and with HPS, which restrikes one minute after extinguishing.

Another sophisticated control employs a small and relatively inexpensive *receiver relay* at each fixture or group of fixtures, with each relay having a unique address. A microprocessor controls a carrier current transmitter whose signal is sent all over the branch circuit power lines within a building or complex to operate each switch as required or as programmed. The program is responsive to sun angle, to time of day, or to any other measurable variable affecting illumination requirements.

The illumination system can be dimmed by controlling the electrical input, to provide stepless attenuation from full output to some very low minimum. In most cases, the reduction in input power is proportional to the reduction in light output.

In the earlier stages, dimmers for incandescent lamps and some fluorescent lamps used variable auto transformers to reduce input voltage. Variable transformers have been replaced by silicon controlled rectifier dimmers, with substantial cost savings. Dimming of fluorescent and HID lamps usually requires special dimming ballasts.

Dimmers can be controlled in a number of ways: man-

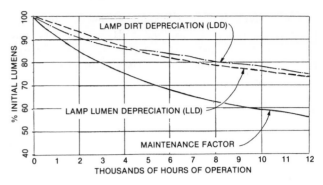

FIG. 5-28.

ually, by means of photocells, and by remote controlled motors. Photocells can automatically lower the level of electric illumination as day lighting increases and vice versa. Generally located in the ceiling, these photocells control the lamps in their zone to maintain constant illumination levels.

A study was conducted for a 720,000 ft² industrial plant using an automatic dimming system to dim 1,000W metal halide lamps to maintain a 50 fc level at all times. Figure 5-28 shows the initial lamp lumen depreciation (LLD) curve, the lamp dirt depreciation (LDD) curve and the maintenance factor curve. A maintenance factor of 0.56 is used. Thus the system was designed to initially deliver 50/.56 = 89.28 fc.

Luminaires were placed in the 300 bays as shown in Figure 5-20. Each 40′ × 60′ bay was equipped with three 1,000W metal halide lamps. Figure 5-29 is a computer printout of the raw fc distribution in each bay. The average initial footcandle level is 106.374. Figure 5-30 shows the light output versus the percent of ballast power input.

Figure 5-31 shows the savings in power output due to automatic dimming. At installation, the average lamp output is 106 fc. The dimmer system reduces this to 50 fc or 47%. From Figure 5-30, the corresponding input power is 60%. Power saved is 40% which becomes the first point on the energy savings curve. At 2000 hours, the lamp output is .85 × 106 = 90.1 fc. The corresponding power saved is 30%. Similarly, all other points of the energy savings curve are obtained by referring to Figures 5-28 and 5-30.

There are 900 lamps, each with a power input of 1100W. The area under the curve in Figure 5-31 represents the total energy saved by using the light dimmer system. The energy saved in three years, the life of the lamp, is

Year	Energy Saved
1	1,346,000 kWh
2	932,400 kWh
3	576,000 kWh
Total	2,854,400 kWh

The capital cost of installing the automatic dimmer system and the dimmer ballasts was $121,520. An economic analysis of the dimmer system was based on a 2¢ energy rate and an 8% escalation factor. The system pays for itself in seven years.

VOLTAGE LEVEL

A 3-phase, 480/277v system has proved to be the most economical for lighting since the wiring cost at this voltage is much lower than at 120v. However, a 480/277v system is not available for incandescent lamps which should be

```
                                   ILLUMINATION
                                   ------------

WORKING PLANE HEIGHT:   2.50


AVERAGE: 106.374   MINIMUM:  95.155   MAXIMUM: 115.198   MEAN DEVIATION:   3.526

ABS. Y      ABSOLUTE X-COORDINATE(S)
 COOR.    40.0  42.0  44.0  46.0  48.0  50.0  52.0  54.0  56.0  58.0  60.0  62.0  64.0  66.0  68.0  70.0  72.0  74.0  76.0
        ***************************************************************************************************************

 117.0  * 110.8 110.8 109.1 106.0 104.6 104.7 104.6 106.5 109.5 111.9 111.7 109.5 107.8 108.3 109.1 109.1 108.5 108.0 107.6

 114.0  * 105.8 106.2 104.3 100.3  98.7  98.7  98.7 100.9 104.7 107.2 106.8 106.9 110.2 113.6 112.8 111.0 112.2 113.3 110.0

 111.0  * 103.2 104.0 102.4  98.1  95.7  95.4  95.6  98.5 102.7 104.8 104.1 106.2 111.4 111.3 105.5 103.0 104.9 111.0 111.2

 108.0  * 103.4 104.0 102.3  98.1  95.9  95.6  95.9  98.5 102.7 104.9 104.3 106.0 111.0 112.0 107.2 104.5 106.7 111.7 110.8

 105.0  * 106.7 107.1 105.1 101.4  93.9 100.0  99.8 101.7 105.4 107.9 107.5 106.5 108.2 111.3 112.2 111.5 111.7 111.0 107.9

 102.0  * 111.7 111.3 108.9 105.7 104.2 104.2 104.0 106.1 109.3 112.0 112.5 110.0 107.0 105.8 105.5 105.4 105.1 105.6 106.8

  99.0  * 112.9 110.7 107.7 105.2 104.1 104.2 103.9 105.6 108.0 111.5 113.6 112.7 109.4 105.9 103.9 103.4 103.5 105.7 109.2

  96.0  * 108.4 106.6 107.0 109.0 110.5 111.0 110.3 109.3 107.3 107.2 109.1 109.3 107.1 103.7 102.3 102.1 101.9 103.4 106.8

  93.0  * 104.1 105.2 110.0 112.5 103.6 107.6 109.4 112.7 110.3 105.8 104.8 105.3 102.9  98.8  96.8  96.4  96.4  98.5 102.7

  90.0  * 103.4 105.6 111.3 110.6 104.5 102.9 104.3 110.8 111.5 106.3 104.1 104.8 102.7  98.5  96.0  95.2  95.6  98.2 102.5

  87.0  * 104.5 105.5 110.4 112.9 110.0 108.0 109.8 113.3 110.9 106.3 105.3 105.8 103.4  99.3  97.3  96.8  96.8  98.9 103.2

  84.0  * 108.8 107.0 107.4 109.5 110.9 111.4 110.8 109.9 107.8 107.8 109.6 110.0 107.7 104.4 103.0 102.5 102.4 103.9 107.3

  81.0  * 113.2 111.1 108.0 105.6 104.5 104.5 104.3 106.1 108.5 112.0 114.2 113.2 109.9 106.5 104.5 103.8 104.0 106.1 109.5

  78.0  * 112.1 111.7 109.3 106.1 104.6 104.5 104.4 106.6 109.8 112.5 113.0 110.6 107.6 106.4 106.1 105.9 105.6 106.0 107.3

  75.0  * 107.0 107.4 105.4 101.8 100.3 100.3 100.1 102.2 105.9 108.4 108.1 107.1 108.8 111.8 112.8 112.0 112.1 111.5 108.5

  72.0  * 103.7 104.4 102.6  98.4  96.1  95.9  95.1  98.9 103.0 105.3 104.7 106.5 111.5 112.5 107.7 104.8 107.2 112.1 111.2

  69.0  * 103.7 104.4 102.8  98.5  96.0  95.7  96.0  99.0 103.2 105.4 104.6 106.7 111.9 111.8 106.1 103.4 105.5 111.4 111.6

  66.0  * 105.8 106.2 104.3 100.3  98.7  98.7  98.6 100.8 104.7 107.2 106.8 106.9 110.2 113.5 112.8 111.0 112.3 113.3 110.0

  63.0  * 110.8 110.9 109.1 106.1 104.6 104.7 104.6 106.5 109.5 111.8 111.7 109.5 107.8 108.2 109.1 109.0 108.6 108.0 107.7

  60.0  * 114.1 112.7 109.7 106.6 104.6 104.4 104.5 107.0 110.2 113.7 115.2 113.8 110.1 106.9 105.0 104.4 104.5 106.6 109.9
```

FIG. 5-29.

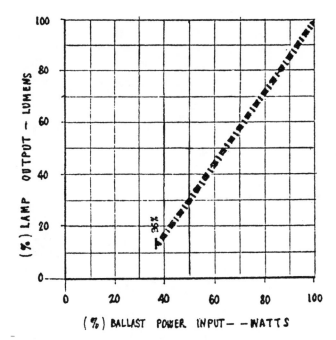

FIG. 5-30. Ratio of lamp output to ballast input.

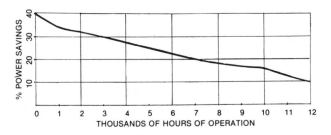

FIG. 5-31. Savings due to automatic dimming.

avoided, as much as possible, for lighting systems.

NONUNIFORM SYSTEM

In the nonuniform lighting system, lighting at the task is greater than in the areas surrounding the work. The Federal Energy Administration and the General Services Administration recommend 10 fc in corridors and lobbies, 30 fc in areas surrounding the task, and 50 fc for the average office task. When tasks involve reasonably good printed or written materials, there is no improvement in performance above 50 fc. Higher footcandle levels should only be considered if there is an appreciable improvement in productivity, and if they can be justified on a cost/benefit basis. Depending on the room layout, visual tasks, and the ap-

proach used, operating cost reductions up to 10 percent can be achieved.[5-13]

A primary limitation of task lighting is that it must be served with 120 volts in order to comply with the National Electrical Code. It is more economical to serve the lighting loads with a 480/277v supply. Large numbers of 120v, furniture-mounted lighting fixtures will increase the number of branch circuits, thus increasing the number of panel boards, large step-down transformers, feeders and distribution equipment.

According to the 1978 National Electrical Code, only 180 volt amperes, at unity power factor, shall be allowed for every receptacle. That means that a 20 ampere, 120v circuit breaker can carry a maximum of ten receptacles. Since integrally-switched fluorescent lamps and mogul base, high-intensity (HID) lamp fixtures that are mounted below eight feet must be served by 120 volts, according to NEC articles *210-6(a)d* and *210-6(a)c* respectively, these task lighting fixtures must be fed from the 120v convenience receptacle circuits. A 400W HID fixture, with ballast, consumes 470 watts, so a 120 volt, 20 ampere branch circuit breaker can only carry four of these at 1880 watts. Since these fixtures will be connected one per receptacle, only four receptacles per circuit can be allowed, due to the higher load per receptacle. With only four receptacles per 20 ampere circuit, rather than the 10 allowed by code (without additional lighting loads), 250% more branch circuits, 250% more branch circuit breakers, and 250% more panelboards are needed.[5-14] This not only increases the first cost of the electrical distribution system but also increases the life cycle costs, due to the greater maintenance required by the additional, larger equipment being used. In addition, lower voltage equipment has larger voltage drops and increases energy wastage. The increased number and sizes of the step-down transformers also introduce a larger energy wastage due to the transformer core and coil losses.

Although a task/ambient lighting design may, under some design conditions, reduce the raw watts used for lighting, when viewed as part of the total electrical system it tends to increase the first cost and the life cycle costs of the project.

While task/ambient lighting produces its light in the working space, it also converts all of of its power consumption into heat within the working space. More air changes are then needed to remove heat from the space, requiring larger, more costly ductwork and air handler fan motors (which consume more energy) to handle the increased ventilation load.

Chapter 6

ENERGY MANAGEMENT OF ELECTRICAL POWER SYSTEMS

Almost 72% of the energy used in the generation of electrical power is lost in boiler and generator losses in the power plant and transmission and distribution losses up to buildings. The amount of power used within the building is dependent upon (1) the power used for lighting, heating, ventilation, etc. (2) power lost while distributing to these loads (3) the operation and selection of electrical equipment. The major causes of electrical power losses are: low power factor, high voltage drops, improper distribution voltage, inefficient operation of motors and transformers.

POWER SYSTEM TERMINOLOGY

Energy efficient electrical design begins with a clear understanding of basic electrical power system terminology.[6-1]

Demand Load (D)   The average rate at which energy is delivered to a building or a piece of equipment during a specified continuous time interval which is usually 15, 30 or 60 minutes.

Maximum Demand (D_{max})   The greatest demand imposed on an energy system by a device or by the simultaneous operation of several devices within some time interval. $D_{max} = D_1 + D_2 + D_3 + - - - + D_n =$ the demand load occuring at the instant when a maximum number of devices are in operation.

Connected Load (L_c)   The sum of the full load power of all the devices connected in the system, whether in use or not. $L_c = L_{c1} + L_{c2} + L_{c3} + - - + L_{cn}$.

Demand Factor (f_{dem})   The ratio of the maximum demand to the connected load.

$$f_{dem} = \frac{D_{max}}{L_c}$$

Diversity Factor (f_{div})   The ratio of the sum of the maximum demands of the subsystems to the maximum demand for the entire system. The diversity factor is always greater than 1.

$$f_{div} = \frac{D_{max_1} + D_{max_2} + D_{max_3} + - + D_{max_n}}{D_{max_s}}$$

Peak Load (L_p)   The highest instantaneous load. Only when the time interval chosen is the same as that for demand load, does the peak load equal maximum demand. The instantaneous peak load, or one averaged during a shorter time interval than the demand load, can far exceed maximum demand.

Load Factor (f_L)   The ratio of average load, during any period of time, to peak load. Average load is the total energy used during a period, divided by the number of hours in the period. Thus, the monthly load factor

$$f_L = \frac{kWh \text{ per } 30 \text{ days}}{720 \text{ hours} \times L_p}$$

ELECTRICAL POWER SYSTEM DESIGN

The design of electrical power systems can be broadly divided into four stages.

- Primary distribution system design
- Building electrical system design
- Specification of equipment
- Design of controls

Primary Distribution System

The considerations that can produce an energy efficient and economical primary system are:

- Combining meters
- Using the highest voltage available
- Purchasing power at the most favorable rate

72

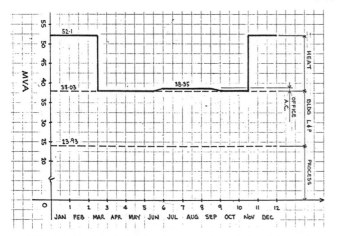

FIG. 6-1. Profile of the electric usage in an industrial complex.

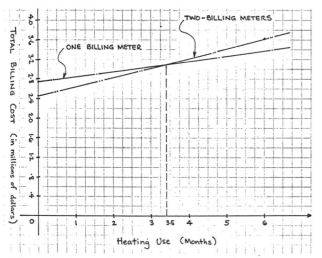

FIG. 6-2. Breakeven point between a one and two-meter system for the industrial complex shown in Figure 6-1.

• Locating utility meters close to the site

Combining Separately Metered Services

Utility rate schedules usually assign the highest rates to initial blocks of monthly kW demand and monthly kWh consumption. With separately metered services, the customer generally pays higher than average demand charges and energy charges for each meter. The utility rate should be analyzed to see if it is more economical to have one meter.

The utility rates for an industrial complex having the profile shown in Figure 6-1 were as follows:

Electric heat rate   =   \$0.0190/kWh
Usage rate (to 100 hours)   =   \$0.0143/kWh
     (over 100 hours)   =   \$0.0089/kWh
Demand rate   =   \$1.95/kW

A general analysis of the above rates showed that if

heating is used less than 3.5 months a year, the total cost of electric power and heat will be less with a two meter system that uses one meter for heat and another for power. On the other hand, if heating is used more than 3.5 months, the one meter system will cost less. Figure 6-2 shows the comparative costs.

Use the Highest Voltage Available

Utilities often offer a cost incentive to customers accepting service at the highest voltage available at the site. To qualify for these reduced rates, the customer is usually required to purchase and maintain all switchgear, transformers, bus and cable on his side of the utilized metering transformers. Additional capital cost, added maintenance costs, and transformer losses offset much of the apparent savings. Thus, each case should be evaluated separately. One study showed that the added costs were paid back in one year.

Purchase Power at an Alternate Rate

Investigation of the utility rate schedules may reveal a schedule that qualifies the customer for a more favorable rate. This is more likely to occur on smaller metered services that are impractical to combine with the major service to a site.

Locate Utility Meters Close to the Site

Utility meters should be located close to the site to decrease line losses and to improve voltage regulation.

Building Electrical System Design

A proper design methodology for building electrical systems produces an energy efficient and economical distribution system. A number of computer programs may be used in the process. Broadly, the methodology can be reduced to the followng steps, with energy conservation being considered at every stage.

Design Step	Energy Conservation Considerations
1. Data collection and design criteria	Selection of proper voltage for motors and electrical distribution
2. Load analysis and preliminary substation design	Use proper demand and diversity factors Use load center concept Select an optimal transformer size
3. Evaluation of distribution system schemes	Select a reliable distribution system
4. Complete system design	Select proper size of feeders Reduce length of cable runs

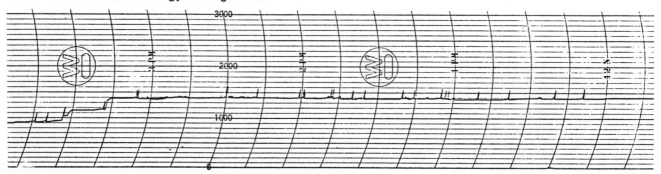

FIG. 6-3. Load profile of an industrial transformer.

Balance the loads on the system

Select proper size of over current and ground fault protective devices

5. Perform load flow analysis of the system — Check the power factor at every node to see if it meets ASHRAE 90 Standard

Check the voltage drop in the feeders to see if the voltage drops meet ASHRAE 90 Standard

5. Perform short-circuit analysis — Check the short-circuit rating of protective devices

6. Protective device coordination — Selectively coordinate the over current and ground fault protective devices

Data Collection

Collecting system design data is usually the first phase of the work. Utility source data, such as available fault MVA and source, and future planning for the source are sought. The design program for the building or site determines the loads and physical factors of the design. Criteria, such as distribution or usable voltages, client's specification, and other factors are accumulated.

If historical data is available, future loads are forecast by using an electrical load forecasting program. Otherwise loads are estimated over the useful life of the project.

Load Analysis & Substation Design

Projections are made as to the location and size of the loads. Estimates are made for the demand and diversity factors. Preliminary size, number and the location of substations is determined. The load center concept is used for electrical distribution; i.e., high voltage power is brought up to the substation, which is located at the center of the loads, thus insuring minimum voltage drop in the feeders.

Transformers are sized close to their demand kva. In sizing transformers, proper consideration is given to no load and full load losses. If the load profile of a building is known, an economical size of a transformer, which has the least operating losses over the life of a transformer, can be found.

Computer programs are available which size the transformers for a given load profile. For example, one such program was used to determine the size of the transformer for the load profile shown in Figure 6-3. The optimal transformer size was 1,500 kva.

Evaluation of Distribution System Schemes

Applicable schemes are then developed for the major components of the distribution system, beginning with the simplest and then adding incremental levels of redundancy and complexity until the most adequate scheme is attained. Engineers must consider various voltages, tie- or network-systems, and emergency or shared generation. Parallel evaluations are made for systems requiring critical or emergency power.

Each scheme is evaluated for its fault-current potential, load flow, power factor, and voltage transient potential and is subjected to a complete reliability analysis of the major elements of the system.

Complete System Design

The major components having been determined, the design is incorporated into the project. At this stage, a computer program is used to design an electrical distribution system. The program input is the riser diagram and individual branch circuit and motor loads. Program output is the optimal sizes of feeders, conduits and starters, protective device ratings and schedules for lighting panels, distribution panels and motor control centers. The program is used iteratively, allowing frequent recalculations of the entire distribution system. During this phase of the work, care should be taken to balance the loads on all three phases.

Load Flow Analysis

Load flow analysis is the most important step in the design of energy efficient electrical distribution systems. Proper design demands rigorous calculations and analyt-

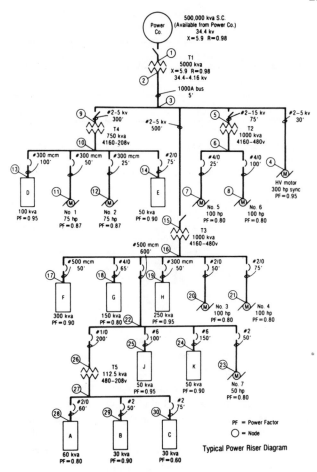

FIG. 6-4. Typical power riser diagram.[6-2]

ical evaluation of system characteristics. A concurrent power flow analysis provides an efficient means of determining a system's operational status.

Load flow analysis can be expedited with a Fortran program developed by the co-author and an associate[6-2]. It is designed to provide the engineer with a clear, concise mathematical record of power flow, line losses, percent voltage drop, power factor, and receiving end voltage. It provides information vital to the verification of optimum systems design in terms of specific component application and the resulting cost benefits.

Development of a complete riser diagram for the system to be analyzed is necessary for determination and accumulation of input data. Each point (node) on the system must be identified with a sequential numeric, as shown in Figure 6-4, a typical power riser diagram. The riser diagram must be completed to a point where component parts (i.e., panels, cables, motors, etc.) are identified by load kilovolt-ampere (kva), cable size and length, voltage, and load power factor as applicable.

Power factor, for a given load, is input as a percent value. It should be determined as accurately as possible. If an appliance type panelboard is one of the end nodes on the system, and is found to contain both power and lighting loads, the resultant power factor for the panel should be used as input data. The same is true for any loads that are input in aggregate form.

Certain data, such as resistance and reactance values for cables, transformer percent resistance and percent reactance are stored in the program data bank. However, these values may be input manually if they differ from the stored values.

Input forms are used to accumulate the data found on the riser diagram. These data then are transferred to key punch cards for processing by the program. The three forms that are used are a title card for job title and number information, a node description card, and a branch description card.

The node description card lists the node number identification, special codes to identify the type of load, node identification (i.e., panel A, main bus, etc.), voltages (if transformer), load kilovolt-amperes, and the power factor of the load. Each line on the form represents a separate key punch card. The number of forms required will depend on the total number of nodes to be computed.

The branch description cards are used to identify the size, type, and length of component parts of the system being analyzed. They further identify the sequential interconnection and relationships between nodes. The first bit of required data is the start node and end node numbers, which describe, for the program, the direction of flow in the system. The code column is used to identify the existence of a transformer. The code name column is used to state the code for a specific feeder or transformer. With the code name given, the program will look for and use the stored data bank information (reactance and resistance) for the given code. If a special code is given, the user may provide X and R information as required. The final entry is length (in feet) of the feeder between the corresponding nodes.

The input data are relatively simple to enter on the forms and subsequently process through the program. Special attention should be given to accurate accumulation of data and preparation of input forms to insure maximum first-run results.

The program output, Figure 6-5, includes a listing of the input data for both node description and branch description that were used by the program during the calculations. These data should be verified as correct before any analysis of the results is attempted.

The program results are an analysis of the power flow,

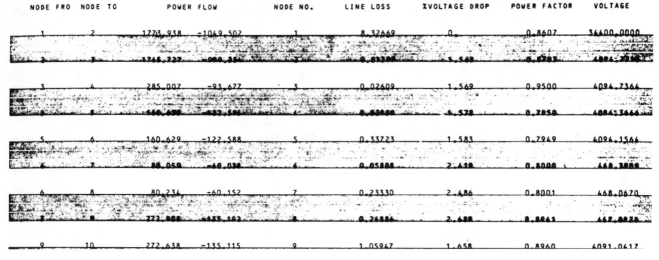

NODE FRO	NODE TO	POWER FLOW		NODE NO.	LINE LOSS	%VOLTAGE DROP	POWER FACTOR	VOLTAGE
1	2	1773.938	-1069.502	1	8.32669	0.	0.8607	36600.0000
3	4	285.002	-93.677	3	0.02609	1.569	0.9500	4094.7364
5	6	160.629	-122.588	5	0.33723	1.583	0.7949	4094.1564
6	8	80.236	-60.152	7	0.23330	2.486	0.8001	468.0670
9	10	272.638	-135.115	9	1.05947	1.658	0.8960	4091.0617

Fig. 6-5. Load flow analysis printout for the system shown in Figure 6-4.

line loss (I^2R), percent voltage drop, power factor, and receiving end voltage occurring at and between each "from-to" node identified. The power flow values are the real (kW) and reactive (kvar) values of power flowing between nodes. The I^2R value represents the power being lost, as heat, between nodes.

Voltage drop is the percent voltage drop occurring at this "to" or end node for a specific branch. The power factor value given is the power factor at the load, taking into account the power factor of the total distribution system. The final piece of information is the receiving end voltage resulting at an "end" node.

The advantage of having an analysis, by branch, of an entire distribution system is clearly recognizable. With the power flow, I^2R losses, power factor, and voltage drop values available, an analysis will determine if optimum operating characteristics exist.

The resultant power factor and voltage drop can be identified for each node in the system. Where values are below the accepted minimum, corrective action can be applied. (ASHRAE 90 requires that power factors lower than 85%, at rated load, be corrected to at least 90% and that voltage drop does not exceed 3% in branch circuits or feeders for a total of 5% to the farthest outlet based on steady state load conditions.)

Percent voltage drop and line losses can be evaluated in terms of the overall effect on the total energy used by the system. If, for example, voltage drop is high and power in the system is assumed constant, then it follows that line losses will be excessive. To limit the losses in the system, it would be necessary, perhaps, to increase the size of certain feeders. This procedure would reduce the voltage drop and effect a corresponding reduction in line losses.

Power flow analysis is an accurate and expeditious way to complete the tedious calculations necessary to define a system in terms of its energy utilization parameters. Power flow analysis can aid the engineer in the search for the proper balance between energy efficient versus cost beneficial design. Knowing the I^2R losses of a line, for example, may require increased capital investment or first cost, because of the increased feeder size required. However, long-term benefits will be realized. The feeder may be "oversized," but it will not needlessly waste energy.

Short Circuit Analysis

Short-circuit analysis of a system is performed by a computer program that gives the values of the fault currents, in case the following faults occur:

- 3-phase bolted fault
- Double line to ground fault
- Line to line fault
- Line to ground fault

The fault current values are used for proper specification of interrupting capacities, i^2t ratings and bracings for all system components.

Protective Device Coordination

Protective devices are coordinated in the system so that, in case of a fault, the protective device closest to the fault opens up, thus maintaining continuity of service to the rest of the system. Protective device coordination is performed by a computer program that gives the setting of each relay, breaker and fuse in the system, for selective coordination of the system. The program also draws the coordination curves for the system.

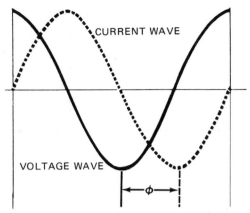

FIG. 6-6. Alternating current and voltage wave forms.

POWER FACTOR IMPROVEMENT

In an AC circuit, having the voltage and current waveforms shown in Figure 6-6, the average power or actual power is given by

$$P = VI \cos \phi \qquad (6\text{-}1)$$

where
P = Actual power in watts
V = rms value of voltage in volts
I = rms value of current in amperes

The value Cos ø is called the power factor of the circuit. The value VI is the apparent power in the circuit. The product VI Sin ø is the reactive power in the circuit. Figure 6-7 shows the power triangle for an inductive load with a lagging power factor. Thus, the power factor of a circuit is defined as the ratio of the actual power to apparent power in a circuit,

$$\text{Power factor} = \frac{\text{Actual Power}}{\text{Apparent Power}} = \cos \phi$$

It is clear, from the definition of power factor, that its value lies between 0 and 1. Power factor is said to be unity (1) when the voltage and current waves are in phase with each other; i.e., in circuits having totally resistive loads. A system is said to have a lagging power factor when it is supplying power to inductive loads; i.e., induction motors, arc furnaces, etc. Capacitive loads, such as synchronous motors, are said to have a

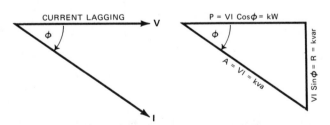

FIG. 6-7. Power triangle for an inductive load with a lagging power factor.

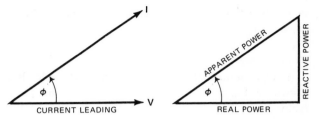

FIG. 6-8. Power triangle for a capacitive load with a leading power factor.

leading power factor. Figure 6-8 is a power triangle for a system having a leading power factor. Table 6-1 lists some commonly used loads having leading, unity and lagging power factors.[6-3]

TABLE 6-1. Typical Power Factors

LOAD	POWER FACTOR
Induction motors (full wound)	.55 to .75
Polyphase squirrel-cage motors	.75 to .90
Wound rotor motors	.80 to .85
Welders	0.50 to .70
Arc furnaces	0.80 to 0.90
Induction furnaces	0.60 to 0.80
Incandescent lamps	1.0
Fluorescent lamps	0.95 to 0.97
Resistor heaters	1.0
Synchronous motors	1.0 or leading
Rotary converters	1.0
Capacitors	0

Disadvantages of Low Power Factor

A low power factor indicates low electrical efficiency, which means that the actual power consumed is less than the apparent power (VI). Transformers and generators are rated on the basis of volt-amperes and, if the volt-ampere value exceeds the real power being used, extra current is being drawn from the service. Thus, poor power factor results in overloaded cables, transformers, etc.; increased copper losses, reduced voltage levels, resulting in poor motor operation and reduced illumination from lighting. Power companies generally use a rate structure based on kWh of energy and kW of demand. This type of rate can lead to a situation where the utility, feeding a poor power factor, is forced to supply a disproportionately large quantity of kilovolt-amperes in relation to the amount of kWh being billed. To counter this possibility, the rates include penalties for any power factor lower than 0.85.

Power Factor of a Group of Loads

Power factor of a single load is generally given or can

be easily estimated. Power factor of a group of loads is calculated.

Example 6-1

A substation supplies three different kinds of loads as shown in Figure 6-9. Calculate the substation power factor.

The following is a step-by-step process used to find the power factor of a substation.

1. Find the kilowatts and kilovars (kva reactive) of each load.

Lighting load
  pf = 1.0
  kva = 50
  kW = kva × pf = 50
  kvar = kva × Sin ϕ = 0

Induction motor load
  pf = 0.8 (lagging)
  kva = 200
  kW = 200 × .8 = 160
  kvar = $\sqrt{(kva^2 - kW^2)} = \sqrt{(200^2 - 160^2)} = 120$

Synchronous motor load
  pf = 0.8 (leading)
  kva = 100
  kW = 100 × .8 = 80
  kvar = $\sqrt{(100^2 - 80^2)} = 60$

2. Find the kilowatts and kilovars that the substation should supply.

Kilowatts
  Lights   =   50 kW
  Induction motor = 160 kW
  Synchronous motor = 80 kW
                    = 290 kW

Kilovars
  Lights   =   0
  Induction motor = 120 (+ lagging)
  Synchronous motor = −60 (− leading)
                    =   60 kvar

3. Find the substation kva and power factor.

$$kva = \sqrt{(kW^2 + kvar^2)} = \sqrt{(290^2 + 60^2)} = 296.1$$

$$pf = \frac{290}{296.1} = 0.979 \text{ (lagging)}$$

Power Factor Improvement

Power factor of a system is improved by installing synchronous motors or capacitors in the system. Synchronous motors, when overexcited (normal operation), act as kilovar generators. When underexcited, synchro-

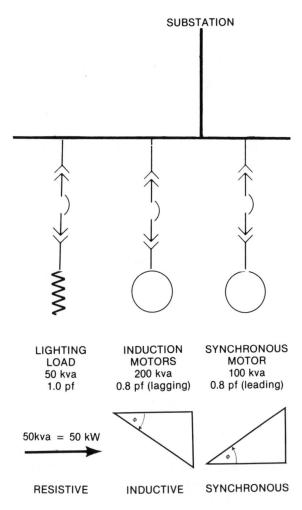

FIG. 6-9. Power factors of a group of loads.

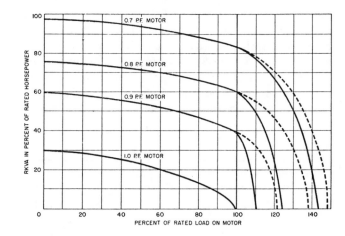

FIG. 6-10. Kilovars supplied by synchronous motors at various power factors. Solid lines indicate under-excitation and dotted lines indicate rated excitation, at overload.

nous motors do not generate enough kilovars to supply their own requirements.[6-4]

Synchronous motors are generally used in industrial plants for power factor improvement. Figure 6-10 shows kilovars supplied by synchronous motors at various power factors.

Power Factor Improvement by Capacitors

Capacitors are used for low power factor correction because the leading current drawn by a capacitive circuit is opposite to the lagging current drawn by an inductive circuit. When two circuits are combined into one, the effect of capacitance reactance tends to cancel the effects of inductance reactance. The net amount of reactive power in the circuit is thus reduced and the power factor is improved. A proper size capacitor gives perfect cancellation, but too little or too much capacitance should be avoided.[6-5]

Capacitors are rated in vars or kilovars, to indicate their power factor correction capabilities. One var is equivalent to one volt-ampere of reactive power, and one kilovar equals 1,000 vars. Thus, one kvar is equivalent to one kva of reactive power. The kvar capacity of the capacitors required to improve the power factor of a system can be calculated by using a power triangle, Figure 6-11.

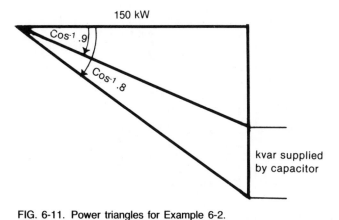

FIG. 6-11. Power triangles for Example 6-2.

Example 6-2

A group of loads, totaling 150 kW, has a power factor of .8. Find the capacitor kvar required to improve the power factor of the substation to .9.

1. Calculate the kvar required from the line at .8 pf.

$$kW = 150$$
$$kva = \frac{150}{.8} = 187.5$$
$$kvar = \sqrt{(187.5^2 - 150^2)} = 112.5$$

2. Calculate the kvar required at .9 pf.

$$kW = 150$$
$$kva = \frac{150}{.9} = 166.67$$
$$kvar = \sqrt{(166.67^2 - 150^2)} = 72.65$$

$$\text{kvar to be provided by the capacitor} = 112.5 - 72.65$$
$$= 39.85 \text{ kvar}$$

Power Factor Correction Factor

From the power triangle, Figure 6-7, we have

$$\text{Cos } \phi = \frac{kW}{kva} = pf$$
$$\text{Sin } \phi = \frac{kvar}{kva}$$
$$\text{Tan } \phi = \frac{kvar}{kW}$$

Assume that we want to improve the power factor from Cos ϕ_1 (.8) to Cos ϕ_2 (.9), then

$$\text{kvar at original pf} = kW \times \text{Tan } \phi_1$$
$$\text{kvar at improved pf} = kW \times \text{Tan } \phi_2$$

The necessary capacitor kvar is the difference in the two tangents

$$\text{Capacitor kvar} = kW(\text{Tan } \phi_1 - \text{Tan } \phi_2)$$
$$= kW \times \Delta\text{Tan } \phi$$

The values for ΔTan ϕ are found in Table 6-2. The original pf is listed in the vertical column, the corrected pf in the horizontal rows. Where they intersect, you will find the ΔTan ϕ factor. To calculate the required kvar, multiply the kW by the ΔTan ϕ factor from the table. Returning to *Example 6-2*,

$$kW = 150$$
$$\text{original pf} = .8$$
$$\text{improved pf} = .9$$
$$\Delta\text{Tan } \phi = 0.266 \text{ (from Table 6-2)}$$
$$\text{capacitor kvar} = 150 \times .266 = 39.9 \text{ kvar}$$

Locating Capacitors

Capacitors can provide either group correction or localized correction, Figure 6-12.

Group correction can be provided at primary or secondary transformers or at a main switchgear or motor control center bus. Group correction is required where loads shift between feeders and where motor voltages are low; such as 230 volts.[6-6]

Here are some of the advantages of group correction.

1. Group capacitors have a lower cost per kilovar than individual capacitors.
2. Group installation improves overall plant power factor, thus reducing power bills.

TABLE 6-2. kW Multipliers to Determine Capacitor Kilovars Required for Power Factor Correction

Orig-inal Power Factor	Corrected Power Factor																			
	0.80	0.81	0.82	0.83	0.84	0.85	0.86	0.87	0.88	0.89	0.90	0.91	0.92	0.93	0.94	0.95	0.96	0.97	0.98	0.99	1.0
0.50	0.982	1.008	1.034	1.060	1.086	1.112	1.139	1.165	1.192	1.220	1.248	1.276	1.306	1.337	1.369	1.403	1.440	1.481	1.529	1.589	1.732
0.51	0.937	0.962	0.989	1.015	1.041	1.067	1.094	1.120	1.147	1.175	1.203	1.231	1.261	1.292	1.324	1.358	1.395	1.436	1.484	1.544	1.687
0.52	0.893	0.919	0.945	0.971	0.997	1.023	1.050	1.076	1.103	1.131	1.159	1.187	1.217	1.248	1.280	1.314	1.351	1.392	1.440	1.500	1.643
0.53	0.850	0.876	0.902	0.928	0.954	0.980	1.007	1.033	1.060	1.088	1.116	1.144	1.174	1.205	1.237	1.271	1.308	1.349	1.397	1.457	1.600
0.54	0.809	0.835	0.861	0.887	0.913	0.939	0.966	0.992	1.019	1.047	1.075	1.103	1.133	1.164	1.196	1.230	1.267	1.308	1.356	1.416	1.559
0.55	0.769	0.795	0.821	0.847	0.873	0.899	0.926	0.952	0.979	1.007	1.035	1.063	1.093	1.124	1.156	1.190	1.227	1.268	1.316	1.376	1.519
0.56	0.730	0.756	0.782	0.808	0.834	0.860	0.887	0.913	0.940	0.968	0.996	1.024	1.054	1.085	1.117	1.151	1.188	1.229	1.277	1.337	1.480
0.57	0.692	0.718	0.744	0.770	0.796	0.822	0.849	0.875	0.902	0.930	0.958	0.986	1.016	1.047	1.079	1.113	1.150	1.191	1.239	1.299	1.442
0.58	0.655	0.681	0.707	0.733	0.759	0.785	0.812	0.838	0.865	0.893	0.921	0.949	0.979	1.010	1.042	1.076	1.113	1.154	1.202	1.262	1.405
0.59	0.619	0.645	0.671	0.697	0.723	0.749	0.776	0.802	0.829	0.857	0.885	0.913	0.943	0.974	1.006	1.040	1.077	1.118	1.166	1.226	1.369
0.60	0.583	0.609	0.635	0.661	0.687	0.713	0.740	0.766	0.793	0.821	0.849	0.877	0.907	0.938	0.970	1.004	1.041	1.082	1.130	1.190	1.333
0.61	0.549	0.575	0.601	0.627	0.653	0.679	0.706	0.732	0.759	0.787	0.815	0.843	0.873	0.904	0.936	0.970	1.007	1.048	1.096	1.156	1.299
0.62	0.516	0.542	0.568	0.594	0.620	0.646	0.673	0.699	0.726	0.754	0.782	0.810	0.840	0.871	0.903	0.937	0.974	1.015	1.063	1.123	1.266
0.63	0.483	0.509	0.535	0.561	0.587	0.613	0.640	0.666	0.693	0.721	0.749	0.777	0.807	0.838	0.870	0.904	0.941	0.982	1.030	1.090	1.233
0.64	0.451	0.474	0.503	0.529	0.555	0.581	0.608	0.634	0.661	0.689	0.717	0.745	0.775	0.806	0.838	0.872	0.909	0.950	0.998	1.068	1.201
0.65	0.419	0.445	0.471	0.497	0.523	0.549	0.576	0.602	0.629	0.657	0.685	0.713	0.743	0.774	0.806	0.840	0.877	0.918	0.966	1.026	1.169
0.66	0.388	0.414	0.440	0.466	0.492	0.518	0.545	0.571	0.598	0.626	0.654	0.682	0.712	0.743	0.775	0.809	0.846	0.887	0.935	0.995	1.138
0.67	0.358	0.384	0.410	0.436	0.462	0.488	0.515	0.541	0.568	0.596	0.624	0.652	0.682	0.713	0.745	0.779	0.816	0.857	0.905	0.965	1.108
0.68	0.328	0.354	0.380	0.406	0.432	0.458	0.485	0.511	0.538	0.566	0.594	0.622	0.652	0.683	0.715	0.749	0.786	0.827	0.875	0.935	1.078
0.69	0.299	0.325	0.351	0.377	0.403	0.429	0.456	0.482	0.509	0.537	0.565	0.593	0.623	0.654	0.686	0.720	0.757	0.798	0.846	0.906	1.049
0.70	0.270	0.296	0.322	0.348	0.374	0.400	0.427	0.453	0.480	0.508	0.536	0.564	0.594	0.625	0.657	0.691	0.728	0.769	0.817	0.877	1.020
0.71	0.242	0.268	0.294	0.320	0.346	0.372	0.399	0.425	0.452	0.480	0.508	0.536	0.566	0.597	0.629	0.663	0.700	0.741	0.789	0.849	0.992
0.72	0.214	0.240	0.266	0.292	0.318	0.344	0.371	0.397	0.424	0.452	0.480	0.508	0.538	0.569	0.601	0.635	0.672	0.713	0.761	0.821	0.964
0.73	0.186	0.212	0.238	0.264	0.290	0.316	0.343	0.369	0.396	0.424	0.452	0.480	0.510	0.541	0.573	0.607	0.644	0.685	0.733	0.793	0.936
0.74	0.159	0.185	0.211	0.237	0.263	0.289	0.316	0.342	0.369	0.397	0.425	0.453	0.483	0.514	0.546	0.580	0.617	0.658	0.706	0.766	0.909
0.75	0.132	0.158	0.184	0.210	0.236	0.262	0.289	0.315	0.342	0.370	0.398	0.426	0.456	0.487	0.519	0.553	0.590	0.631	0.679	0.739	0.882
0.76	0.105	0.131	0.157	0.183	0.209	0.235	0.262	0.288	0.315	0.343	0.371	0.399	0.429	0.460	0.492	0.526	0.563	0.604	0.652	0.712	0.855
0.77	0.079	0.105	0.131	0.157	0.183	0.209	0.236	0.262	0.289	0.317	0.345	0.373	0.403	0.434	0.466	0.500	0.537	0.578	0.626	0.685	0.829
0.78	0.052	0.078	0.104	0.130	0.156	0.182	0.209	0.235	0.262	0.290	0.318	0.346	0.376	0.407	0.439	0.473	0.510	0.551	0.599	0.659	0.802
0.79	0.026	0.052	0.078	0.104	0.130	0.156	0.183	0.209	0.236	0.264	0.292	0.320	0.350	0.381	0.413	0.447	0.484	0.525	0.573	0.633	0.776
0.80	0.000	0.026	0.052	0.078	0.104	0.130	0.157	0.183	0.210	0.238	0.266	0.294	0.324	0.355	0.387	0.421	0.458	0.499	0.547	0.609	0.750
0.81		0.000	0.026	0.052	0.078	0.104	0.131	0.157	0.184	0.212	0.240	0.268	0.298	0.329	0.361	0.395	0.432	0.473	0.521	0.581	0.724
0.82			0.000	0.026	0.052	0.078	0.105	0.131	0.158	0.186	0.214	0.242	0.272	0.303	0.335	0.369	0.406	0.447	0.495	0.555	0.698
0.83				0.000	0.026	0.052	0.079	0.105	0.132	0.160	0.188	0.216	0.246	0.277	0.309	0.343	0.380	0.421	0.469	0.529	0.672
0.84					0.000	0.026	0.053	0.079	0.106	0.134	0.162	0.190	0.220	0.251	0.283	0.317	0.354	0.395	0.443	0.503	0.646
0.85						0.000	0.027	0.053	0.080	0.108	0.136	0.164	0.194	0.225	0.257	0.291	0.328	0.369	0.417	0.477	0.620
0.86							0.000	0.026	0.053	0.081	0.109	0.137	0.167	0.198	0.230	0.264	0.301	0.342	0.390	0.450	0.593
0.87								0.000	0.027	0.055	0.083	0.111	0.141	0.172	0.204	0.238	0.275	0.316	0.364	0.424	0.567
0.88									0.000	0.028	0.056	0.084	0.114	0.145	0.177	0.211	0.248	0.289	0.337	0.397	0.540
0.89										0.000	0.028	0.056	0.086	0.117	0.149	0.183	0.220	0.261	0.309	0.369	0.512
0.90											0.000	0.028	0.058	0.089	0.121	0.155	0.192	0.233	0.281	0.341	0.484
0.91												0.000	0.030	0.061	0.093	0.127	0.164	0.205	0.253	0.313	0.456
0.92													0.000	0.031	0.063	0.097	0.134	0.175	0.223	0.283	0.426
0.93														0.000	0.032	0.066	0.103	0.144	0.192	0.252	0.395
0.94															0.000	0.034	0.071	0.112	0.160	0.220	0.363
0.95																0.000	0.037	0.079	0.126	0.186	0.329
0.96																	0.000	0.041	0.089	0.149	0.292
0.97																		0.000	0.048	0.108	0.251
0.98																			0.000	0.060	0.203
0.99																				0.000	0.143

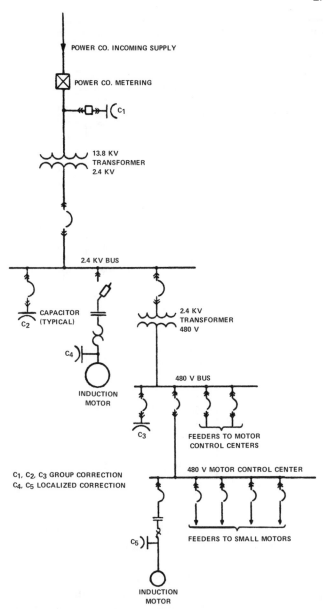

FIG. 6-12. Capacitors can provide either a group or localized correction.

3. Labor and material costs for installing group capacitors are lower per kvar than the cost of individual capacitors for each motor.
4. All or part of the group capacitors can be switched on or off manually or automatically, depending upon load requirements and power factor correction requirement.

Individual power factors can be corrected by installing capacitors on small feeders, on branch motor circuits, or directly on motors or a group of motors and switching them with the motors. For greater benefits, the capacitors should be connected as near as possible to the load or near the end of feeders. When capacitors are used directly at the motor terminals, the size for each motor may be determined from Table 6-3. The table shows capacitor ratings in accordance with AIEE recommendations; these values will improve power factor to about 95%.

The main advantages of individual capacitor installations are:
1. Capacitors are switched in and out of the circuit with the motors, thus the power factor is adjusted whenever needed.
2. Voltage drop to the individual motors is reduced, resulting in better motor performance.
3. More flexibility is provided by installing capacitors on individual motors. In the case of plant alterations, the motor and capacitor are moved together.

Since small capacitors cost more per kvar than larger units, it is more economical to use individual capacitors with motors larger than 10 hp and to correct smaller motors in groups.

Advantages of Power Factor Correction

Reduced Power Costs   The utility rate consists of an energy charge plus a demand charge. By improving power factor, the energy consumed is the same but there is a reduction in kva demand.

For example: a building had 2,000 kW and 2,700 kva demand for power. It was decided to install capacitors to improve its power factor to .95. The following simple analysis shows the savings in utility costs by improving the power factor.

$$\text{Original pf of the system} = \frac{2,000}{2,700} = 0.74$$
$$\text{Corrected pf} = 0.95$$

From Table 6-2, the multiplier for power factor
$$\text{correction} = 0.58$$
$$\text{Capacitor kvar required} = 2,000 \times 0.58$$
$$= 1160.0$$

The installed cost of capacitors for a 480v system is about $10 per kvar.

Installed cost of 1160.0 kvar capacitors = $11,600

New kva demand $= \frac{2,000}{.95} = 2105$ kva

Savings in the demand charge per month $= (2700 - 2105) \times \$1.65 = \$982$.

Annual savings $= \$982 \times 12 = \$11,784$. Thus, the cost of the installed capacitors is repaid in less than one year.

Releasing Existing System Capacity System capacity can be released by adding capacitors to a sub-

TABLE 6-3. Maximum Capacitor Rating When Motor* and Capacitor are Switched as a Unit

Induction Motor Horse-Power Rating	Nominal Motor Speed in RPM											
	3600		1800		1200		900		720		600	
	Capacitor Rating KVAR	Line Current Reduction %	Capacitor Rating KVAR	Line Current Reduction %	Capacitor Rating KVAR	Line Current Reduction %	Capacitor Rating KVAR	Line Current Reduction %	Capacitor Rating KVAR	Line Current Reduction %	Capacitor Rating KVAR	Line Current Reduction %
3	1.5	14	1.5	15	1.5	20	2	27	2.5	35	3.5	41
5	2	12	2	13	2	17	3	25	4	32	4.5	37
7½	2.5	11	2.5	12	3	15	4	22	5.5	30	6	34
10	3	10	3	11	3.5	14	5	21	6.5	27	7.5	31
15	4	9	4	10	5	13	6.5	18	8	23	9.5	27
20	5	9	5	10	6.5	12	7.5	16	9	21	12	25
25	6	9	6	10	7.5	11	9	15	11	20	14	23
30	7	8	7	9	9	11	10	14	12	18	16	22
40	9	8	9	9	11	10	12	13	15	16	20	20
50	12	8	11	9	13	10	15	12	19	15	24	19
60	14	8	14	8	15	10	18	11	22	15	27	19
75	17	8	16	8	18	10	21	10	26	14	32.5	18
100	22	8	21	8	25	9	27	10	32.5	13	40	17
125	27	8	26	8	30	9	32.5	10	40	13	47.5	16
150	32.5	8	30	8	35	9	37.5	10	47.5	12	52.5	15
200	40	8	37.5	8	42.5	9	47.5	10	60	12	65	14
250	50	8	45	7	52.5	8	57.5	9	70	11	77.5	13
300	57.5	8	52.5	7	60	8	65	9	80	11	87.5	12
350	65	8	60	7	67.5	8	75	9	87.5	10	95	11
400	70	8	65	6	75	8	85	9	95	10	105	11
450	75	8	67.5	6	80	8	92.5		100	9	110	11
500	77.5	8	72.5	6	82.5	8	97.5	9	107.5	9	115	10

*For use with 3-phase, 60 cycle NEMA Classification B Motors to raise full load power factor to approximately 95%.

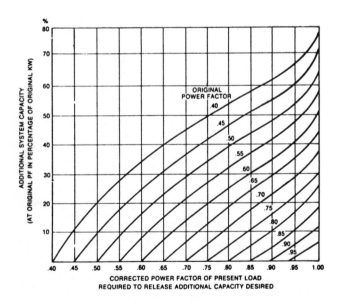

FIG. 6-13. Additional system capacity released by power factor correction.

station. For a given kW load, the higher the power factor, the lower the kva. Thus, adding capacitors to a substation is the cheapest way to release extra capacity. Figure 6-13 gives the curves for finding extra system capacity released by the addition of capacitors.

For example, a fully loaded substation operates at .70 power factor and requires additional capacity to serve 20 percent more load. Figure 6-13 shows that 20 percent more capacity can be released by correcting the power factor to .88. Table 6-4 can be used to find how many kvar of capacitors are required per kva of released capacity.

Power System Losses   As described earlier, the higher the power factor, the lower the kva or, in other words, the lower the current. The losses in a power distribution system are given by I^2R; i.e., they are directly proportional to the square of the current or they are inversely proportional to the square of the power factor. Capacitors are effective in reducing only that portion of losses due to the kva current.

TABLE 6-4. KVAR/KVA of Released Capacity

Original PF	Final PF of original load										
	.90	.91	.92	.93	.94	.95	.96	.97	.98	.99	1.00
.95							3.374	3.587	3.873	4.314	5.947
.94						3.058	3.210	3.397	3.644	3.816	5.366
.93					2.824	2.934	3.110	3.233	3.451	3.779	4.909
.92				2.633	2.723	2.825	2.946	3.094	3.287	3.575	4.541
.91			2.487	2.553	2.635	2.727	2.835	2.968	3.143	3.395	4.229
.90		2.346	2.407	2.479	2.553	2.638	2.739	2.858	3.011	3.240	3.970
.89	2.243	2.296	2.332	2.413	2.475	2.558	2.649	2.757	2.896	3.100	3.742
.88	2.195	2.241	2.293	2.350	2.413	2.484	2.566	2.665	2.792	2.974	3.543
.85	2.053	2.098	2.140	2.186	2.238	2.293	2.354	2.431	2.524	2.661	3.068
.80	1.878	1.909	1.939	1.969	2.002	2.041	2.083	2.133	2.196	2.283	2.528
.75	1.735	1.753	1.775	1.797	1.820	1.846	1.876	1.910	1.952	2.008	2.164

$$\text{kW losses} \propto \left(\frac{\text{original pf}}{\text{improved pf}} \right)^2 \qquad (6\text{-}2)$$

Loss reduction due to improved pf

$$= 1 - \left(\frac{\text{original pf}}{\text{improved pf}} \right)^2 \qquad (6\text{-}3)$$

Voltage Improvement   Voltage improvement is an additional benefit of installing capacitors and will be discussed next.

VOLTAGE DROP

Voltage drop is due to the flow of current through an impedence such as that of a transformer, bus, reactor, cable, etc. Voltage drop in the power system within any commercial or industrial building creates a difference in voltage in various parts of the power system.

In designing electrical power systems, voltage drop should be kept to a minimum. This assures normal life and efficient operation of electrical equipment. According to ASHRAE 90, voltage drop should not be allowed to exceed 3% in branch feeders and circuits, for a total of 5% to the farthest outlet based on steady state conditions.

With large voltage drops in the feeders, the voltage at the terminals of the electrical equipment will be low, thus reducing the life of the equipment and impairing its performance.

Here is a brief description of the effect of low voltage on the electrical equipment most commonly used in industrial and commercial buildings.

Motors   Low voltage at the terminals of induction or synchronous motors reduces starting torque, reduces maximum running torque, decreases full load efficiency and increases full load temperature rise.

Luminaires   Low voltage affects the output of most luminaires. For incandescent units, there is approximately a three percent change in lumens for each one percent of voltage deviation. For mercury luminaires with high reactance ballasts, there is a change of approximately three percent in lamp lumens for each one percent deviation from the rated ballast voltage. Fluorescent luminaire output changes approximately one percent for each two and a half percent change in primary voltage. Figure 6-14 shows these variations in lumen output due to changes in voltage.

Resistance Heating   The heat output of a resistive device varies approximately with the square of the impressed voltage. Thus, a 10% drop in voltage will cause a 19% drop in heat output, provided the resistance remains constant.

Capacitors   The power corrective capacity of capacitors varies with the square of the impressed voltage. A drop of 10% in the supply voltage therefore reduces the corrective capacity by almost 20%.

Calculation of Voltage Drop

Steady state voltage drops are due to current flowing through impedance: impedance being the value, in

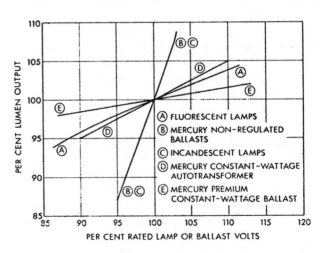

FIG. 6-14. Effect of voltage variations on lamp lumen output.

ohms, of the rms of the resistance and the inductive resistance. To calculate the voltage drop, the circuit impedance, circuit current and power factor should be known. In practice, voltage drop is determined either by formula or from charts.

Voltage Drop by Formula   A number of approximate and exact formulas are available for calculating the voltage drop in the power system.[6-3]

If e_r, the receiving end voltage is known, then the line-to-neutral voltage drop

$$= \sqrt{(e_r \cos \theta + IR)^2 + (e_r \sin \theta + IX)^2} - e_r \quad (6\text{-}4)$$

If e_s, the sending end voltage is known, the line-to-neutral voltage drop

$$= e_s + IR \cos \theta + IX \sin \theta$$
$$- \sqrt{e_s^2 - (IX \cos \theta - IR \sin \theta)} \quad (6\text{-}5)$$

When either e_r or e_s is known, then the approximate line-to-neutral voltage drop

$$= I(R \cos \theta + X \sin \theta)$$

$$\text{Percentage voltage drop} = \frac{kva (R \cos \theta + X \sin \theta)}{10 \, kv^2} \quad (6\text{-}6)$$

where

    I = line current in amperes
    R = resistance of the circuit in ohms
    X = reactance of the circuit in ohms
    $\cos \theta$ = load power factor
    $\sin \theta$ = load reactive power factor
    kva = three-phase kva
    kv = line-to-line kilovolts
    e_r = line-to-neutral voltage at load end
    e_s = line-to-neutral voltage at source end

The vector diagram in Figure 6-15 shows that the error between actual voltage drop and approximate voltage drop is very small.

Voltage Drop by Charts   Figure 6-16 is used to find the voltage drop in 3-phase, 60 cycle, liquid-filled, self-cooled transformers. The chart can also be used for single phase transformers by entering the chart at three times the single phase rating.

Example 6-3

Find the voltage drop in a 1,500 kva, 13,200 to 480v transformer carrying a full load of 1,500 kva at a pf of .85.

Enter Figure 6-16 at 1,500 kva. Move up vertically until you intersect the pf = .85 line. Move horizontally to the percent voltage drop scale. The voltage drop is 3.7%.

$$\text{Actual voltage drop} = 480 \times \frac{3.70}{100} = 17.76 \text{ volts.}$$

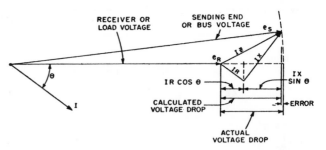

FIG. 6-15. Variation between actual and approximated voltage drop.

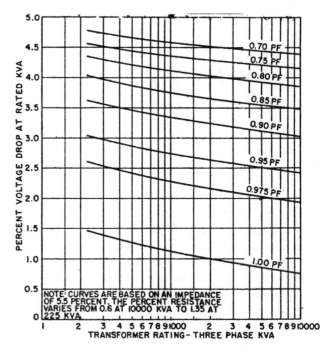

FIG. 6-16. Transformer voltage drop chart.

If the transformer was only carrying 1,000 kva at .85 pf

$$\text{Actual voltage drop} = \frac{1,000}{1,500} \times \frac{3.7}{100} \times 480 = 11.84 \text{ volts.}$$

Example 6-4

Given the power system shown in Figure 6-17, calculate the voltage at the 480v bus and at the motor terminals.

$$\text{Cable resistance (13.2 kv)} = .0577\Omega/1000 \text{ ft} \times 1.5$$
$$= .0865\Omega$$

$$\text{Cable reactance (13.2 kv)} = .0436\Omega/1000 \text{ ft} \times 1.5$$
$$= .0654\Omega$$

$$\text{Resistance of transformer} = 1.0\% \text{ on its base}$$
$$= 1.0 \times \frac{(13.2)^2 \times 10}{2,000} \times .871\Omega$$

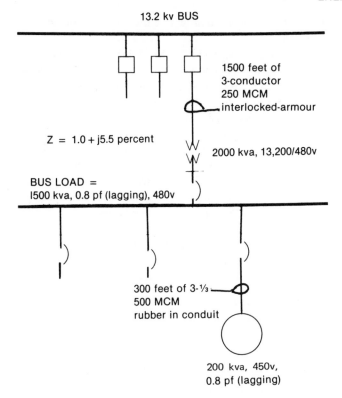

13.2 kv BUS

1500 feet of
3-conductor
250 MCM
interlocked-armour

Z = 1.0 + j5.5 percent

2000 kva, 13,200/480v

BUS LOAD =
1500 kva, 0.8 pf (lagging), 480v

300 feet of 3-⅓
500 MCM
rubber in conduit

200 kva, 450v,
0.8 pf (lagging)

13.2 kv cable resistance = .0577 per 1000 feet
13.2 kv cable reactance = .0436 per 1000 feet
600v cable resistance = .0294 per 1000 feet
600v cable reactance = .0466 per 1000 feet

FIG. 6-17. Example 6-4 power system.

Reactance of transformer = 5.5% on its base

$$= 5.5 \times \frac{(13.2)^2 \times 10}{2,000} \times 4.79\Omega$$

Total resistance = .871 + .0865 = .9575Ω
Total reactance = 4.79 + .0654 = 4.855Ω

Assuming 450 volts on bus B,

Current drawn by the load = $\frac{1,500 \text{ kva}}{\sqrt{3} \times .45 \text{ kv}}$ = 1924A

Current flowing through 13.2 kv cable =

$$1924 \times \frac{480}{13,200} = 69.9\text{A}$$

Using the approximate formula, voltage drop,

$$u = \sqrt{3}I(R \text{ Cos } \theta + \text{Sin } \theta)$$
Cos θ = 0.8
Sin θ = 0.6
$$u = \sqrt{3} \times 69.9 \times (.9575 \times .8 + 4.855 \times .6)$$
$$= 445.4 \text{ volts}$$

Bus B voltage = (bus A voltage − u) × transformer ratio

$$= (13,200 - 445.4) \times \frac{480}{13200} = 463.8 \text{ volts}$$

Using 463.8 volts as the bus voltage and recalculating u,

15 kv cable amperes = $69.9 \times \frac{450}{463.8}$ = 67.8 amps

$$u = \sqrt{3} \times 67.8 \times (.9575 \times .8 + 4.855 \times .6)$$
$$= 432 \text{ volts}$$

Bus B voltage = $(13,200 - 432) \times \frac{480}{13200}$ = 464.3 volts

To calculate the secondary load voltage, assuming 450 volts at the load:

Load Current = $\frac{200 \text{ kva}}{\sqrt{3} \times .45 \text{ kv}}$ = 256.6 amps

Cable resistance (.6 kv) = .0294Ω/1000 ft × .3
$$= .00882\Omega$$
Cable reactance (.6 kv) = .0466Ω/1000 ft × .3
$$= .0139\Omega$$
$$u = \sqrt{3} \times 256.6 \times (.00882 \times .8 + .0139 \times .6)$$
$$= 6.83 \text{ volts}$$
Load voltage = 464.3 − 6.83 = 457.4 volts.

Recalculating u, using 457.4v as the load voltage

$$u = 6.72 \text{ volts}$$

load voltage = 464.3 − 6.72 = 457.6 volts

Reducing Voltage Drop

There are four commonly used methods for reducing voltage drop.

1. Reduce the impedance of the system.
2. Use the load center concept.
3. Use switched capacitors.
4. Use regulating equipment.

Reduce Impedance   From Equation 6-6, it is clear that a reduction in circuit impedance, which is the vector sum of the resistance and the reactance (Z=R+jX), will reduce the voltage drop. Here are some of the methods used to reduce circuit impedance:

- Reduce the length of low voltage feeders.
- Use series capacitors to partially neutralize inductive reactance with capacitive reactance.
- Use two small cables in parallel, rather than one large one.
- Use normal reactance transformers instead of high reactance transformers. (High reactance transformers are used for limiting short circuit currents.)

Load Center Concept   Figure 6-18 shows an electrical distribution system employing a load center. As the power is carried to the load center at high voltage, the length of the secondary feeders is minimized, thus reducing voltage drop.

Switched Capacitors   From equation 6-6, we have
voltage drop e = RI Cos θ + XI Sin θ
$$e = R(\text{kW current}) + X(\text{kvar current})$$

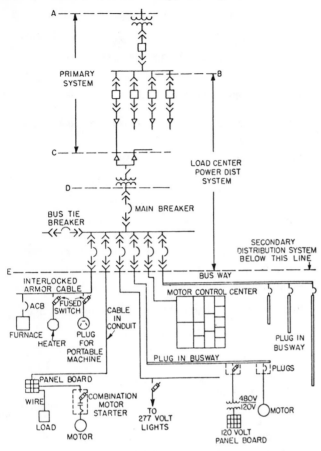

FIG. 6-18. Distribution system employing a load center.

the improvement is small where there is low voltage distribution.

Regulating Equipment   Most industrial plants are supplied with high voltage that is stepped down below 15 kv. In such cases, voltage regulation can be built into the transformer. Regulation is obtained by automatic tap changing, under load. There are generally 2 to 2.5% taps above and below the normal value.

When power is supplied to a building below 15 kv, the only transformation is in the load center. In such cases, the individual primary lines are equipped with voltage regulators. These can be 3-phase, step voltage regulators or induction voltage regulators.

SPECIFICATION OF EQUIPMENT
Energy efficiency in new buildings starts with the equipment specifications. To minimize energy costs

- Select energy efficient motors
- Use synchronous motors
- Select the proper transformers

Motor Selection   It is estimated that 35% of all the electricity generated in the U.S. is consumed by motors. By improving their efficiency, a large portion of that energy can be saved.

Induction motors are most commonly used in industrial and commercial buildings. Load variation has little effect on the efficiency of induction motors, but it has a significant impact on the power factor, as shown in Figure 6-19. Efficiency is essentially constant over a range of 50 to 125% of full load, but the power factor is dramatically reduced as the motor is unloaded. It is essential, therefore, to specify a motor that will operate close to full load.

This expression shows that kvar current operates on reactance. Since capacitors reduce kvar current, they also reduce voltage drop by a value equal to the capacitor current multiplied by the reactance. However,

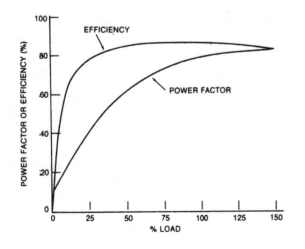

FIG. 6-19. The effect of load variation on efficiency and power factor.[6-7]

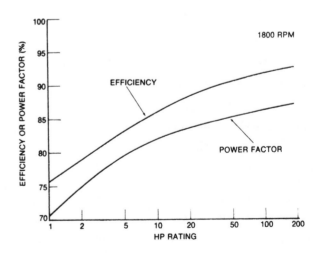

FIG. 6-20. Relationship between hp rating, efficiency and power factor.[6-7]

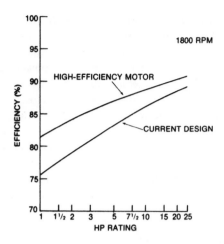

FIG. 6-21. Efficiency vs. horsepower of high efficiency and traditional motors.[6-7]

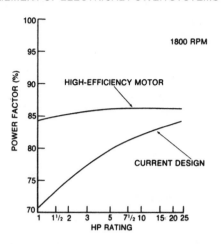

FIG. 6-22. Power factor vs. horsepower of high efficiency and traditional motors.[6-7]

Figure 6-20 shows the relationship between the efficiency, power factor and horsepower rating of induction motors. It is obvious that there is an increase in efficiency and power factor as the horsepower rating increases; i.e., a 50 horsepower induction motor is more efficient than a 25 hp motor. In addition, two-pole motors (3,600 rpm at 60 hertz) have higher efficiencies and power factors than four-pole motors (1,800 rpm at 60 hertz). This trend continues as the number of poles increases.

A recent development is energy efficient motors having a minimum .85 power factor. Internal losses in these motors have been reduced by adding more material in the magnetic core structure.[6-7]

Figure 6-21 shows the efficiency versus horsepower rating of the new high efficiency motors and traditional designs. Figure 6-22 shows the power factor versus horsepower rating of these motors. The curves indicate that, compared to traditional designs, the new motors offer significantly improved efficiency and power factor. The first cost of these new motors is higher than that of conventional motors, but their higher efficiency often repays the extra initial cost in about a year.

High efficiency motors also offer performance benefits. Operating temperatures are reduced by 10°C, because the motors dissipate considerably lower losses, which increases insulation life. Another major advantage is that these motors are less sensitive to voltage variations.

The size of the motors should be the smallest size that will do the job. Engineers sometimes specify a larger motor than is needed, overlooking the fact that most integral horsepower motors have a 15% service factor.

Synchronous Motors Synchronous motors, that are operating at either leading or unity power factor, are capable of correcting power factor. A synchronous motor operates at a lower leading power factor, and delivers more reactive leading kva, at part load than it does at full load, if the excitation remains unchanged. Figure 6-10 illustrates this characteristic. For instance, a 500 hp, 0.8 pf motor delivers $500 \times .6 = 300$ kvar to the system at full load. At half load, this value increases to $500 \times .72 = 360$ kvar. The general formula for reactive kva is

$$kvar = \frac{.746 \times hp \times 100}{Eff \times pf} \times \sqrt{1 - (pf)^2}$$

where

hp = horsepower output
Eff = efficiency expressed in percent
pf = power factor expressed as a decimal

Synchronous motors can be applied to any load driven by a NEMA B design squirrel cage motor. As a rule of thumb, the first cost of synchronous motors is less than that of squirrel cage motors, if the rating exceeds one horsepower per rpm. Besides cost, there are some other attractive features of synchronous motors.

- High efficiency
- Better power factor
- Adaptability to large frame, low speed applications
- Low kva in-rush capability for low torque applications

Here are some of the rules that govern the selection of synchronous versus induction motors.

1. At 3,600 rpm, synchronous motors may be used at 2,000 hp or above and are the first choice above 7,000 hp where tandem drive must be used on a squirrel cage motor.

2. At 1,800 rpm, cylindrical rotor motors are avail-

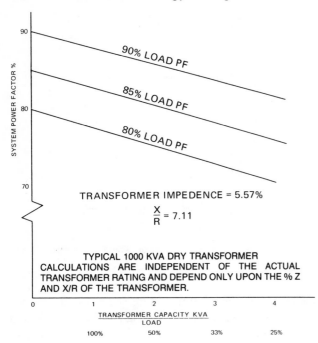

FIG. 6-23. Power factor of a transformer increases with system load.

able above 1,500 hp. Their capital cost is very high. Thus, in most cases, a squirrel cage induction motor will be used.

3. At speeds from 500 rpm to 1,200 rpm inclusive, any motor above 700 hp should normally be synchronous. From 201 to 700 hp, the choice should be based on life-cycle costs instead of just the first cost. Through 200 hp, the motors used are generally induction motors.

4. For speeds below 500 rpm, induction motors should rarely be used because of their low efficiency and low power factor. Synchronous motors can be built with a unity or leading power factor and at high efficiency, at speeds as low as 72 rpm. For direct connection, with ratings above 200 hp and speeds below 500 rpm, the synchronous motor must be the first choice for pumps, compressors, grinders, mixers, mill lines, etc.

Transformers   A considerable amount of energy can be saved by the proper selection of the transformers that are used in all industrial and commercial buildings. Transformers should be sized close to their demand kva and consideration should be given to no-load and full-load losses. If the load profile of a building is known, an economical transformer, that has the least operating losses over the life of the transformer, can be selected. Available computer programs can be used to size the transformers for a given load profile.

The efficiency of transformers varies from about 96% at light loads to nearly 99% at full load. Since the power factor of transformers varies with the loading, a lightly loaded transformer operates at a low power factor. Figure 6-23 shows the relationship between the loading of transformers and the system power factor.

Dry-type transformers have higher total losses than silicon or oil-filled transformers. Among the dry-type transformers, the energy loss is greatest with 150°C rise transformers, followed by the 115° and 80°C rise models in the order of descending energy loss. Selection of transformers should always be based on life-cycle cost analysis.[6-8]

Example 6-5

Two manufacturers have quoted on a 1,000 kva, standard impedance, 13,200/480v transformer.

	Transformer A	Transformer B
Price	$8,000	$8,500
No-load losses	2,700 W	3,100 W
Full-load losses	18,000 W	15,000 W

If the selection is based on capital cost, transformer A is a better buy, but if losses are considered, the results are different.

Let's assume that the transformer is to provide power for a plant working five days per week, two shifts per day. During working hours, the transformers will be loaded to 90% capacity, At all other times the transformers will be loaded to only about 5% capacity. The operating loss for the transformers will be no load loss plus the I^2R loss, as I^2R loss varies as the square of any change in the load. The following table gives the losses of transformers A and B.

	Transformer A	Transformer B
Full load loss, watts	18,000	15,000
No load loss, watts	2,700	3,100
Full load I^2R loss, watts	15,300	11,900
I^2R loss at 90% load, watts	12,393	9,639
I^2R loss at 5% load, watts	38	30
Total losses at 90% load	15,093	12,739
Total loss at 5% load, watts	2,738	3,130

During peak operation of the plant, the transformer A's loss is 2,354 watts more than the loss of transformer B. But during off duty hours, transformer A has the advantage by 392 watts. The annual cost advantage of each transformer is

Transformer A during off duty hours
$$= \frac{392 \text{ W}}{1000} \times \frac{88 \text{ hr}}{\text{wk}} \times \frac{52 \text{ wk}}{\text{yr}} = 1794 \text{ kWh/yr}$$

Transformer B during working hours

$$= \frac{2354 \text{ W}}{1000} \times \frac{80 \text{ hr}}{\text{wk}} \times \frac{52 \text{ wk}}{\text{yr}} = 9793 \text{ kWh/yr}$$

Therefore, transformer B has a net annual advantage of 7,999 kWh. If the power costs 2¢ per kWh, the additional cost of operating transformer A is $160 per year. Assuming an 8% increase in the power cost, the extra capital cost for transformer B pays itself within three years.

ELECTRICAL DEMAND CONTROL SYSTEMS

Utilities charge their customers for kWh consumed and for kW of maximum demand. Demand is the measurement of total energy usage over some fixed interval of time. The utilities generally compute demand by utilizing the following relationship over each demand interval (15, 30 or 60 minutes),

$$D = \int_0^t f(kW_t)dt \qquad (6-8)$$

where
  D = demand
  $f(kW_t)$ = equation of kilowatt usage curve
  t = demand period
  dt = time differential

The greatest value of D computed during any demand interval of the month, determines the maximum demand. Demand control produces savings by limiting the demand charge. This is generally achieved by turning off or shedding certain loads when it is determined, by the demand controller, that the preset demand may be exceeded. For this reason, it is important that portions of the facility's total load be controllable. Figure 6-24 shows a typical demand chart. Each line represents the demand during a demand interval and the longest line represents maximum demand over the billing period. With a demand controller installed, the demand chart looks like Figure 6-25.

Electrical loads in most of the facilities can be divided into two categories: essential and sheddable. Essential loads include lighting, elevators, computer room power, escalators and production machinery. Electrical power to these loads cannot be deferred for any period of time. Such loads as heating, air conditioning, exhaust fans, pumps, compressors and water heaters are classified as sheddable (controllable) because they can be turned off for a short period without causing any inconvenience. It is these sheddable loads that are properly controlled by demand controllers and duty cycling equipment to control the facility's electrical demand.

Various types of demand controllers are available. All of them measure the rate at which electrical energy is

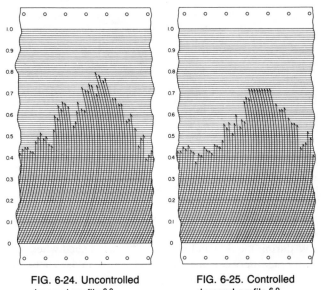

FIG. 6-24. Uncontrolled demand profile.[6-9]   FIG. 6-25. Controlled demand profile.[6-9]

consumed, compare this rate with a predetermined allowable value, and turn off specified sheddable loads, when the preset maximum value is reached.

Simple demand controllers control one or group of loads and turn them off during the demand interval. They cannot turn them back on, during the demand interval, even if the rate of use falls off. Thus, the controllable loads are not being used optimally.

Smart demand controllers react to the rate of change of energy use and, if the use decreases during the interval in which loads are shed, restore a portion or all of the dropped loads during the demand interval. Loads may be shed or restored on a priority schedule or, with some devices, the order in which loads are dropped and restored can be rotated.

Computer-based demand controllers have extremely large capabilities. They control the operation of all primary energy using systems—boilers, chillers, motors, pumps, ventilating systems, lighting and power. In some cases building security, fire protection and communication are also part of this system.[6-9]

Computer-based demand controllers generally consist of CPU modules, consoles, CRT units and remote terminal units. A one line diagram of a demand control system is shown in Figure 6-26.

Demand control systems use one of three methods to compute demand.[6-10]

1. Ideal rate
2. Instantaneous rate
3. Predictive

The ideal rate method measures the rate at which energy is consumed during the demand interval, Figure

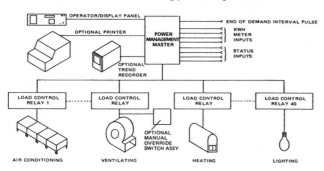

FIG. 6-26. Demand control system.

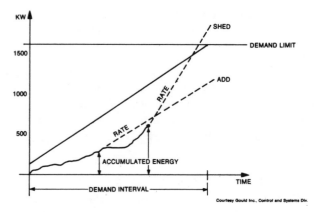

FIG. 6-28. Instantaneous rate method of demand control.[6-10]

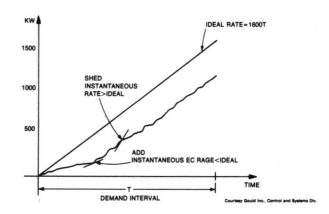

FIG. 6-29. Predictive load control.[6-10]

6-27. This rate is compared with a predetermined ideal rate of consumption, based on the maximum allowable kW during the demand interval. A load is shed when the difference between the ideal and actual rates reaches a preset minimum. Loads are usually restored at the beginning of the next demand interval and the process is repeated.

In operation, the ideal rate line extends from the demand limit point to an offset value at the beginning of the demand period. The offset value is usually equal to the maximum demand limit minus the kW of the non-shedable loads.

The signal for *kW used* is generated by the rotor in the watthour meter. Pulse initiators within the meter convert a specific number of turns of the rotor into a pulse. Since the rotor measures the amount of energy being used, each pulse represents a number of kW. The number of pulses generated is compared to the preset value of the counter to determine whether the kW consumed exceeds the *shed line* of the program. Loads are shed if the actual usage exceeds the shed line at any point and are restored when the count falls below the preset rate. The end-of-interval signal is obtained from the utility's demand meter whereupon the count is restarted.

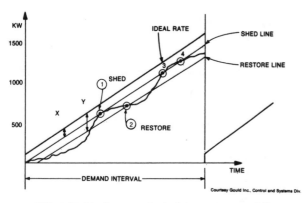

FIG. 6-27. Ideal rate method of demand control.[6-10]

This method is used where only one or two loads are controlled or the controlled leads have the same rating. This method has one major disadvantage. All the loads are put on line at the start of each demand interval, thus causing heavy transients in the system.

The instantaneous rate method evaluates the rate of power consumption at short time intervals and compares it with the preset limit. With this type of controller, loads are shed whenever the instantaneous rate exceeds the preset rate, Figure 6-28. When the instantaneous rate is less than the preset rate, loads are added. Because this method is based on the instantaneous rate of energy consumption, it is independent of demand patterns and does not need synchronization with the utility's demand interval.

Predictive load control measures the instantaneous rate of energy consumption of all loads at specific points within the demand interval. The controller then predicts whether or not the preset demand limit will be exceeded. This forecast is made by adding the accumulated energy to the rate of energy consumption multiplied by remaining time in the demand interval, Figure 6-29. If the total indicates demand in excess of the limit, the unit calculates the excess kW and the number of kW that must be shed to ensure that the demand limit is not exceeded. When the predicted demand is less than the limit, loads may be restored. In some controllers, the loads are only restored before the start of the next demand interval. Predictive controllers are more complex and, therefore, more expensive.

The predictive principle requires synchronization with the utility's demand interval. In many cases, this is not possible so the most modern of these controllers use the sliding window method to calculate demand.

Controllers using the sliding window method receive the kW meter inputs from the pulse initiators, accumulate the kW values over a demand interval of say 15 minutes, then average these values to determine the average power consumed. The controller receives new kW values at 3 second intervals and continuously updates demand.

The method is best explained by envisioning a histogram, Figure 6-30, showing the past 18 seconds of power consumption. At position A, the system gathers the data from time sequences 10 to 25. Three seconds later, it

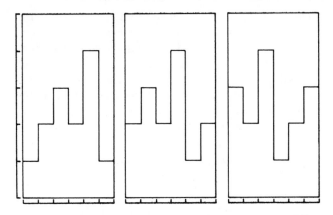

FIG. 6-30. Sliding window method of demand control.[6-10]

slides over and reads the data from sequences 13 to 28. In another three seconds, it reads sequences 16 to 31. In this way, it goes on calculating average power over all possible window positions during the 15 minute demand interval which is divided into 300 segments of 3 seconds each. Obviously, demand controllers using this method do not require synchronization with the utility's demand interval.

TABLE 6-5. Duty Cycling Schedule

UNIT	LOAD SCHEDULES						
	Time On	Cycle Start	Time Off	Cycle Interval Minutes	% Off Time	Motor Nameplate HP	Motor Usage HP
AC-D1	0700	0900	1800	60	10	25	16.8
RF-D1	0700	0900	1800	60	10	7.5	5.1
AC-D2	0700	0900	1800	60	10	25	16.8
RF-D2	0700	0900	1800	60	10	7.5	6.3
AC-D3	0700	0900	1800	60	10	30	22.3
RF-D3	0700	0900	1800	60	10	15	9.9
AC-D4	0700	0900	1800	60	10	20	13.9
RF-D4	0700	0900	1800	60	10	10	9.2
AC-D5	0700	0900	1800	60	15	25	16.4
RF-D5	0700	0900	1800	60	15	10	7.2
AC-D6	0700	0900	1800	60	15	25	16.4
RF-D6	0700	0900	1800	60	15	7.5	5.1
AC-D7	0700	0900	1800	60	15	30	18.4
RF-D7	0700	0900	1800	60	15	7.5	4.7
AC-D8	0700	0900	1800	60	15	25	18.6
RF-D8	0700	0900	1800	60	15	10	8.0
AC-D9	0700	0900	1800	60	15	15	9.6
RF-D9	0700	0900	1800	60	15	5	3.7
AC-D10	0700	0900	1800	60	15	20	12.2
RC-D10	0700	0900	1800	60	15	5	4.1
AC-D11	0700	0900	1800	60	15	10	6.8

Duty cycling is a computer-based system that turns nonessential loads off and on according to a predetermined schedule. Each load can be assigned one or more cycle intervals, ranging from 15 to 120 minutes, to reduce kilowatt consumption. For instance, with 20% off time during a 60 minute interval, the load will be on for 48 minutes and off for 12 minutes out of each hour. Table 6-5 shows the schedule for cycling nonessential loads in a typical plant.

Great care should be taken in specifying the equipment that is to be duty cycled, especially motors that are vulnerable to short-cycling operations. Higher insulation levels should be specified for these motors.

Cost savings result from the controller's ability to limit both utility demand charges (by shedding loads during high demand periods) and energy charges by duty cycling the equipment. The savings depend upon the user's power profile but, on average, the payback period for a demand controller is 6 to 18 months, with annual savings of 10 to 20 percent.

Chapter 7

ECONOMICS OF ENERGY MANAGEMENT

Making energy-using systems more efficient invariably involves expenditures. It is, therefore, necessary to evaluate the benefits received in terms of annual savings in energy costs.

Since natural fuel resources are finite, the possibility of future escalations of energy costs should be considered in evaluating present investment decisions.

Other factors to be considered include the effect of income taxes on potential annual savings; the effect of depreciation on income taxes; carrying costs, such as property taxes and insurance, on capital assets; maintenance and administrative expenses; the cost of capital and the economic life of the projected investment.

Let us review some of the basic tools of financial analysis that are useful in economic evaluation of alternatives considered for energy management.

TIME VALUE OF MONEY

The value of money varies with time because it can grow if profitably invested or earn interest if placed in a savings account. A dollar today is equivalent to $1.08 owed a year from today, if it can be invested at 8% per year. The concept of equivalence is important in economic studies because it allows values that occur at different points in time to be compared on a common time basis.

Time scales are often used for visualizing the cash flows that occur at different points in time. A time scale is illustrated in Figure 7-1. A downward arrow indicates an expense or an investment and an upward arrow represents an income.

Interest is defined as money paid for the use of borrowed money. The rate of interest is the ratio, expressed as a percentage of the interest payable at the end of the period of time, usually a year, and the money owed at the beginning of the year. Thus, if $8 is payable annually on a debt of $100, the interest is $8 and the interest rate is $8/\$100 = .08$ or 8 percent.

SYMBOLS AND TERMS

The following symbols are used in discussing interest;

I = total interest
i = the annual interest rate
n = the number of annual interest periods
P = present principal sum
A = a uniform series of equal end-of-period payments
F = sum of money at a specified future date

SIMPLE INTEREST

With simple interest, no interest is paid on the interest earned. The interest to be paid is proportional to the length of time the principal sum is borrowed. The interest that will be earned is given by the following formula.

$$I = Pni$$

Suppose that $500 is borrowed at a simple interest

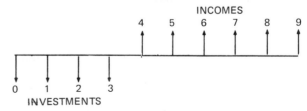
FIG. 7-1. Cash flow time scale.

93

rate of 8% per annum. At the end of one year, the interest would be

$$I = \$500 \times 1 \times 0.08 = \$40$$

and at the end of two years the interest would be

$$I = \$500 \times 2 \times 0.08 = \$80$$

COMPOUND INTEREST

If money is deposited in a bank and the interest is left on deposit, the bank generally pays interest on the interest earned. This reinvestment of interest and the payment of interest on the original investment, as well as on the earned interest, is called compounding and is designated by the term compound interest. For example, Table 7-1 gives the interest and the amount at the end of each year, when $1000 is deposited at a compounded rate of 5% annually.

TABLE 7-1. Application of Compound Interest

Year	Amount at start of year	Interest at end of year	Total at end of year
1	$1000.00	$1000 + .05 = $50	$1050.00
2	$1050.00	$1050 + .05 = $52.50	$1102.50
3	$1102.50	$1102.50 + .05 = $55.12	$1157.62
4	$1157.62	$1157.62 + .05 = $57.88	$1215.50
5	$1215.50	$1215.50 + .05 = $60.77	$1276.27

SINGLE PAYMENT COMPOUND-AMOUNT FACTOR

The final compound amount shown in Table 7-1 may be computed using the single payment compound-amount factor. This factor is used to find the future worth, F, at the end of n periods for a present principal amount, P. The time scale is shown in Figure 7-2 and the factor is derived as follows:

At the end of the first period,
$$F = P + Pi = P(1 + i)$$
At the end of the second period,
$$F = P(1+i) + P(1+i)i = P(1 + i + i + i^2) = P(1+i)^2$$
At the end of the nth period, by induction,

$$F = P(1+i)^n \tag{7-1}$$

The term $(1+i)^n$ is called the single payment compound-amount factor and is represented by the symbol

$$\left(\frac{F}{P}\right)_i^n$$

Thus
$$F = P\left(\frac{F}{P}\right)_i^n \tag{7-2}$$

To calculate the future value, multiply the present value by the single payment compound-amount factor for n periods at i rate of interest.

The values of $\left(\frac{F}{P}\right)_i^n$ are given in column two of Appendix A.

Example 7-1:
P = $1000
n = 5
i = .05
$$F = \$1000(1 + .05)^5$$
$$= \$1276.28$$

Also, from Appendix A, the value of $\left(\frac{F}{P}\right)_5^5 = 1.2763$ and F = $1000 \times 1.276 = $1276.30

SINGLE PAYMENT PRESENT WORTH FACTOR

The single payment present worth factor is used to find the present worth, P, of a future amount, F. The value of P is derived from Equation 7-1.

$$P = F\left[\frac{1}{(1+i)^n}\right] \tag{7-3}$$

The factor within the brackets is called the single-payment present worth factor. It is designated by $\left(\frac{P}{F}\right)_i^n$ and its values are given in Appendix A.

From Example 7-1, F = $1276.28, n = 5, i = 5%.

$$P = \frac{1276.28}{(1+.05)^5} = \$1000$$

Also from Appendix A
$$\left(\frac{P}{F}\right)_5^5 = .7835$$
and
$$P = 1276.28 \times .7835 = \$1000$$

EQUAL PAYMENT SERIES, COMPOUND-AMOUNT FACTOR

For a given series of equal payments, A, occurring at the end of each annual period, the sum of compound amount can be found by using the equal payment series, compound-amount factor.

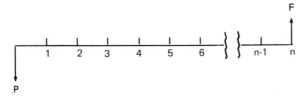

FIG. 7-2. Single payment, compound-amount time scale.

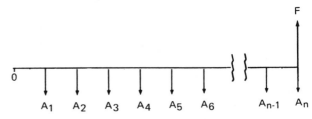

FIG. 7-3. Equal payment series, compound amount time scale.

It is clear from the cash flow diagram, Figure 7-3, that each payment, A, has a different compound amount sum. The first one is at compound interest for $(n-1)$ years, the second one for $(n-2)$ years and the last one does not draw any interest. The total future amount is given by

$$F = A(1+i)^{n-1} + A(1+i)^{n-2} + A(1+i)^{n-3} + ---- + A(1+i) + A$$

Multiply the above equation by $(1+i)$:

$$F(1+i) = A(1+i)^n + A(1+i)^{n-1} + A(1+i)^{n-2} + ---- + A(1+i)^2 + A(1+i)$$

Subtracting the first equation from the second gives

$$F(1+i) - F = -A + A(1+i)^n$$

$$F = A\left[\frac{(1+i)^n - 1}{i}\right] \qquad (7-4)$$

The factor within the brackets is called the equal payment series, compound-amount factor and is denoted by $\left(\dfrac{F}{A}\right)_i^n$. The values of this factor are given in column four of Appendix A.

Example 7-2

Find the future amount for a series of five $1000 payments made at the end of each year at 5% annual compound interest.

$$A = \$1000$$
$$n = 5$$
$$i = 5\%$$
$$F = \$1000\left[\frac{(1+.05)^5 - 1}{.05}\right]$$
$$= \$1000(5.5256) = \$5525.60$$

From Appendix A,

$$\left(\frac{F}{A}\right)_5^5 = 5.5256$$
$$F = \$1000(5.5256) = \$5525.60$$

SINKING FUND DEPOSIT FACTOR

The sinking fund deposit factor is used to find a uniform series of year-end payments, A, to provide a future amount, F. From Equation 7-4, the value of A is given by

$$A = F\left[\frac{i}{(1+i)^n - 1}\right] \qquad (7-5)$$

The factor within the brackets is called the sinking fund deposit factor and is denoted by $\left(\dfrac{A}{F}\right)_i^n$

The values of this factor are given in Appendix A. From Example 7-2,

$$F = \$5525.60$$
$$i = 5\%$$
$$n = 5$$
$$A = \$5525.60\left[\frac{.05}{(1+.05)^5 - 1}\right]$$
$$= \$5525.60(.18097) = \$1000$$

Also from Appendix A,

$$\left(\frac{A}{F}\right)_5^5 = 0.181$$
$$A = \$5525.60(.181) = \$1000.13$$

CAPITAL RECOVERY FACTOR

The capital recovery factor is used to find the end of period payments, A, that will be provided by a present amount, P, as illustrated in Figure 7-4. From Equations 7-5 and 7-1,

$$A = F\left[\frac{i}{(1+i)^n - 1}\right]$$

and

$$F = P(1+i)^n$$

Substituting for F

$$A = P\left[\frac{i(1+i)^n}{(1+i)^n - 1}\right] \qquad (7-6)$$

The factor within the brackets is called the capital recovery factor and is denoted by $\left(\dfrac{A}{P}\right)_i^n$. Its values are given in Appendix A.

Example 7-3

$5000 is invested at 5% annual compound interest for 5 years. Find five year-end payments provided by the investment.

$$P = \$5000$$
$$i = 5\%$$
$$n = 5$$
$$A = \$5000.00\left[\frac{.05(1+.05)^5}{(1+.05)^5 - 1}\right]$$
$$= \$5000(.23097) = \$1154.88$$

From Appendix A,

$$\left(\frac{A}{P}\right)_5^5 = 0.231$$
$$A = \$5000 \times .231 = \$1155.0$$

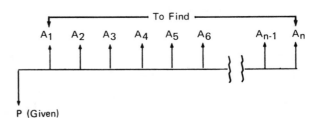

FIG. 7-4. Capital recovery time scale.

EQUAL PAYMENT SERIES, PRESENT WORTH FACTOR

The equal payment series, present worth factor is used to find the present worth, P, of a series of end-of-period payments, A. Equation 7-6 can be restated as follows:

$$P = A\left[\frac{(1+i)^n - 1}{i(1+i)^n}\right] \quad\quad (7-7)$$

The factor within the brackets is the equal payment series present worth factor and is denoted by $\left(\frac{P}{A}\right)_i^n$.

From the previous example,

A = $1155
n = 5
i = 5%

$$P = \$1155\left[\frac{(1+.05)^5 - 1}{.05(1+.05)^5}\right]$$
$$= \$1155(4.32946) = \$5000.52$$

From Appendix A,

$$\left(\frac{P}{A}\right)_5^5 = 4.3295$$
$$P = \$1155(4.3295) = \$5000.57$$

LIFE CYCLE COSTING

The decision to isntall energy-saving equipment should be based on life cycle costing (L.C.C.) and not on the capital cost of the equipment. Life cycle costing determines the total cost of the equipment over its useful life and inforporates the following costs:

Investment cost includes the cost of purchasing and installing the equipment. If the component is a transformer, for example, the true cost includes the price of the transformer, freight, and the contract cost of installation.

Operating costs generally include energy costs, utility demand charges, maintenance, service, insurance and taxes. Utility demand charges and energy costs can be easily estimated. There are computer programs, discussed in Chapter 11, that perform dynamic simulation of equipmet and systems to calculate thier monthly energy and power needs. Maintenance and service costs are generally based on the performance of similar equipment already in use. Taxes and insurance are the least significanty part of operating costs. Insurance costs can be obtained from underwriters and tax costs are obtained from taxing agencies. It is sometimes important to consider tax and insurance costs in the analysis.

For analysis purposes, inflationary increases in the cost of operation are calculated. The formula used for computing the yearly escalation of operating costs is as follows:

$$D_n = D_1\left(1+\frac{x}{100}\right)^{n-1} \quad\quad (7-8)$$

where

D_n = operating cost for one year that is *n* years in the future

D_1 = operating cost for the current year, n = 1

x = percent escalation in the operating cost

Example 7-4

The current operating cost is $2000. What will be the operating cost at the end of the second and tenth years at an escalation rate of 6%?

D_1 = $2000
x = 6%
n = 2

At the end of the second year, $D_2 = 2000\left(1+\frac{6}{100}\right)^{2-1}$
$$= \$2000(1.06)$$
$$= \$2120$$

At the end of the 10th year, $D_{10} = 2000\left(1+\frac{6}{100}\right)^{10-1}$
$$= 2000(1.06)^9$$
$$= \$3378.96$$

L.C.C. studies involve the principles of investment analysis and the translating or converting of accrued costs at one time to equivalent costs at other times. These methods are also used to compare different energy conservation alternatives.[7-1]

PRESENT WORTH AMOUNT

Present worth is the value of money at time zero (t = 0). Any cash receipts or disbursements that occur at time zero are taken at their current or stated values. Any receipts or disbursements that will occur in the future will have the anticipated worth adjusted to present worth and included in the first cost.

If the future payments are in unequal amounts, their present worth is obtained by using the single payment, present worth factor $\left(\frac{P}{F}\right)_i^n$.

If the future payments are in the form of an annuity, or in equal amounts, the present worth is obtained by using the equal payment present worth factor $\left(\frac{P}{A}\right)_i^n$.

Thus, the present worth of an energy conservation proposal, having a life of *n* years, at interest rate *i* and with the cash flows shown in Figure 7-5 is given by

$$P = -F_0 + F_1\left(\frac{P}{F}\right)_i^1 + F_2\left(\frac{P}{F}\right)_i^2 + ---$$
$$+ F_n\left(\frac{P}{F}\right)_i^n \quad\quad (7-9)$$

Present worth amount is often used for comparison of alternatives since

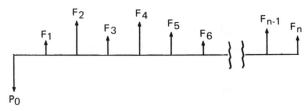

FIG. 7-5. Present worth of a series of unequal payments.

(a) It concentrates the equivalent value of a cash flow at a particular point in time; i.e., $t = 0$

(b) It considers the time value of money by considering the yearly interest rate.

(c) The value of present worth amount is unique, no matter what the cash flow patterns are.

Example 7-5

The installed cost of an energy demand controller is $11,000. The annual savings in energy and power cost are $2000. The interest rate is 8% and the escalation rate for the energy cost is 6%. Annual savings in energy cost due to escalation in the cost of energy are give by

$$SE_n = SE_1 \left(1 + \frac{x}{100} \right)^{n-1} \qquad (7\text{-}10)$$

At the end of the second year, the savings in energy cost are given by

$$SE_2 = \$2000 \left(1 + \frac{6}{100} \right)^{2-1}$$
$$= \$2120$$

The present worth amount for the energy savings is calculated by the factor $\left(\dfrac{P}{F} \right)^n_i$

For example, the present worth amount for the savings at the end of the second year is

$$2120 \left(\frac{P}{F} \right)^2_8 = \$2120(.8573) = \$1817.47.$$

Table 7-2 gives the cash flow and the present worth of the cash flow which is $6047.30.

Thus, it is seen that the cost of the demand controller is recovered out of the future energy savings discounted at 8% plus a present sum of $6047.30.

ANNUAL EQUIVALENT AMOUNT

The annual equivalent amount is another basis for comparing alternatives. With this method, all receipts and disbursements occuring over a period of time are converted to an equivalent uniform yearly charge as visualized in Figures 7-6 and 7-7. The annual equivalent amount for an investment having nonuniform cash flows is determined by

(a) Calculating the present worth amount using Equation 7-9.

TABLE 7-2. Present Worth Calculations

End of Year	Cash Flow	Present Worth Factor	Present Worth Amount
0	−$11,000	1	−$11,000
1	2,000	0.9259	1,851.80
2	2,120	0.8573	1,817.47
3	2,247	0.7938	1,783.66
4	2,382	0.7350	1,750.77
5	2,524	0.6806	1,717.83
6	2,676	0.6302	1,686.41
7	2,837	0.5835	1,655.39
8	3,007	0.5403	1,624.68
9	3,187	0.5002	1,594.14
10	3,379	0.4632	1,565.15
			6,047.30

(b) Multiplying the present worth amount by the capital recovery factor $\left(\dfrac{A}{P} \right)^n_i$.

Thus, the annual equivalent amount for interest rate i and n years is defined as

$$AE = P \left(\frac{A}{P} \right)^n_i \qquad (7\text{-}11)$$

The annual equivalent amount basis of comparison of alternatives is preferred over the present worth amount method for a series of cash flows.

Example 7-6

The annual equivalent amount of the cash flows listed in Table 7-2 is

$$AE = P \left(\frac{A}{P} \right)^{10}_8$$
$$= \$6,047.30(0.149) = \$901.05$$

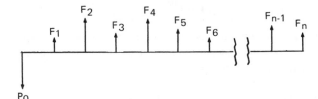

FIG. 7-6. Original time line for annual equivalent amount.

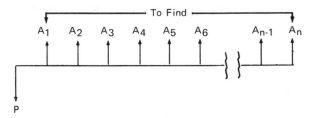

FIG. 7-7. Equalized restatement of Figure 7-6.

ANNUAL EQUIVALENT COST OF AN ASSET

Any equipment installed for energy management is a unit of capital. Such an asset loses its value over a period of time and this loss represents an expenditure of capital. The amount spent on the procurement of an asset is called first cost and the amount received on its retirement is called salvage cost. From the first cost and the salvage value, it is possible to derive an annual equivalent cost for an asset. Figure 7-8 shows the cash flow time scale for an asset with salvage value.

Let

P = First cost of an asset
S = Estimated salvage value
n = Estimated years of service life

The annual equivalent cost of the asset may be expressed as the annual equivalent salvage value.

$$P\left(\frac{A}{P}\right)_i^n - S\left(\frac{A}{F}\right)_i^n \tag{7-12}$$

but since

$$\left(\frac{A}{P}\right)_i^n - i = \frac{i(1+i)^n}{(1+i)^n - 1} - i = \frac{i(1+i)^n - i(1+i)^n + i}{(1+i)^n - 1}$$

$$= \frac{i}{(1+i)^n - 1} = \left(\frac{A}{F}\right)_i^n$$

by substitution

$$P\left(\frac{A}{P}\right)_i^n - S\left[\left(\frac{A}{P}\right)_i^n - i\right] =$$

$$(P - S)\left(\frac{A}{P}\right)_i^n + Si \tag{7-13}$$

Example 7-7

A machine with a first cost of $10,000 has an estimated economic life of 10 years and an estimated salvage value of $2,000. For an interest rate of 8%, the annual equivalent cost is:

$$(\$10,000 - \$2,000)\left(\frac{A}{P}\right)_8^{10} + \$2,000(.08) = \$1,352.00$$

This calculation is also useful in evaluating the true cost of leasing when compared to an outright purchase of the needed equipment.

FUTURE WORTH AMOUNT

Sometimes future worth amount is used for comparison of alternatives. Future worth amount is the equivalent amount of cash flow calculated at some future time. The future worth amount for a proposal at some future time *n* years from the present at an interest rate *i* is given by

$$FW = F_0\left(\frac{F}{P}\right)_i^n + F_1\left(\frac{F}{P}\right)_i^{n-1} + - - + F_n\left(\frac{F}{P}\right)_i^0$$

$$= \sum_{t=0}^n F_t\left(\frac{F}{P}\right)_i^{n-t} \tag{7-14}$$

Since

$$\left(\frac{F}{P}\right)_i^{n-t} = (1+i)^{n-t}; \quad FW = \sum_{t=0}^n F_t(1+i)^{n-t} \tag{7-15}$$

The future worth amount of a given cash flow can also be calculated by first calculating the present worth of the cash flow and then multiplying by the single payment compound amount factor, i.e.,

$$FW = P\left(\frac{F}{P}\right)_i^n$$

The future worth amount, annual equivalent amount and the present worth amount are consistent bases of comparison. For fixed values of i and n, the following relation will hold true for two alternatives, A and B.

$$\frac{P_A}{P_B} = \frac{AE_A}{AE_B} = \frac{FW_A}{FW_B}$$

Example 7-8

The future worth amount for the cash flow for the demand controller is given by:

Future worth amount of the first cost

$$= -\$11,000\left(\frac{F}{P}\right)_8^{10}$$

$$= -\$11,000(2.159)$$

$$= -\$23,749$$

Table 7-3 gives the future worth amount for Example 7-8.

INTERNAL RATE OF RETURN

The internal rate of return is a widely accepted method for comparison of alternatives. It is defined as the interest rate that reduces the present worth amount of a cash flow to zero. Alternatively, the present rate of return for an investment is the interest rate that satisfies the equation:

$$P(i^*) = \sum_{t=0}^n F_t(1+i^*)^{n-t} = 0 \tag{7-16}$$

where n is the life of the investment proposal.

In comparing alternatives, the rate of return measures the return from a larger investment against those obtained from smaller investments to ascertain if the incremental investment earns a stipulated minimum acceptable rate. This is accomplished by equating the present worths of two investments. The rate of return is the in-

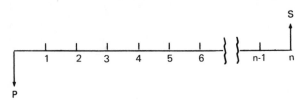

FIG. 7-8. Annual equivalent amount for an asset with salvage value.

TABLE 7-3. Future Worth Amount Calculations

End of Year	Cash Flow	Future Worth Factor $(F/P)_i^n$	Future Worth Amount
0	−$11,000	2.159	−$23,749.00
1	2,000	1.999	3,998.00
2	2,120	1.851	3,924.12
3	2,247	1.714	3,851.36
4	2,382	1.587	3,780.23
5	2,524	1.469	3,707.75
6	2,676	1.360	3,639.36
7	2,837	1.260	3,574.62
8	3,007	1.166	3,506.16
9	3,187	1.080	3,441.96
10	3,379	1.000	3,379.00
		Total =	$13,053.56

Future worth amount FW = $13,053.56.

terest rate that balances the equation. If the rate is greater than the minimum required return, the larger investment is preferred.

For a single investment, Equation 7-16 is solved for i* by the trial and error method or by using Newton Rapson's method for finding the roots of a polynomial. Figure 7-9 shows the variation of present worth with the interest rate for an investment.

Example 7-9

Find the rate of return on energy demand controller whose cash flow is given in Table 7-2.

First cost = $11,000

We have to find the value of i that satisfies

$$0 = P(i)$$

$$= -\$11,000 + \$2,000\left(\frac{P}{F}\right)_i^1 = \$2,120\left(\frac{P}{F}\right)_i^2$$

$$+ \$2,247\left(\frac{P}{F}\right)_i^3 + \$2,382\left(\frac{P}{F}\right)_i^4$$

$$+ \$2,524\left(\frac{P}{F}\right)_i^5 + \$2,676\left(\frac{P}{F}\right)_i^6$$

$$+ \$2,837\left(\frac{P}{F}\right)_i^7 + \$3,007\left(\frac{P}{F}\right)_i^8$$

$$+ \$3,187\left(\frac{P}{F}\right)_i^9 + \$3,379\left(\frac{P}{F}\right)_i^{10}$$

In general, where there are negative and positive cash flows, the value of i cannot be estimated. With all positive cash flows, the value of i can be estimated by

$$\frac{F_1}{F_0} \times 100$$

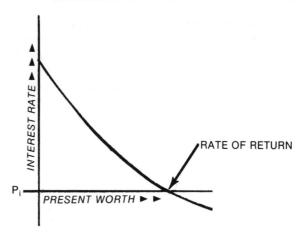

FIG. 7-9. Present worth decreases with increasing internal rate of return.

For example, the estimated value of i is

$$\frac{2,000}{11,000} \times 100 = 18.18\%$$

Let us use 18% as our first trial value of i in Equation 7-16 and, for clarity's sake, list the steps in Table 7-4.

As seen in Table 7-4, $i = 18\%$ produces a negative result, indicating that the trial value of i is too great. The next trial, $i = 17\%$, produces a positive result. This indicates that 17% is too low and that i lies between 17% and 18%. Interpolating,

$$i = 17\% + 1\% \times \frac{406.89}{406.89 - (-36.78)} = 17.92\%$$

PAY-BACK PERIOD

Pay-back period is one of the most common methods of evaluating different investment alternatives. It is defined as the length of time required to recover the first

TABLE 7-4

n	F_t	$\left(\frac{P}{F}\right)_{18}^n$	P_{18}	$\left(\frac{P}{F}\right)_{17}^n$	P_{17}	$\frac{1}{(1+.1792)^n}$	$P_{17.92}$
0	−11,000	1.0000	−11,000	1.0000	−11,000	1.0000	−11,000
1	2,000	.8475	1,695.00	.8547	1,709.40	.8480	1,696.00
2	2,120	.7182	1,522.84	.7305	1,548.66	.7192	1,524.70
3	2,247	.6086	1,367.52	.6244	1,403.03	.6099	1,370.45
4	2,382	.5158	1,228.64	.5337	1,271.27	.5172	1,231.97
5	2,524	.4371	1,103.24	.4561	1,151.20	.4386	1,107.93
6	2,676	.3704	991.19	.3898	1,043.10	.3719	995.20
7	2,837	.3139	890.53	.3332	945.29	.3154	894.79
8	3,007	.2660	799.86	.2848	856.39	.2675	804.37
9	3,187	.2255	718.67	.2434	775.72	.2268	722.81
10	3,379	.1911	645.73	.2080	702.83	.1924	650.12
			−36.78		406.89		−2.56

cost of an investment from the net cash flow produced by that investment for an interest rate equal to zero. That is, if F_0 = first cost of the investment, and if F_t = net cash flow in period t, then the pay-back period is defined as the value of n that satisfies the equation,

$$0 = -F_0 + \sum_{t=1}^{n} F_t \qquad (7\text{-}17)$$

Example 7-10

The pay-back period for the demand controller is given by

Sum of savings for 4 years = \$8,749

Sum of savings for 5 years = \$11,273

$$\$11,000 = \$8749 + \frac{x}{12}(2524)$$

$$x = 10.7$$

Thus, the pay-back period equals 4 years and 11 months.

DEPRECIATION

Depreciation is defined as the decrease in the value of a physical asset with the passage of time.[7-2] The common types of depreciation are: (1) physical depreciation, (2) functional depreciation or obsolescence, and (3) accidents. In financial accounting, the main objectives in charging a depreciation cost are: (1) to recover capital invested in production assets, and (2) to include the cost of depreciation in the operating expenses for tax purposes.

BOOK VALUE

Book value of an asset is defined at the original value of an asset, less its accumulated depreciation at any point in time.

Let

$$
\begin{aligned}
P &= \text{First cost of an asset} \\
S &= \text{Salvage value} \\
B_t &= \text{Book value at the end of year t} \\
n &= \text{Estimated life of the asset} \\
D_t &= \text{Depreciation during year t}
\end{aligned}
$$

then

$$B_t = B_{t-1} - D_t \qquad (7\text{-}18)$$

where $B_0 = P$

The following are several methods for determining D_t.

STRAIGHT LINE DEPRECIATION

The straight line method of depreciation assumes a constant rate of depreciation, Figure 7-10. The rate of depreciation is equal to the reciprocal of the depreciation life of the asset in years or $1/n$. The depreciation for any year is

$$D_t = \frac{P-S}{n}$$

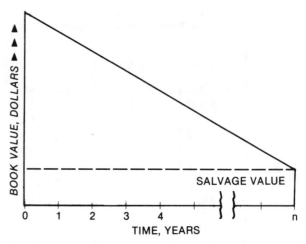

FIG. 7-10. Straight line depreciation of an asset with salvage value.

The book value is:

$$B_t = P - t\frac{P-S}{n} \qquad (7\text{-}19)$$

Example 7-11

Energy conserving equipment has a first cost of \$10,000 and an estimated salvage value of \$1,000. If the estimated life is 10 years, the depreciation rate is 1/10 and the charge per year will be $0.1(10,000 - 1,000) =$ \$900. Table 7-5 gives the yearly depreciation charge and book value.

DECLINING BALANCE METHOD

With the declining balance method, a fixed depreciation rate, R, is assumed throughout the life of the asset. Depreciation for a given year is determined by multiplying the book value of the asset by the depreciation rate. That is:

$$D_t = RB_{t-1} \qquad (7\text{-}20)$$

$$B_t = B_{t-1} - D_t$$

This process is continued until the economic life of the

TABLE 7-5. Straight Line Method of Depreciation

End of Year	Depreciation Charge	Book Value
0	——	\$10,000
1	\$900	\$ 9,100
2	\$900	\$ 8,200
3	\$900	\$ 7,300
4	\$900	\$ 6,400
5	\$900	\$ 5,500
6	\$900	\$ 4,600
7	\$900	\$ 3,700
8	\$900	\$ 2,800
9	\$900	\$ 1,900
10	\$900	\$ 1,000

TABLE 7-6. The Declining Balance Method (30% rate)

End of Year	Depreciation Charge	Book Value
0	——	$10,000
1	.3($10,000) = $3,000	$ 7,000
2	.3($ 7,000) = $2,100	$ 4,900
3	.3($ 4,900) = $1,470	$ 3,430
4	.3($ 3,430) = $1,029	$ 2,401
5	.3($ 2,401) = $ 720	$ 1,631

TABLE 7-7. The Declining Balance Method (27.5% rate)

End of Year	Depreciation Charge	Book Value
0	——	$10,000
1	.275($10,000) = $2,750	$ 7,250
2	.275($ 7,250) = $1,994	$ 5,256
3	.275($ 5,256) = $1,445	$ 3,811
4	.275($ 3,811) = $1,048	$ 2,763
5	.275($ 2,763) = $ 760	$ 2,003

asset is expended or the total depreciation is equal to the first cost minus the salvage value of the asset.

The maximum rate that can be used for the declining balance method is equal to twice the rate used in the straight line method. That is:

$$R \leq \frac{2}{n}$$

When the maximum depreciation rate of 2/n is applied, the method is called the double-declining balance method. However, any value between 1/n and 2/n may be elected. Table 7-6 is a tabulation of the depreciation per year and net book value of a machine with a $10,000 first cost and an estimated salvage value of $2,000 after 5 years of useful life. The depreciation rate that has been chosen is 1.5/n or 30% of the declining balance.

In general, the depreciation rate is given by[7-3]

$$R = 1 - \left(\frac{S}{P}\right)^{1/n}$$

If the depreciation rate is rate is not specified, it can be calculated by using

$$R = 1 - \left(\frac{2,000}{10,000}\right)^{1/5}$$
$$= .275 \text{ or } 27.5\%$$

Table 7-7 gives the depreciation charge and the book value using the depreciation rate of 27.5%.

SUM OF THE DIGITS METHOD

The sum of the digits method of depreciation may be used on property qualifying for the double-declining bal-

TABLE 7-8. Sum of Years' Digits Method

End of Year	Depreciation Charge	Book Value
0	——	$10,000
1	5/15($8,000) = $2,667	$ 7,333
2	4/15($8,000) = $2,133	$ 5,200
3	3/15($8,000) = $1,600	$ 3,600
4	2/15($8,000) = $1,067	$ 2,533
5	1/15($8,000) = $ 533	$ 2,000

ance method. Sum of the digits produces larger depreciation charges in the early life of the asset and smaller changes near the end of the asset's life. This method is easily explained by considering an asset with 5 years of life, having a first cost of $10,000 and a salvage value of $2,000. The sum of the years is $1+2+3+4+5=15$ and the depreciation during the first year will be $5/15 \times (10,000-2,000)=\$2,666$. The annual depreciation charge and the book value are shown in Table 7-8.

SINKING FUND METHOD

With the sinking fund method, depreciation has two components. The first is the annual equivalent amount, over the life of the asset, whose accumulated value is the first cost of the asset minus the salvage value. The second component is the amount of interest earned on the accumulated value of the sinking fund at the beginning of each year. The sinking fund depreciation for any year can be determined by the general expression,

$$D_t = (P-F)\left(\frac{A}{F}\right)_i^n \left(\frac{F}{P}\right)_i^{t-1} \tag{7-21}$$

Example 7-12

A machine has a first cost of $10,000, an estimated life of 5 years, an estimated salvage value of $2,000, and the interest rate is 8%. The first component of the depreciation is given by:

$$(\$10,000-\$2,000)\left(\frac{A}{F}\right)_8^5 \left(\frac{F}{P}\right)_{.8}^0 = \$1364$$

The depreciation charge for the second year is
$1364 \times 1.08 = \$1473$
For the third year, the charge is
$1364 \times 1.1664 = \$1591$

The depreciation charge for each year and the book value at the end of the year are given in Table 7-9.

Figure 7-11 compares the four depreciation methods, without factoring in taxes or profit on investment. Each method has its own unique advantages. The sinking fund method has the slowest rate of capital recovery. The straight-line method is also slow, but is simplifies the accounting procedure. Both the sum-of-the-digits and declining balance methods recover a larger share of the

TABLE 7-9. Sinking Fund Method

End of Year	Depreciation Charge	Book Value
0	—	$10,000
1	$1,364	$ 8,636
2	$1,473	$ 7,163
3	$1,591	$ 5,572
4	$1,718	$ 3,854
5	$1,855	$ 1,999

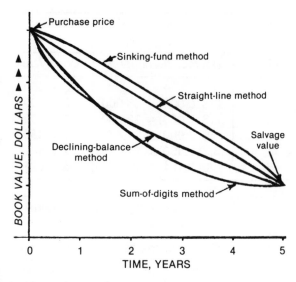

FIG. 7-11. Book values of an asset as determined by various depreciation methods.

initial investment in the first few years of the asset's life. This early recovery of investment is perceived as an advantage by the most conservative managements since it eliminates some of the uncertainty about the future cost of replacing the asset.

The various depreciation methods do not change the total depreciation charge of the total after-tax returns, but the net cash flow pattern is changed. Rapid capital recovery, by accelerated depreciation, allows a slightly higher rate of return on the investment because the present worth of early cash flows is greater than the same amount of cash flows spread evenly over the life of the asset.

TAXES IN ECONOMIC CALCULATIONS

In order to calculate present worth or rate of return of net gains from an investment, one has to calculate net income after taxes. The following equation gives a general formula for calculation of profits after taxes. Let

P_b = Profits before income taxes
P_a = Profits after taxes
D = Annual depreciation
GI = Gross income from the investment
NC = Annual costs
i = Interest rate
Rt = Income tax rate
T = Tax payable
FI = Interest paid

Profits before taxes:

$$P_b = GI - (NC + D + FI)$$
$$T = P_b \times Rt \qquad (7\text{-}22)$$

Profits after taxes

$$P_a = P_b - (P_b \times Rt)$$
$$= P_b(1 - Rt) \qquad (7\text{-}23)$$

Example 7-13

Capacitor banks are installed on substations to improve the power factor. The installed cost of the equipment is $27,200. The value of the energy saved is $5,939 annually and increases at the rate of 8% annu-

ally. Using straight line depreciation and a 50% tax rate, the cash flow, after taxes, is shown in Table 7-10.

First cost of capacitor banks = $27,200
Net energy savings at the end of year one = $5,939
Net energy savings at the end of the second year

$$= \$5,939\left(1 + \frac{8}{100}\right)^{2-1}$$
$$= \$6,414$$

Depreciation charge every year $= \frac{\$27,200}{10}$
$$= \$2,720$$

Income before taxes at the end of the first year
= Net energy savings − Depreciation charge
= $5,939 − $2,720
= $3,219

Income after taxes at the end of the first year

$$= \text{Income before taxes} \times (1 - \tfrac{\text{tax rate}}{100})$$
$$= \$3,219(1 - \tfrac{50}{100})$$
$$= \$1,610$$

Cash flow at the end of the first year
= Income after taxes + Depreciation charge
= $1,610 + $2,720
= $4,330.00

From the cash flows shown in Table 7-10 the pay back period for the capacitor banks is 5 years and 6 months. Based on the cash flows, the internal rate of return of the investment is 14.5%.

Example 7-14

Three types of chiller systems are proposed for a

TABLE 7-10. Cash Flow for Example 7-13

Year	Net Energy Savings (1)	Depreciation Charge (2)	Income Before Taxes (1)-(2) (3)	Income After Taxes (3)-.5(3) (4)	Cash Flow (4)+(2)
1	$ 5,939	$2,720	$3,219	$1,610	$4,330
2	$ 6,414	$2,720	$3,694	$1,847	$4,567
3	$ 6,927	$2,720	$4,207	$2,104	$4,824
4	$ 7,481	$2,720	$4,761	$2,381	$5,101
5	$ 8,080	$2,720	$5,360	$2,680	$5,400
6	$ 8,726	$2,720	$6,006	$3,003	$5,723
7	$ 9,424	$2,720	$6,704	$3,352	$6,072
8	$10,178	$2,720	$7,458	$3,729	$6,449
9	$10,993	$2,720	$8,273	$4,136	$6,856
10	$11,872	$2,720	$9,152	$4,576	$7,296

TABLE 7-11

ENERGY MAINTENANCE AND OPERATING COSTS
AIR COOLED CHILLER

YEAR	ENERGY COST	MAINTENANCE AND OPERATING COST	TOTAL COST
1	43920.00	1600.00	45520.00
2	47565.40	1732.80	49298.20
3	51513.30	1876.60	53389.90
4	55788.90	2032.40	57821.30
5	60419.40	2201.10	62620.40
6	65434.20	2383.80	67817.90
7	70865.20	2581.60	73446.80
8	76747.00	2795.90	79542.90
9	83117.00	3027.90	86145.00
10	90815.70	3279.30	93295.00
11	97487.00	3551.40	101038.50
12	105578.50	3846.20	109424.70
13	114341.50	4165.40	118506.90
14	123831.80	4511.20	128343.00
15	134109.90	4885.60	138995.50
16	145241.00	5291.10	150532.10
17	157296.00	5730.30	163026.30
18	170351.60	6205.90	176557.40
19	184490.70	6721.00	191211.70
20	199803.50	7278.80	207082.30

TABLE 7-12

ENERGY MAINTENANCE AND OPERATING COSTS
ABSORPTION MACHINE

YEAR	ENERGY COST	MAINTENANCE AND OPERATING COST	TOTAL COST
1	33512.60	1700.00	35212.00
2	35782.10	1841.10	37623.20
3	38239.90	1993.90	40233.80
4	40901.60	2159.40	43061.00
5	43784.30	2338.60	46122.90
6	46906.20	2532.70	49439.00
7	50297.30	2743.00	53030.20
8	53948.90	2970.60	56919.60
9	57914.50	3217.20	61131.70
10	62209.20	3484.20	65693.50
11	66860.40	3773.40	70633.80
12	71897.70	4086.60	75984.30
13	77353.00	4425.80	81788.80
14	83261.10	4793.10	88054.30
15	89698.70	5191.00	94850.60
16	96589.20	5621.80	102211.00
17	104094.00	6088.40	110182.40
18	112221.60	6593.80	118815.30
19	121023.80	7141.00	128164.80
20	130556.80	7733.70	138290.40

140,000 ft^2 facility. Energy costs and maintenance costs are both assumed to increase by 8% annually and the expected life of chiller is assumed to be 20 years. Compare the life cycle costing of the three alternatives.

No.	Type	Capital Investment	Energy Cost	Maintenance and Operating Cost
1	Air-Cooled Chiller	$126,524	$43,920	$1,600
2	Absorption Machine	$165,244	$33,512	$1,700
3	Water-Cooled Chiller	$ 88,562	$38,869	$1,700

Here is the step by step method for comparing the alternatives.

1. For each alternative, calculate the energy costs, maintenance and operating costs, and total costs. The total costs are the sum of the energy costs and maintenance and operating costs. Tables 7-11, 7-12, and 7-13 show these costs for the air-cooled chiller system, the absorption machine, and the water-cooled chiller system, respectively.

2. To compare alternatives 1 and 2, calculate the extra first cost for alternative 2 over alternative 1, i.e., $165,244 − $126,524 = $38,720. Calculate the net savings in the total cost by using alternative 2 over alternative 1, i.e., for first year net savings = $45,520 − $35,213 = $10,307.

3. Table 7-14 shows the cash flow for the comparison of alternatives 1 and 2. From Table 7-14, the pay-back period for the differential investment is 5 years and 1 month and the rate of return on the differential investment is 24.8 percent. Thus, alternative 2 (absorption machine) is better than alternative 1 (air-cooled chiller).

4. Similarly, compare alternative 2 with alternative 3.

The differential investment of alternative 2 over alternative 3 is $165,244 − $88,562 = $76,682. The net savings at the end of the first year in total cost by using alternative 2 over alternative 3 is $40,569 − $35,213 = $5,356. Table 7-15 shows the cash flow for the comparison of alternatives 2 and 3. The diffe-

TABLE 7-13

ENERGY MAINTENANCE AND OPERATING COSTS
WATER-COOLED CHILLER

YEAR	ENERGY COST	MAINTENANCE AND OPERATING COST	TOTAL COST
1	38859.20	1700.00	40569.20
2	42095.20	1841.10	43936.40
3	45589.30	1993.90	47583.20
4	49373.20	2159.40	51532.60
5	53471.10	2338.60	55809.80
6	57909.20	2532.70	60442.00
7	62715.70	2743.00	65458.70
8	67921.10	2970.60	70891.70
9	73558.60	3217.20	76775.80
10	79663.90	3484.20	83148.10
11	86276.00	3773.40	90049.40
12	93436.90	4086.60	97523.50
13	101192.20	4425.80	105618.00
14	109591.20	4793.10	114383.30
15	118687.20	5191.00	123878.20
16	128538.30	5621.80	134160.10
17	139207.00	6088.40	145295.40
18	150761.10	6593.80	157354.90
19	163274.30	7141.00	170415.30
20	176826.70	7733.70	104559.80

rential investment pays itself back in 10 years and 4 months and the rate of return on the differential investment is 10.6 percent. Thus, alternative 2 (absorption machine) is better than alternative 3 (water-cooled chiller).

ECONOMICS OF CONSERVATION

Energy conservation proposals for old and new facilities vary widely in their initial costs. Some alternatives can be economically justified, while others cannot[7-4]. Tables 7-16 and 7-17 list some of the alternatives that can be justified on the basis of statistical data collected through years of work in energy conservation.

COST/BENEFIT RELATIONSHIPS

An existing, university-owned indoor athletic facility in the northeast was used to equate the costs and benefits of a series of energy conservation methods. The solid curve in Figure 7-12 shows the cumulative savings in annual energy costs plotted as a function of the cumulative cost of successive energy conservation modifications, which are expressed as a percent of annual energy

TABLE 7-14. Cash Flow Comparison of Alternatives 1 and 2

DIFFERENTIAL CAPITAL INVESTMENT = 38720.0

YEAR	NET SAVINGS	DEPRECIATION	INCOME BEFORE TAXES	INCOME AFTER TAXES	CASH FLOW
1	10307	1936	8371	4186	6122
2	11675	1936	9739	4869	6805
3	13156	1936	11220	5610	7546
4	14760	1936	12824	6412	8348
5	16498	1936	14562	7281	9217
6	18379	1936	16443	8221	10157
7	20417	1936	18481	9240	11176
8	22623	1936	20687	10344	12280
9	25013	1936	23077	11539	13475
10	27002	1936	25666	12833	14769
11	30409	1936	28469	14234	16170
12	33440	1936	31504	15752	17688
13	36728	1936	34792	17396	19332
14	40299	1936	38353	19176	21112
15	44145	1936	42209	21104	23040
16	40321	1936	46385	23193	25129
17	52644	1936	50908	25454	27390
18	57742	1936	55806	27903	29839
19	63047	1936	61111	30555	32491
20	68792	1936	66856	33428	35364

THE PAY BACK PERIOD = 5 YEARS 1 MONTHS

RATE OF RETURN ON INVESTMENT = 24.8 PERCENT

TABLE 7-15. Cash Flow Comparison of Alternatives 2 and 3

DIFFERENTIAL CAPITAL INVESTMENT =   76682.0

YEAR	NET SAVINGS	DEPRECIATION	INCOME BEFORE TAXES	INCOME AFTER TAXES	CASH FLOW
1	5356	3834	1522	761	4595
2	6313	3834	2479	1240	5074
3	7349	3834	3515	1758	5592
4	8472	3834	4637	2319	6153
5	9687	3834	5853	2926	6760
6	11003	3834	7169	3504	7419
7	12423	3834	8594	4297	8131
8	13972	3834	10138	5069	8903
9	15644	3834	11810	5905	9739
10	17455	3834	13621	6810	10644
11	19416	3834	15581	7791	11625
12	21538	3834	17705	8853	12607
13	23839	3834	20005	10003	13837
14	26330	3834	22496	11248	15062
15	29028	3834	25193	12597	18431
16	31949	3834	28115	14057	17892
17	35112	3834	31279	15639	19474
18	34540	3834	34705	17353	21167
19	42251	3834	38416	19208	23042
20	46269	3834	42435	21218	25052

THE PAY BACK PERIOD = 10 YEARS 4 MONTHS

RATE OF RETURN ON INVESTMENT = 10.6 PERCENT

TABLE 7-16. New Buildings

1. Increase wall and roof insulation.
2. Reduce glass area.
3. Install storm windows.
4. Site orientation.
5. Change maintained temperature.
6. Reduce lighting levels.
7. Reduce ventilation rates.
8. Move internally generated heat between zones.
9. Electrical demand management control.
10. Power factor improvement.
11. High efficiency pumps, motors, etc.
12. Heat pumps.
13. Use economizer system.
14. Heat recovery schemes.
15. Variable air volume system.
16. Periodic, night or weekend shutdown of HVAC systems.
17. Nighttime set-back of HVAC systems.

TABLE 7-17. Existing Buildings

1. Shift work hour schedule.
2. Monitor housekeeping.
3. Establish energy-use budgets.
4. Increase wall and roof insulation.
5. Change maintained temperature.
6. Reduce lighting levels.
7. Reduce ventilation rates.
8. Increase HVAC equipment efficiency.
9. Use economizer system
10. Periodic, night or weekend shutdown of HVAC systems.
11. Nighttime set-back of HVAC systems.
12. Power factor improvement.
13. Electrical demand management control.

costs. The dashed curves show Spiegel's estimate of the limits of energy savings with similar buildings in this geographic area. Evaluation of Figure 7-12 shows that

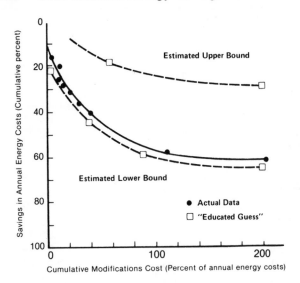

FIG. 7-12. Decreasing cost effectiveness of cumulative energy conservation modifications.

(a) The incremental cost effectiveness of each succeeding modification decreases and the curve flattens out.

(b) Most of the initial cost savings are both significant and relatively inexpensive. For instance, a cumulative expenditure of slighty more than 40% of the base-year energy costs produces a savings on the order of 40-45%.

(c) There is a point beyond which the slope of the curve is zero, indicating that no future energy savings can rationally be realized at any investment. This limit will occur at different points for different buildings and the objective should be not to exceed this limit under any financial criteria.

Figure 7-13 shows four additional curves resulting

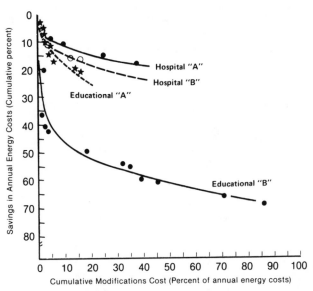

FIG. 7-13. Cumulative energy savings varies with the size and function of the building.

from studies undertaken by Spiegel. Hospital "A" is a metropolitan hospital with a 103,000 square-foot area. Hospital "B" is a 170,000 square-foot addition to a large metropolitan hospital. Educational "A" is a 109,000 square-foot university classroom building with a lab/teaching wing and Educational "B" is a 92,000 square-foot combination university classroom and administration building. Each of the curves is slightly different, representing both the individual nature of each building and energy conservation economics association with each structure.

It is always imperative to analyze every energy management consideration with respect to its economics.

Chapter 8

ENERGY INFORMATION
SYSTEMS AND ENERGY AUDITS

Energy Information Systems (EIS) are developed to determine how much energy is used by an organization. The EIS survey inventories energy using devices and determines the quantities of energy they use and the purpose for which they are used in order to evaluate the energy use trends and patterns throughout the organization.

An Energy Information System should be more than a mere record of energy quantities and costs. It must be dynamic: it must cause people to modify the information input, hopefully for the better. Energy Information Systems are established with the following objectives in mind:[8-1]

1. Stimulate management interest and control

A good Energy Information System informs management about the quantity and cost of each fuel used, thus stimulating their interest in controlling these costs. As a result, management tends to be more receptive to energy conserving ideas.

2. Stimulate employee interest and cooperation

A widespread knowledge of the energy usage of the organization stimulates employees to do their share in conserving energy. Appreciation by management for each employee's contribution to energy conservation is likely to keep this interest intact.

3. Develop energy audit systems

An energy Information System generally generates a good energy flow data base which can be used for the development of detailed energy audits. Energy Information Systems can also be used for evaluation of the potential costs and benefits of energy management opportunities.

ENERGY DATA BASE

The first step in the development of an Energy Information System is to compile the energy use history of the organization. This data base identifies the major factors affecting energy use and indicates the important usage trends. The basic information needed for establishment of a data base is as follows.
1. Fuel bills.
2. Quantities of major internally-generated energy carriers such as steam and compressed air.
3. Quantities of captive fuels consumed and quantities of other materials such as wood chips and waste oil that are used as energy sources and not otherwise counted.
4. Some measure of production or activity levels.

Tables 8-1 through 8-15 illustrate the data collection process for electricity and steam.[8-2]

Electrical Power

The following tables can be used for recording the amounts of electrical power purchased or generated on-site. Table 8-1 is used to record the readings of utility meters at each incoming service. It is important to note the day of the month when the meters are read and to attach copies of the bills for the 12-month period under study.

Table 8-2 is used to calculate the Btu equivalant of the annual kWh consumption.

Figure 8-1 is used to plot the monthly demand in kW and the monthly usage in kWh. A separate graph is prepared for each service.

Table 8-3 is a record of electricity purchases, in kWh,

TABLE 8-1. Utility Meter Readings and Date

Service No.	Building or Facility	Meter Type			Reading Date
		Total kWh	Demand kW	Power Factor	
1					
2					
3					
4					
5					
6					

TABLE 8-2. Annual kWh and Equivalent Btu

Service No.	Annual kWh $\times 10^6$	Annual Btu $\times 10^9$ kWh = 3412 Btu
1		
2		
3		
4		
5		
6		
Combined		

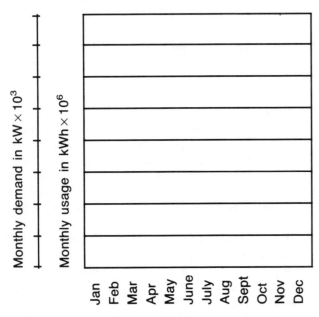

FIG. 8-1. Profile of monthly demand and usage.

for every quarter and for the year. For conversion to equivalent Btu, use 3412 Btu per kWh.

Table 8-4 records onsite electrical generation. Under *type,* list whether steam turbine, diesel or gas turbine. Under *heat recovery,* indicate the form of heat in its physical units: Btu for hot water, lb/hr for steam. Indicate the rated capacity of the heat recovery device in terms of Btu or lb/hr of steam.

Table 8-5 is used to record the quantities supplied to the site from each central power plant. Demand kW and kWh can be obtained from the meters. Site distributed kWh represents the difference between generated kWh and the kWh consumed by auxiliaries in the central power plant. Columns 5 through 8 record the thermal recovery.

Table 8-6 is a summary of purchased and generated power. To estimate the combined peak kW, add 90% of the peak kW generated to the peak kW purchased.

Steam

The following tables are used to record the amount of

TABLE 8-3. Total Purchased Electricity

	kWh × 10³	Btu × 10⁶	Cost	Cost per Btu × 10⁶
January				
February				
March				
1st QUARTER				
April				
May				
June				
2nd QUARTER				
July				
August				
September				
3rd QUARTER				
October				
November				
December				
4th QUARTER				
YEAR TOTAL				

TABLE 8-4. Onsite Generation

Generator No.	Building or Facility	kW Rating		Prime mover			Hours of Operation	Remarks
		Contin.	Emerg.	Type	Heat Recovery			
					Units	Cap.		

TABLE 8-5. Quantities Supplied from Central Power Plant to Site

1	2	3	4	5	6	7	8
	Electrical Quantities			Thermal Recovery Quantities			
	Generated		Site Dist. kWh	Generated	Wasted	Utilized	Maximum Flow
	kW	kWh					
January							
February							
March							
1st QUARTER							
April							
May							
June							
2nd QUARTER							
July							
August							
September							
3rd QUARTER							
October							
November							
December							
4th QUARTER							
YEAR TOTAL							

TABLE 8-6. Summary of Purchased and Generated Power

	Purchased (All Metered Services)		Generated		Combined	
	Peak kW	kWh	Peak kW	kWh	Peak kW	kWh
January						
February						
March						
1st QUARTER						
April						
May						
June						
2nd QUARTER						
July						
August						
September						
3rd QUARTER						
October						
November						
December						
4th QUARTER						
YEAR TOTAL						

TABLE 8-7. Site Service Characteristics

Service Entry		Steam Conditions			Service Fluctuation psig	Maximum Capacity lb/hr	Remarks
Location	Meter No.	Press psig	Temp °F	Quality %			

TABLE 8-8. District Steam Meters

Meter No.	Type	Scale Ranges			Areas Served
		Indicator	Recorder	Totalizer	

TABLE 8-9. Total Purchased Steam

-	Quantities 10^6 lb	Flow (lbs/hr)		Remarks
		Maximum	Minimum	
January				
February				
March				
1st QUARTER				
April				
May				
June				
2nd QUARTER				
July				
August				
September				
3rd QUARTER				
October				
November				
December				
4th QUARTER				
YEAR TOTAL				

steam purchased and generated at the site.

Table 8-7 records the characteristics of the steam delivered to the site. Under *service fluctuations* note the minimum pressure in psig.

Table 8-8 is a record of all the steam meters. The ranges should be specified in lb/hr and the after applying multiplier.

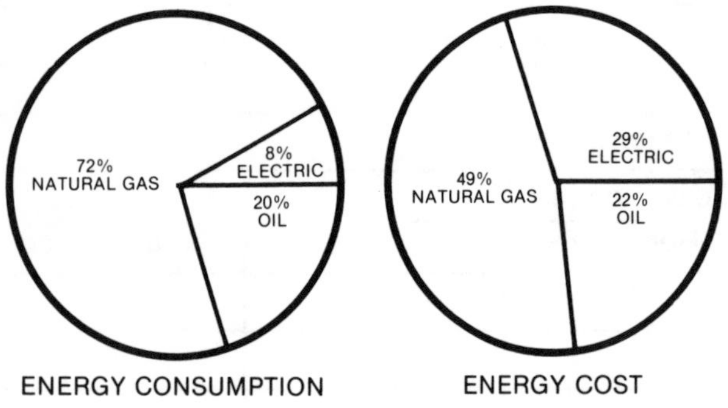

ENERGY CONSUMPTION   ENERGY COST

FIG. 8-2.

Table 8-9 is used to record monthly purchases of steam and totals for the year.

Table 8-10 records the equivalent Btu's of the total pounds of steam purchased each month. Use the following values for the heat content of steam at the district steam meters. For other steam conditions, refer to a standard saturated steam table.

PSIG	Heat Content, Btu/lb
25	1170
50	1180
75	1186
100	1190
150	1197
200	1200
250	1202
300	1204

Tables 8-11 through 8-14 record the steam generated onsite.

Table 8-15 summarizes the data on fossil fuel usage. After one year of operation, the data can be presented to

TABLE 8-10. Equivalent Btu of Purchased Steam

	Quantities		Cost	Cost/10^6 Btu
	10^3 lb	10^6 Btu		
January				
February				
March				
1st QUARTER				
April				
May				
June				
2nd QUARTER				
July				
August				
September				
3rd QUARTER				
October				
November				
December				
4th QUARTER				
YEAR TOTAL				

TABLE 8-11. Central Heating Plant (CHP) Output

CHP Bldg No.	CHP Location	Fuel	Steam Generated		CHP Hours of Operation	Buildings Served
			Pressure psig	Temperature °F		

TABLE 8-12. Central Heating Plant Meters

CHP No.	Meter No.	Type	Scale Ranges Indicator	Scale Ranges Recorder	Scale Ranges Totalizer	Items Metered

TABLE 8-13. Quantities Supplied from Each CHP to Site

	Steam Quantities – 10^6 LBS Production	Steam Quantities – 10^6 LBS CHP Usage	Steam Quantities – 10^6 LBS Net Supply	Flow lb/hr Max.	Flow lb/hr Min.	Other Metered Media Makeup Water	Other Metered Media Conden-sate	Other Metered Media Feed Water
January								
February								
March								
1st QUARTER								
April								
May								
June								
2nd QUARTER								
July								
August								
September								
3rd QUARTER								
October								
November								
December								
4th QUARTER								
YEAR TOTAL								

TABLE 8-14. Btu Equivalent of Steam Supplied to Distribution System

	Steam Supplied 10⁶ Btu	REMARKS
January		
February		
March		
1st QUARTER		
April		
May		
June		
2nd QUARTER		
July		
August		
September		
3rd QUARTER		
October		
November		
December		
4th QUARTER		
YEAR TOTAL		

management. A typical year's results are shown in Figure 8-2. It is evident that, although natural gas fulfills nearly three-fourths of the plant's energy requirements, it accounts for only half the energy cost.

FACTORS AFFECTING ENERGY USE

Energy consumed in an industrial or commercial building must be related to one or more of the major parameters such as the size of the building, the weather and the level of activity. Table 8-16 is used to summarize these factors.

Data is entered into the Table as follows:

Column 1: List the consecutive months of the year.
Column 2: Weather data is available from local offices of the weather service. Heating degree days for one day = 65° − (high temperature + low temperature)/2. Calculate the total heating degree days for each month and enter in Column 2.
Column 3: Cooling degree days for one day = (high temperature + low temperature)/2 − 65°. Cal-

culate the total heating degree days for each month and enter in Column 3.
Column 4: Estimate the floor space area in use and enter in Column 4.
Column 5: Estimate the volume of space in use and enter in Column 5.
Column 6: Enter the production level in presently used physical units (number of items, weight, volume, etc.) in Column 6.
Column 7: Enter operating cost in Column 7.
Column 8: Enter a measure of operating level (man hours, hours of operation, etc.) in Column 8.

Figure 8-3 illustrates the impact of one of these parameters, production level, on energy use per unit of production.

Utility Bills

The quantities used in all the tables are for a calendar month. However, most utility bills are not computed on an end-of-month basis. For example, find the consump-

TABLE 8-15. Summary of All Fuel Usage and Cost

	OIL			COAL			GAS			OTHER			TOTAL	
	Gals.	Total Cost	10^6 Btu	Tons	Total Cost	10^6 Btu	Mcf	Total Cost	10^6 Btu	Quan.	Total Cost	10^6 Btu	10^6 Btu	Btu/ ft^2
January														
February														
March														
1st QUARTER														
April														
May														
June														
2nd QUARTER														
July														
August														
September														
3rd QUARTER														
October														
November														
December														
4th QUARTER														
YEAR TOTAL														

TABLE 8-16. Factors Affecting Energy Use

Month (1)	Degree Days		Size of Facility		Level of Activity		
	Heat (2)	Cool (3)	ft² (4)	ft³ (5)	Production Units (6)	Operating Cost (7)	Operating Level (8)
January							
February							
March							
April							
May							
June							
July							
August							
September							
October							
November							
December							

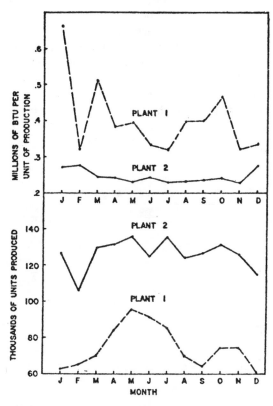

FIG. 8-3. Energy use related to production levels.

tion of electricity during February from the following bills:

Billing Period	kWh Consumed
1/20/80 - 2/6/80	10,000
2/7/80 - 3/25/80	24,000

The January/February billing is for 18 days, 6 of which are in February. Prorating, consumption during the first 6 days of February is

$$6/18 \times 10,000 = 3,333 \text{ kWh}$$

The February/March billing is for 47 days, 22 of which are in February. Prorating, consumption during the last 22 days of February is

$$22/47 \times 24,000 = 11,234 \text{ kWh}$$

Total consumption for the 28 days of February is

$$3,333 + 11,234 = 14,567 \text{ kWh}$$

ENERGY AUDITS

An energy audit is a method of determining the energy efficiency of an operation, plant or building. It performs two functions. First, it measures the total energy input of the system so that it can be compared with the useful energy output in the form of finished products or completed processes. The difference between the input and the output is the energy lost. Losses must be identified and, as far as possible, reduced or converted into useful energy. Second, an ongoing program of periodic audits tracks the effectiveness of the conservation program.

An energy audit should reveal the following facts:
- Energy consumption of each fuel-using unit such as furnaces and boilers
- Energy losses
- Energy consumption by special users

There are many levels of energy audits. At the most general level, there is the audit of the facility as a whole, called the overall energy audit.

At the next level, there are four areas of interest corresponding to the four categories into which the building and its mechanical/electrical systems can be divided: building skin, control equipment, distribution systems, and utilization equipment.[8-3]

Beyond this, there are an extremely large number of areas into which an energy audit can be expanded. In nonindustrial sectors, it is reasonable to do an audit on a department by department basis. In a commercial building, the audit may spotlight energy usage by tenants or by lighting, ventilation, and air conditioning.

Overall Audits

The overall audit summarizes the energy usage of a building as a whole. This information is useful in comparing the energy used by the building with the usage of other buildings of the same type. All forms of energy used in a building should be converted to the common denominator of raw source energy used per ft^2 of building area. Table 8-17 lists the energy content of various energy sources.[8-4]

Detailed Audits

There are three methods generally used for detailed energy audits:
1. Measure and calculate
2. Flow graph
3. Computer simulation

Before any of these methods is used, certain factors must be delineated. First of all, the level of detail must be established, then the season of greatest interest and finally the various energy consuming processes. As far as level of detail is concerned, audits should focus on processes rather than departments. A wintertime audit might cover energy conversation by boilers; energy losses through glass, walls, roof, infiltration and ventilation; power supply, motors and lighting; and special uses such as domestic water, computers, food services and the like.

The Department of Housing and Urban Development has compiled a statistical analysis of the amount of energy being used by buildings to establish energy

TABLE 8-17. Approximate Energy Content of Various Energy Sources.

FUEL OIL	
No. 2	140×10^3 Btu/gal
No. 6 (high sulphur)	152×10^3 Btu/gal
No. 6 (low sulphur)	144×10^3 Btu/gal
GAS	
Natural	1×10^3 Btu/ft³
COAL	
1% sulphur	14×10^3 Btu/lb
.3% sulphur	9×10^3 Btu/lb
COKE	13×10^3 Btu/lb
STEAM	1×10^3 Btu/lb
ELECTRICITY	3412 Btu/kW
OTHER	
Wood	9×10^3 Btu/lb
Refuse (20% moisture) (20% noncombustible)	5×10^3 Btu/lb

budgets for future buildings. The data was collected from 1,661 nonresidential buildings, in 37 randomly selected metropolitan areas, and include existing data on 230,000 housing units. This summary, Figure 8-4, includes all climatic regions and shows the energy consumption of the most efficient building, the most wasteful building and the average in each category.

Overall energy audits will show how the facility ranks relative to similar facilities and indicate the potential for a cost effective energy conservation program or the need for a detailed energy audit.

Infrared thermography has been successfully used for pinpointing excessive skin heat leaks and distribution leaks.

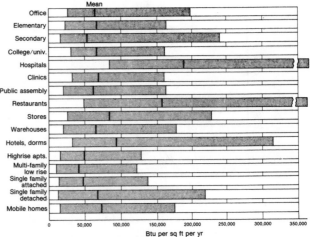

FIG. 8-4. High, low, and mean energy consumption of 1,661 buildings arranged by occupancy categories.

TABLE 8-18.

Energy Source	Measurement	Where to Measure	Method Used	Purpose of Measurement
Fuel	Maximum & minimum consumption rate.	Total plant & each combustion unit in the plant.	Flow meter & recorder, thermometer and gauges.	To calculate heat input to the plant or combustion unit.
Steam & hot water	Peak and off-peak production & consumption, temperature & pressure.	At each boiler & at each major process.	Flow meter & recorder, thermocouple, thermometer & gauges.	To calculate heat consumption of the heating process. To optimize factors.
Combustion Products	Analysis of CO, CO_2 and O_2.	Boiler or furnace flue before & after economizer.	Continuous electrochemical or infrared or manual absorption kit.	To calculate heat combustion efficiency of boiler or furnace.
Electrical power	Peak and off-peak consumption and power factor.	All lighting, heating & motors.	Ammeters, watt hour meter & power factor meter.	To calculate the electrical energy consumed by each source.
Compressed Air	Storage pressure vs. required pressure & leakage.	At receiver & ring main system.	Pressure gauge.	To check for right pressure.
Feed Water	Peak and off-peak feed water rate, temperature.	At each boiler, make-up feed & condensate return.	Flow meter and recorder, thermometer and gauges.	To estimate heat recovered via condensate & heat cost.
Lighting	Lighting levels.	All lighting systems.	Footcandle meter.	To calculate the lighting levels.

Generally, winter, summer, spring/fall, or all three, are selected for an energy audit. If the audit is by computer simulation, every day of the year can be included in the audit.

Finally, the time slice covered by the audit must be established. An audit can develop energy use information in the form of an instantaneous load break or can integrate use over several days or months. Generally, a period of one month is used for an audit. The total energy data is then divided by the days of the month to establish average daily consumption.

Measure and Calculate

The ideal situation is a facility with a meter at each energy user. This would permit instantaneous auditing by reading each meter. But such a situation is far from realistic. Most of the time, the measure and calculate method is used for an audit, using the methods listed in Table 8-18.

Besides taking all the measurements outlined in Table

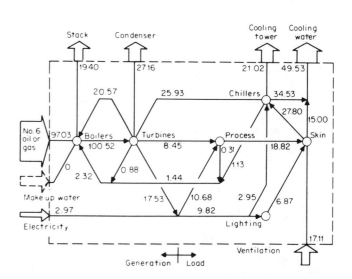

FIG. 8-5. Energy flow graph.[8-5]

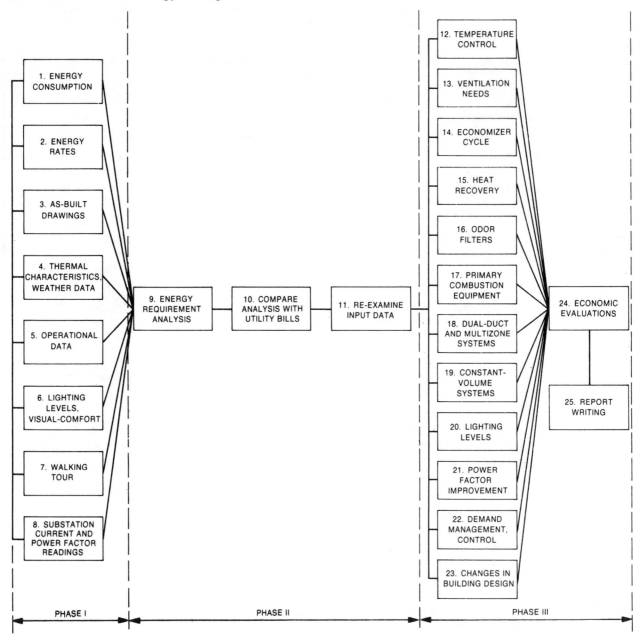

FIG. 8-6. Steps in an energy audit and subsequent energy conservation evaluations.[8-6]

8-18, occupancy profiles must be determined to correlate the data. For example, analysis of the efficacy of night temperature setback requires a knowledge of when the occupants leave the building.

Flow Graph Method

After the measure and calculate procedure, the data is used to draw an energy flow graph, Figure 8-5, in order to gain greater insight into the energy audit. The energy flow graph is a pictorial representation of the energy flow throughout a facility. It consists of a series of arrows connecting various nodes with the magnitude of

energy flow indicated along each arrow. Each node represents a process or a special use.

According to the first law of thermodynamics, the magnitude of the energy flowing into a particular node must equal that leaving the node. This relationship is used to iteratively balance the entire flow network by recalculating, remeasuring and adjusting, as necessary, until each node and then the entire facility is balanced and consistent with the metering records.

Figure 8-5 is a typical flow graph. It shows the major energy sources, electricity, gas, and fuel oil, entering on the left. There are incidental gains, shown along the right side and the bottom, which come into the control volume from the sun, in the ventilation air and from people inside the building. Shown along the top is the low grade energy being lost to the atmosphere. The width of arrows entering and leaving the facility is somewhat proportional to the energy flow. The left half of the graph represents energy input and conversion, the right half represents end use and load.

Thus, an energy flow graph is used to:

1. Organize the available data
2. Correlate sketchy data
3. Cross-check data and increase confidence in the survey's completeness
4. Determine where energy is used: the major flows, the major losses
5. Aid in selecting from among various combinations of energy conserving possibilities

Computer Simulation

With this energy audit method, the facility is simulated in the computer using its mathematical model. A dynamic simulation of the facility provides a very detailed energy audit.

It is frequently advantageous to create an energy audit by computer simulation because, once a dynamic model of a facility is working, proposed energy conservation modifications can be easily evaluated. A number of computer programs are available for detailed energy audits and they are discussed in Chapter 11.

Figure 8-6 is a flow chart showing the steps in an energy audit and the subsequent energy conservation studies using computer simulation.

Chapter 9

ENERGY CONSERVATION THROUGH HEAT RECOVERY

Recovering some of the heat that would otherwise be rejected into the exhaust air, condenser steam and water, or flue gases helps reduce operating costs and peak load demand. Because heat recovery devices require higher capital investment in HVAC equipment, they should be carefully analyzed for their cost effectiveness. Energy conservation also requires a careful analysis of the building energy requirements and use. A heat recovery system should be designed to avoid excessive energy consumption in one part of the system or it will limit the availability of heating/cooling demanded by another zone.

Farsighted engineers like Olivieri* were looking at the concept of heat recovery systems even before the crunch of the 1973 oil embargo. In 1971, he reviewed the state of the art in *Air Conditioning, Heating and Refrigeration News.*[9-1]

Heat can be conserved by recovering energy from the exhaust air and water condensate. We shall first discuss the recovery of heat from exhaust air in HVAC systems.

HEAT RECOVERY FROM EXHAUST AIR

Many buildings, such as hospitals, hotels, schools and factories, use a large proportion of outside air for ventilation. This is to satisfy the codes safeguarding the health and safety of the occupants and to meet air requirements of a process.

Waste energy from exhausted inside air may be recovered and used to heat or cool the makeup air or to preheat water. This helps reduce the overall energy consumption for heating and cooling. The various systems for recovering the energy from exhaust air are as follows:

Runaround Coils

Runaround coils recover waste energy from the exhaust air and supply it to the makeup air. In a typical layout, Figure 9-1, two finned tube coils are linked together by a loop of pipe. One coil is placed in the exhaust air stream and the second coil is placed in the makeup stream. A pump continuously circulates a fluid, such as ethylene glycol and water, between the two air streams. In winter, the exhaust air releases its heat at the exhaust coil. The working fluid transfers this energy to heat the incoming air. In summer, the heat flow is reversed.

A typical winter and summer operation, showing temperature changes for the incoming air and the exhaust air[9-2] is shown in Figure 9-2. Winter operation supplies up to 70% of the makeup air heating load and summer operation supplies up to 20% of the makeup air cooling load.

The runaround coil has two important advantages over other air-to-air heat recovery equipment:

1. The exhaust air outlet and makeup air inlet can be located apart from each other. This eliminates the cost of bringing large volumes of exhaust air back into the process room.
2. The freezing of exhaust air condensate can be avoided by:
    a. controlling the flow rate of fluid in the runaround coils to maintain the tube wall temperature at or above 32°F.

*The authors wish to thank Mr. Olivieri for his contributions to this chapter.

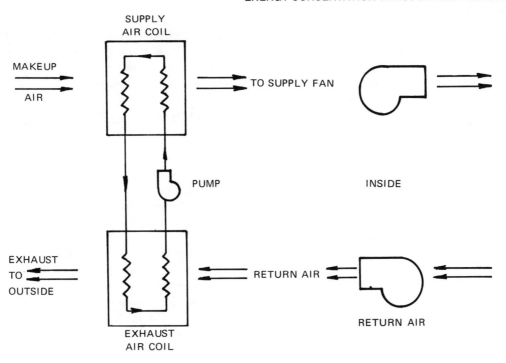

FIG. 9-1. Schematic of a heat recovery system employing runaround coils.

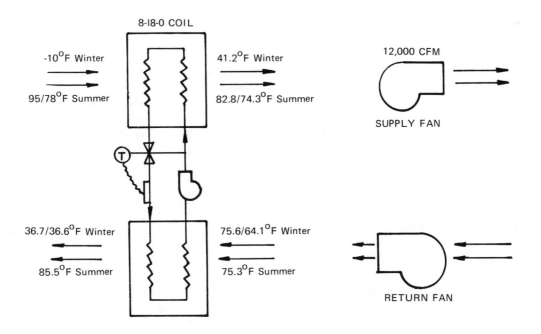

FIG. 9-2. Typical temperature ranges of runaround coils during winter and summer operation.[9-2]

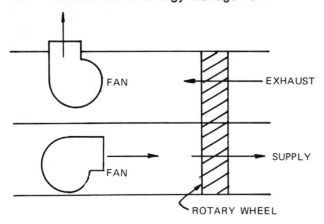

FIG. 9-3. Schematic of a heat recovery wheel.

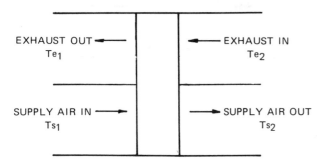

FIG. 9-4. Heat exchanges in a heat recovery wheel.

b. using a temperature-controlled mixing valve to maintain the ethylene glycol-water mixture entering the exhaust air coil at a minimum temperature of 30°F.

Heat Recovery Wheels

Heat recovery wheels are rotary air-to-air heat exchangers of which there are two common types:

1. The metallic wheel, 16 to 22 inches in diameter, constructed of knitted aluminum or stainless steel heat transfer media.
2. The desiccant wheel that has corrugated air passages composed of silicate-backed asbestos impregnated with lithium chloride.

The metallic wheel readily transfers sensible heat from one air stream to another, with virtually no moisture transfer. The desiccant wheel transfers both sensible and latent heat between air streams. Moisture is also transferred, in the vapor phase, through the hygroscopic nature of lithium chloride.

With both of these air-to-air heat exchangers, the wheel rotates slowly (approximately one revolution each 5 to 10 seconds) in a casing between the outside air intake duct and the exhaust air duct, Figure 9-3. As it rotates through the counterflowing air streams, the wheel absorbs energy from the higher-energy air stream and transfers it to the lower-energy air stream, Figure 9-4. The efficiency of the heat wheel is as follows:

$$\text{Efficiency} = \frac{\text{energy recovered from exhaust}}{\text{maximum possible heat recovery}}$$

$$= \frac{\dot{m}C_p(T_{e2} - T_{e1})}{\dot{m}C_p(T_{e2} - T_{s1})}$$

$$= \frac{T_{e2} - T_{e1}}{T_{e2} - T_{s1}}$$

The efficiency of heat wheels varies from 60% to 90% depending upon the face velocity through the unit. For example, some units have 90% efficiency at 200 fpm but only 70% at 800 fpm. The lower face velocity implies a larger unit. One has to balance the higher initial cost of the more efficient unit against a lower operating cost.

To minimize carry-over or cross-contamination between air streams, a purge section is built into the desiccant wheel housing, Figure 9-5. This allows a continuous purging of the exhaust air from the wheel by the outside air before the wheel is exposed to the supply air side. Cross contamination of the supply air stream is held to less than one tenth of one percent, and particle carry-over is less than 0.2% whereas, without a purge section, contamination is 1% to 8%.

Heat recovery wheels are usually justifiable with air handling systems requiring very large quantities of makeup air. Additional duct work is required because the exhaust air and intake air streams must be brought together.

Example 9-1

A sensible heat wheel unit recovers up to 80% of the heat in the exhaust. On a 0°F day, the outside air is heated from 0 to 56°F while the exhaust air is cooled from 70 to 14°F. This heat wheel handles 10,000 cfm and requires a total of about 6.75 hp for each hour of operation. Estimate the net reduction in energy consumption.

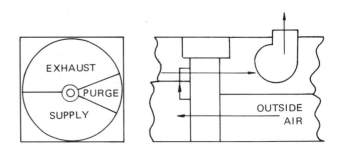

FIG. 9-5. Purge section in a dessicant wheel housing.

Solution 9-1

Heat gained by the
outside air

$= $ sp. heat $\times$ mass $\times$ temp. change
$= 1.08 \times$ cfm $\times (T_i - T_o)$
$= 1.08 \times 10,000 \times (56-0)$
$= 604,800$ Btuh

Additional fan
power

$= 6.75$ hp
$= 6.75 \times .746 \times 3412$
$= 17,180$ Btuh

Net energy recovery $= 604,800 - 17,180 = 587,620$ Btuh

This net recovery decreases as the outside temperature increases. On a day with a 30°F outside air temperature,

Net energy recovery $= 1.08 \times 10,000 \times (56-30) - 17,180$
                    $= 263,620$ Btuh

Heat Pipe Air-to-Air Exchangers

The heat pipe was developed by G. M. Grover at Los Almos, N.M. in 1963. Its outstanding advantage is that it has no moving parts.

Figure 9-6 shows the principles of the heat pipe cycle. The heat pipe is constructed of finned tubes which are sealed at both ends. Each tube is actually a heat pipe consisting of an envelope (the tube), a wick and a working fluid (generally a refrigerant). Hot air streams flow over one end of the tubes, transferring heat energy to the working fluid which evaporates. The vapors flow to the cold end, condense, and transfer energy to the cold stream. The liquid returns to the hot end through the wick and the cycle is complete.

Some of the disadvantages of this system are reported to be as follows:

1. The first cost of the units are 50–100% greater than those of the runaround coils.
2. Expensive duct work is needed to bring the exhaust stream back to the equipment room because side-by-side discharge and intake is required.
3. A preheat coil is required for outside temperatures

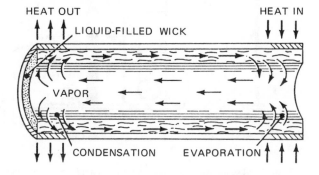

HEAT OUT   HEAT IN

LIQUID-FILLED WICK

VAPOR

CONDENSATION   EVAPORATION

FIG. 9-6. Heat pipe cycle.

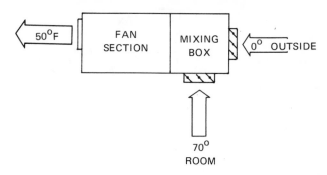

50°F   FAN SECTION   MIXING BOX   0° OUTSIDE

70°
ROOM

FIG. 9-7. Recirculated air can be used to preheat incoming outside air.

below 18°F to prevent the freezing of the exhaust air condensate.

Mixed Air Systems

A method not generally thought of as recovering heat mixes room air with the incoming outside air to preheat the new air, Figure 9-7. Mixed air systems have limited application and can only be used in areas with large internal heat gains. This method should be termed heat conservation because heat is not recovered as it is in other systems.

Buildings with high heat gains need supply air cooler than room temperature, therefore room and outside air are mixed to reach the required temperature. The amount of makeup air needed and the net internal heat gain determine if this method is practical.

Evaluating Exhaust Air Systems

Heat recovery systems are available, but are not always feasible for a given situation. Several variables must be considered when evaluating exhaust air heat recovery systems.

1. Amount of air being exhausted.
2. Temperature of exhaust and relative humidity if total heat recovery is considered.
3. Cost of fuel.
4. Cost of power.
5. Capitalization period.
6. Internal heat gain from lights, equipment, and people.
7. Outside temperature range.

A typical building will now be examined to see how the various heat recovery systems can be used.

Industrial buildings require large amounts of exhaust air which is discharged locally at a process, Figure 9-8, in an attempt to capture contaminants and keep them out of the plant. General exhaust, shown in Figure 9-9, is also used to dilute any contaminant that is not, or cannot be, captured by the local exhaust.

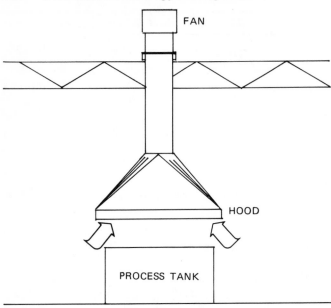

FIG. 9-8. Local exhaust.

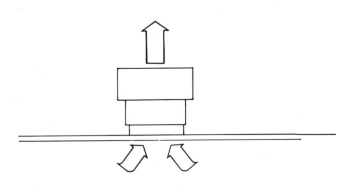

FIG. 9-9. General exhaust.

Exhaust also removes heat. Although this is desirable in summer, heat is a problem for many plants in both summer and winter. In fact, many industrial plants produce so much heat that a heating system is not needed. Why then are some of these plant engineers deluged with complaints that the plant is cold? Oddly enough, the excess heat is the cause of the cold.

Most plants have sufficient exhaust but little makeup air. During the exhausting of excess heat, makeup air is drawn through open doors, cracks and stacks, causing cold drafts.

The answer is to air-temper the makeup air by heating the makeup air or by attempting to recover the heat being thrown away in the exhaust. It may be more dramatic to

think of the Btu's thrown away in a 500°F stack, or even a 150°F stack, but why not try to recover the heat from 70°F air using the methods described earlier?

Example 9-2

Consider a plant with a heat gain of 40 Btuh/ft². Rather than deal with the whole plant, examine a 10,800 ft² interior section. In summer, add another 10 Btuh/ft² for the roof heat gain. In winter, subtract 0.2 Btuh/ft² for every degree difference between indoors and outdoors for roof heat loss.

The amount of ventilation air required is set by the summer requirements. To keep the plant only 10°F warmer than outdoors requires a circulation of 4.63 cfm/ft² as shown below.

Let Q_s = Heat gain/loss of the space
         = 50 Btuh for summer
   cfm = Supply air flow rate in cubic feet per minute
   T_s = Supply air temperature
   T_{sp} = Space air temperature

Taking an energy balance on the space

$$Q_s = 1.08 \times cfm \times (T_{sp} - T_s)$$
$$50 = 1.08 \times cfm \times (10)$$
$$cfm = \frac{50}{1.08 \times 10}$$
$$= 4.63 \ cfm/ft^2$$

This air must be introduced to cover every square foot of the space. This is a lot of air to move, but remember that at 90°F outdoors the space will be at 100°F, while the air conditioned offices are at 70° to 75°.

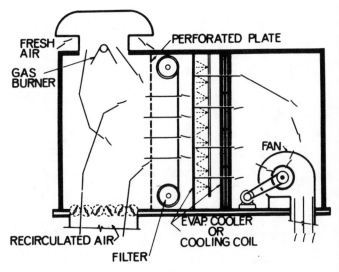

FIG. 9-10. Makeup air unit mixes warmed outside air with recirculated room air.

TABLE 9-1. Mixed Air System

T_o	Q_2	Q_3	Q_4	T_5	T_6
70	+40	0	+40	51	70
60	+40	−2	+38	52	67.5
50	+40	−4	+36	53	65
40	+40	−6	+34	54	62.5
30	+40	−8	+32	55	60
20	+40	−10	+30	56	57.5
10	+40	−12	+28	57	55
0	+40	−14	+26	58	52.5

T_o = Outside air temperature, °F
Q_2 = Heat gain, Btuh/ft^2
Q_3 = Heat loss, Btuh/ft^2
Q_4 = Net heat gain, Btuh/ft^2
T_5 = Required air temperature for 70°F space

$$= T_o - \frac{Q_4}{1.08 \times cfm} = T_o - \frac{Q_4}{1.08 \times 2}$$

T_6 = Mixed air temperature with minimum outside
air

$$= (0.5 \times T_o + 1.5 \times 70)/2.$$

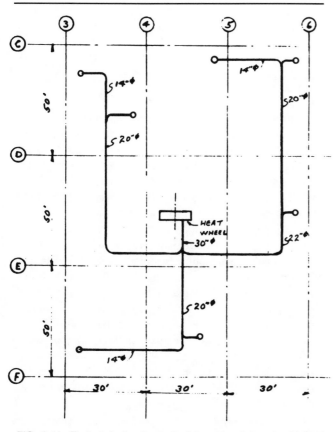

FIG. 9-11. Typical ductwork required to route air to a heat wheel.

TABLE 9-2. Heat Wheel System

T_o	T_w	T_5	T_m
Outside temp.	Temp. off wheel	Required air supply temp.	Mixed air temp.
70	70	51	70
60	68	52	69.5
50	66	53	69
40	64	54	68.5
30	62	55	68
20	60	56	67.5
10	58	57	67
0	56	58	66.5

$$T_m = (0.5\ T_w + 1.5 \times 70)/2$$

The air quantity can be reduced to 2 cfm/ft^2 if evaporative cooling is used along with spot cooling. So, for the following example, assume 2 cfm/ft^2 of supply air.

The plant under examination has a makeup air requirement of 0.5 cfm/ft^2 for local and general exhaust, which is mixed with room air as shown in Figure 9-10.

Table 9-1 shows the air temperature required to keep the plant at 70°F and the mixed air temperature, Column 6, that will result by mixing 0.5 cfm outside air with 1.5 cfm room air. No heat is required until it is 10°F outdoors. This system is the mixing method of heat recovery, for which no special equipment is needed.

Comparing Heat Wheels and Runaround Coils

When a heat wheel is used, it is necessary to run the exhaust to the heat wheel so that the heat can be recovered. In some cases this requires extensive ductwork. Figure 9-11 shows a typical ductwork arrangement.

Assuming that all the exhaust is ducted to the heat

TABLE 9-3. Runaround Coil System

T_o	T_c	T_5	T_m
Outside temp.	Temp. off coil	Required air supply temp.	Mixed air temp.
70	70	51	70
60	65.9	52	69.1
50	61.6	53	68
40	57.2	54	66.9
30	52.9	55	65.8
20	48.6	56	64.7
10	44.3	57	63.6
0	40	58	62.5

$$T_m = (0.5\ T_c + 1.5 \times 70)/2$$

wheel, Table 9-2 lists the temperature off the heat wheel and the resulting air temperature when mixed with room air. At no time is any supplementary heat added because the mixed air temperature is greater than the required supply air temperature.

Next, look at the runaround coil (the circulating liquid system of heat recovery). The supply air temperatures we can expect by mixing room air and air off the runaround coils are shown in Table 9-3. Again, no heat is required.

Economic Analysis of Heat Recovery Systems

Now to compare the three systems on the basis of economy.

The mixing system requires some heat, as seen in Table 9-1. The amount of heat required at each outdoor temperature:

$$Q = 1.08 \times cfm \times (T_s - T_m) \times n_h$$

$\quad$ = Heat supplied in Btu/season at outdoor temperature

cfm = Air circulated, cubic feet per minute

n_h = Hours of outdoor temperature

T_s = Supply air temperature = T_5 (Table 9-1)

T_m = Mixed air temperature = T_6 (Table 9-1)

The amount of heat required at each outdoor temperature is summarized in Table 9-4. Even operating three shifts a day, seven days a week, we only require 11,200,000 Btu. If the internal heat generated in the plant is not considered, then we need 4,200,000,000 Btu or 375 times as much.

We must put a price tag on this heat. Assuming 95% efficiency for a direct-fired unit, the 11,200,000 Btu requires 11,790 ft^3 of gas. At $2.50 per 1000 ft^3, the gas costs approximately $29.50.

The heat wheel and circulating liquid require no heat, as shown in Tables 9-2 and 9-3. However, the supply and exhaust fans need additional power because of pressure

TABLE 9-4. Additional Heat Required by Mixed Air System

T_o	cfm	T_s	T_m	n_h	Q
70	21,600	51	70		0
60	21,600	52	67.5	1,433	0
50	21,600	53	65	1,247	0
40	21,600	54	62.5	1,367	0
30	21,600	55	60	1,353	0
20	21,600	56	57.5	644	0
10	21,600	57	55	185	− 8,631,360
0	21,600	58	52.5	20	− 2,566,080
					− 11,197,440

TABLE 9-5. Yearly Auxiliary Power Costs @ $.05/kWh

System	Supply fan	Exhaust fan	Motor	Total kWh	Cost $
Mixing	22,872	4,901		27,773	1,389
Wheel	40,844	22,872	230	63,946	3,197
Circul. Liquid	40,844	22,872	230	63,946	3,197

TABLE 9-6. Proportional Auxiliary Power Costs

System	Continuous	One shift	Two shift	Three shift
Mixing	1389	331	662	992
Wheel	3197	761	1522	2284
Circul. Liquid	3197	761	1522	2284

TABLE 9-7. Total Owning and Operating Costs

	Mixing	Heat wheel	Runaround coil
First cost	$40,000	$52,000	$46,000
Capitalization	10,280	13,364	11,822
Fuel	30	0	0
Power	331	761	761
Maintenance	200	600	600
Total cost/year	10,841	14,725	13,183

drop caused by the resistance of the heat wheel or coils.

The supply fan brake horsepower will increase from 3.5 to 6.25 and the exhaust fan horsepower from 0.75 to 3.5. Approximately 1.5 bhp is required by the pump and by the heat wheel rotor motor. The heat recovery equipment does not need to operate continuously because we only need heat recovery at 10° and 0°F outdoors, so the wheel or pump motor can be shut off above 10° outdoors.

The yearly power requirements are shown in Table 9-5. The fan power consumption is multiplied by 8,760 hours and the pump and rotor motors by the 205 hours they operate to get the yearly kilowatt hours.

The figures shown in Table 9-5 are for continuous plant operation. The cost figures can be modified, as shown in Table 9-6, for one, two or three 40-hour shift operations.

Maintenance, also an operating cost, is slight because of the short period of operation. A conservative estimate would be $200 per season per shift over the cost of a conventional system.

With the high interest rates presently being charged,

capitalization or owning costs is a very important factor. For example, if the rate of interest is 9%, in order to repay the money in ten years, one must pay back 15% per year. Table 9-7 summarizes the owning and operating costs for the three systems, assuming 9% interest and 5-year capitalization. Thus, we see that mixing is more cost effective when exhaust requirements are only 0.5 cfm/ft² and the internal heat gain is 40 Btu/ft². The annual owning and operating cost for the mixing system is $10,841, the least of the three.

This type of analysis can be used with other applications to determine the cost effectiveness of a given alternative.

ENERGY CONSERVING VENTILATING SYSTEMS

Several companies are developing ventilation systems which can capture up to 95% of the total heating or cooling energy from an exhaust system. The working principles of this type of system are shown in Figure 9-12.

HEAT RECOVERY FROM WASTEWATER

In many industrial installations, considerable heat is lost every day by discharging hot water and oil into drains. As one example, a plant was using city water to cool annealing furnaces. They were throwing away 5 million Btuh in hot water and, at the same time, were purchasing gas to heat 50,000 cfm of air from 0° to 80°F. The layout shown in Figure 9-13 was designed to recover this heat.

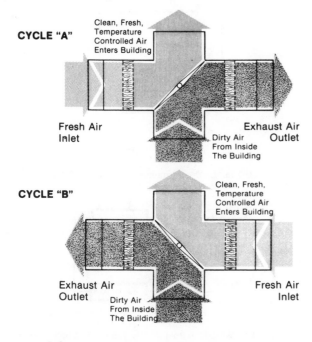

FIG. 9-12. Energy conserving ventilation system.

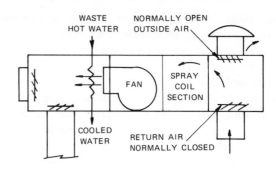

FIG. 9-13. System for recovering heat from industrial wastewater.

Through the use of dampers, air can be discharged to the atmosphere in the summer and into the plant in winter. The unit is really a multicell arrangement divided into three parts. Two parts always discharge to the atmosphere, and one unit can discharge either into the plant or to the atmosphere.

The discharge air temperature is controlled by starting and stopping the two fan sections. If the air is too cold, the sprays are turned off and the unit is protected against freezing by a flow switch and an aquastat. If the water gets too cold (50°F or lower) or the flow stops, the fan stops, the outside damper closes, and a room damper opens. If the water temperature drops to 40°F, a dump valve opens and drains the coil.

Another source of heat in industry is hot oil. Oil must be cooled with either city water, which is costly and wasteful, or cooling tower water that can cause corrosion and scaling problems.

In addition, with some of the oils being used in modern heat treating plants, a water leak will permit water to mix with the oil and cause disastrous results. The same unit used in Figure 9-13 is used for oil cooling, except that an air washer and finned coil are used in series so that the water and oil will not mix in the event of a leak. This is arranged for heating makeup air in winter.

HEAT RECOVERY FROM AIR CONDITIONING SYSTEMS/HEAT PUMPS

Years ago we had an air conditioning system and a heating season. With the use of higher illumination levels, air conditioning has become necessary all year. The most critical times are during the in-between seasons.

In many buildings it is necessary to run both the boiler and the refrigeration unit at the same time. We are burning a fuel to heat a building while, at the same time, we are extracting heat and throwing it away at the cooling tower.

It is possible to use the condenser heat to warm the building by making use of heat pumps with or without storage tanks. These systems are generally of two types:

1. Semicentral systems
2. Central systems

Semicentral Systems

A semicentral system is a year-round heating and cooling system. Decentralized, unitary, water-to-air heat pumps are served by a two-pipe, closed-loop water circuit acting as a heat source or a heat sink for the system. Several companies manufacture unitary heat pumps that may range from 6,100 to 49,000 Btuh in cooling capacity and from 6,800 to 45,000 Btuh in heating capacity. Figure 9-14 shows the heating cycle and cooling cycle of a water-to-air heat pump.

In the heating cycle, the compressor pumps hot compressed refrigerant vapor to the air side coil. Here the refrigerant vapor is condensed to a liquid by transferring heat to the room supply air. The liquid refrigerant is then directed to the refrigerant-to-water exchanger where it evaporates by absorbing heat from the circulating water loop. The refrigerant vapor returns to the suction side of

the compressor and the cycle starts over again.

In the cooling cycle, the reversing valve changes and directs the hot compressed refrigerant vapor to the condenser. The refrigerant vapor condenses to the liquid state by rejecting heat to the water loop. Liquid refrigerant then evaporates in the air side coil by absorbing heat from the supply air. The refrigerant vapor is directed back to the suction side of the compressor and the cycle is repeated.

With many units in a system, those in the heating mode will be extracting heat from the water circulating loop, while those in the cooling mode will be replacing it. In effect, the heat rejected from areas being cooled will be transported and become the heat source for those areas requiring heat. Figure 9-15a represents a balanced condition where the amount of heating required is equal to the heat rejected by the cooling cycle, but this is an ideal condition that rarely exists.

Most often, more heating or more cooling will be required, creating an unbalanced condition. For example, full cooling may be required from one unit, but only half the heating from another. In a short time, the water temperature will become too high and the units will shut off for safety. We must, therefore, add a heat rejection device, shown in Figure 9-15b, to cool the water and keep it from exceeding a maximum temperature. If more heat-

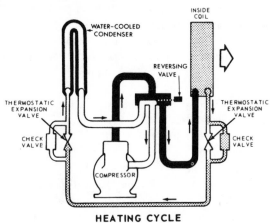

HEATING CYCLE

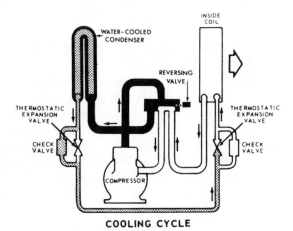

COOLING CYCLE

FIG. 9-14. Water-to-air heat pump.

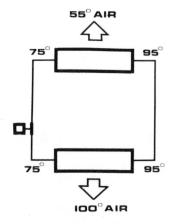

FIG. 9-15a. Balanced multiunit heat pump loop.

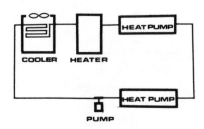

FIG. 9-15b. Unbalanced demand may require the inclusion of a heater and/or cooler in the heat pump loop.

ing than cooling is required, a water heater or boiler-heat exchanger combination is used to heat the water.

The heat rejection device shown diagramatically in Figure 9-15b is an evaporative cooler, sometimes called an industrial or closed circuit cooler. The heated water flows through a pipe coil, and outside air is used to cool it. To increase the efficiency of the unit, water is sprayed on the coil in the manner of a refrigeration evaporative condenser. If this system is used in areas where the outdoor temperatures can drop below freezing, some precautions must be taken. Pumps can and do fail and with no water flow a freeze-up will take place in just a few minutes.

A flow switch or other flow detection device should be installed in the line. When no flow is detected, this switch should close the outside air and discharge air dampers, and stop the fan motor. In addition, a dump valve should open, draining the coil and all outdoor piping.

This is nearly foolproof, but even protective devices can fail. The arrangement shown in Figure 9-16 is preferable. In place of the evaporative cooler, a conventional cooling tower with an inside sump and a heat exchanger is used. An aquastat will energize the cooling tower sump whenever the high limit is reached.

Water from the heat exchanger will be pumped to the tower and drained to the inside sump and then returned to the heat exchanger. Another aquastat in the sump is used to start and stop the tower fan as required.

These types of systems are well adapted for use in buildings with high internal heat gains, particularly those with large internal areas. Several years ago, a feasibility study was done for a proposed office building that was 50 ft wide by 280 ft long and five stories high. The building's perimeter was 37.5% single unshaded glass and the long dimensions faced east and west. The lighting load was 6 watts/ft^2.

It was found that no external heat needed to be added to the water loop until the outdoor temperature dropped

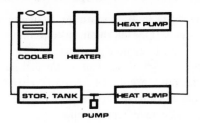

FIG. 9-17. Heat pump loop with a heat storage tank.

below 2°F. Much of the time, enough heat was being rejected during the day to heat the building at night, which suggested the use of a heat storage tank, Figure 9-17.

Table 9-8 shows the number of gallons of water heated from rejected heat during 10 hours/day operation at various outdoor temperatures, along with the gallons required to meet the heating loads during 14 hours of night operation. As you can see, one can store enough heat, down to nearly 22°F, to handle the night load.

How important is this? It depends on the amount of cold weather. At night, 1.75 billion Btu per year will be required. To provide this amount of heat, approximately 1.2 billion Btu will be needed to heat the water furnished to the heat pumps. At 70% efficiency, 1.7 million ft^3 of gas are required. If an electric water heater is used, then 350,000 kWh will be required. If a tank is used, the nighttime requirements are reduced to 800,000 ft^3 of gas or 170,000 kWh.

The dollar savings varies with the gas or electric rate. Assuming $2.50/1,000 ft^3 and 5 cents/kWh, the savings will be $2,250 per year for gas and $9,000 per year for electricity. The cost of the tank installation is $10,000, so it is a good investment with either fuel.

This system is best suited for any building that has high internal heat gains and/or large interior areas, such as office buildings, schools, and hospitals. An office that is part of a plant is also well suited for this system. If the plant has high internal loads, this heat can be used as a nighttime heat source.

The system has several advantages. First, each unit permits the occupants to select heating or cooling any

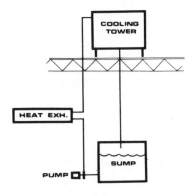

FIG. 9-16. Cooling tower with an indoor sump.

	TABLE 9-3.	
Outside temp °F	Gallons heated 20° in 10 hrs.	Gallons required at night
62	92,840	6,312
52	57,181	14,202
42	50,429	22,092
32	39,668	29,982
22	33,127	37,872

time they wish, at the same cost as heating-only or cooling-only systems. One of the most appealing advantages is the decentralization of equipment. Unlike central systems, the breakdown of one unit means that only that unit is down, and not the complete system. If spare units are kept on hand, an out-of-service unit can be quickly replaced and taken to a repair station.

Central Systems

Buildings having central chillers for cooling and boilers for heating can also incorporate the principles of energy conservation discussed for semicentral heat pumps. Buildings that require simultaneous heating of perimeter zones and cooling of interior zones can find a heat source in the energy rejected by the condenser while the chillers are operating.

This can be done in a variety of ways. Figure 9-18 shows a flow diagram for a 3-pipe terminal mix system. A double-tube condenser is used and its heat can be absorbed by the building hot line or a cooling tower line, depending on the load.

Another method is shown in Figure 9-19. Here the hot line is connected to the water-cooled condenser. The refrigerant discharge line goes first to an air-cooled condenser, and then to the water-cooled condenser. A head pressure control modulates the air-cooled condenser, as required.

The air-cooled condenser arrangement shown in Figure 9-19 will have a higher efficiency than the other systems shown. Whenever the air-cooled condenser operates, the water-cooled condenser acts as a subcooler.

At 95° outdoors, an air-cooled condenser generally condenses at 115° and subcools to 105°. If the water entering the condenser is at 65°, we will obviously further

subcool to some point between 65° and 105°, depending on the quantity of water circulated through the condenser.

A typical 3-pipe system will have 10% of the total water flow in the hot line on a hot summer day. This flow will subcool the refrigerant from 105° to 80°, while heating the water from 65° to 90°. A 10% increase in capacity, without an increase in horsepower, will result.

Central Heat Pump with Storage

Very often, large commercial buildings have high internal loads during the day resulting in high cooling loads during occupied hours. At night, the building may require heating to maintain even the set-back temperature. In such cases, a storage tank can store the excess heat rejected by the chillers during the day and use it during off hours. Figure 9-20 is a typical layout for heat recovery using a heat storage tank.

During the day a cold line (C) sensor loads and unloads the chiller to maintain the desired cold line temperature. Another sensor in the hot line (H) modulates a 3-way valve to send condenser water (CW) through or around a heat exchanger (HEAT-X) to maintain the desired hot line temperature. The temperature of the condenser water is maintained at 100°F by mixing tank water with either return water or cooling tower water and by modulating the valves shown.

Figure 9-21 shows the condenser water flow when the tank is below maximum storage temperature. Some hot return water flows through the tank, and some bypasses and mixes with cold water withdrawn from the tank as

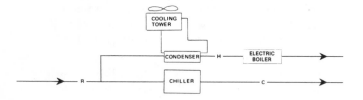

FIG. 9-18. 3-pipe terminal mix system.

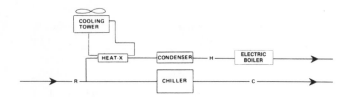

FIG. 9-19. 3-pipe system with an air-cooled and water-cooled condenser.

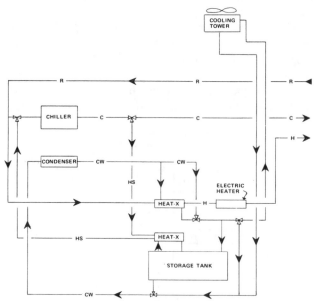

FIG. 9-20. Heat recovery system employing a large storage tank.

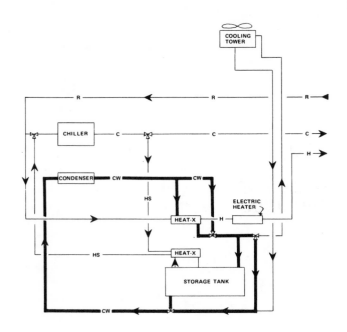

FIG. 9-21. Condenser water flow when storage tank is below maximum storage temperature.

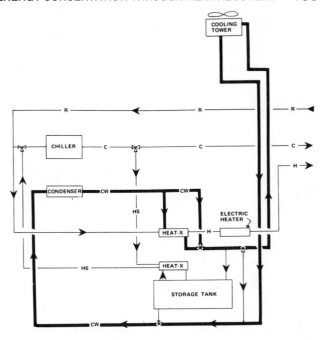

FIG. 9-22. Cooling tower operates when storage tank is above maximum temperature.

long as it is below maximum temperature.

When the tank reaches maximum temperature, it is locked out of the circuit and the water bypasses the tank completely. If we are unable to maintain the 100°F condenser water supply temperature, some water is diverted to the cooling tower to be cooled, as shown in Figure 9-22.

Figure 9-23 shows the night operation of the system. The storage tank water heats the exchanger directly above the tank. This heated water is circulated through the chiller; no cold water is sent to the building. The hot line is heated by the condenser water. When the tank temperature reaches 40°F, the chiller is shut down and the water is heated by the electric heater.

Figures 9-20 through 9-23 show a 3-pipe system, but there is also a 4-pipe system. The storage tank will generally have to be furnished, although sometimes a building has a fire storage tank that can be used. The 3-pipe and 4-pipe systems operate in conjunction with fancoil units, dual duct systems, induction units, or radiant panels.

With proper controls and design, a good fraction of energy can be recovered from exhaust gases and waste fluids in order to reduce energy consumption.

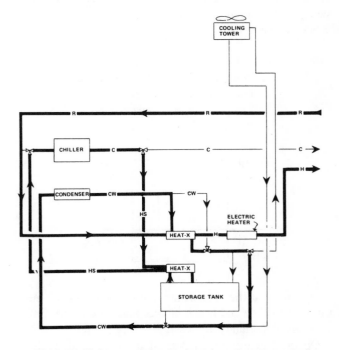

FIG. 9-23. Night operation using heat from the storage tank.

Chapter 10

CONTROL SYSTEMS AND
OPERATING PRACTICES

Chapters One through Six discussed new designs that use energy conservation techniques. This chapter covers control systems, operating practices, and features of mechanical and electrical systems that result in energy savings, starting with the control concepts of air-type distribution systems.

MIXING OUTSIDE AND RETURN AIR

Air section controls can provide three options:
1. Fixed amount of outside air
2. Temperature economizer control
3. Enthalpy economizer control

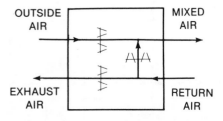

Option 1 satisfies mandatory ventilation codes for ensuring the safety and health of occupants; it does not provide any opportunity for energy savings.

Option 2 adjusts the amount of outside air (above the minimum code requirement) so that the mixed air temperature will approach the temperature required by the space heating/cooling load.

Option 3, the enthalpy economizer control, senses the total enthalpy of the air and adjusts the intake of outside air (above the minimum code requirements) so that the mixed air enthalpy approaches the desired air enthalpy

at the space entrance. The controls for Option 3 lead to greater energy savings than those for Option 2.

Limitations of Economizer Cycles

There are a limited number of hours during the cooling season when an economizer cycle can save energy. The cooling energy saved should be compared with the additional cost of operating a return air fan and the extra humidification that may be needed when using dry, cool outside air.

The economizer cycle should definitely be investigated when there is a 24-hour occupancy. In Atlanta, a temperature economizer cycle can use the outside air (when it is 55°F or lower) 42% of the time to cool a 24-hour occupancy.[10-1] However, with a 12-hour occupancy, cooling outside air can be used only 20% of the time.

Dual Duct Air Systems

Three options are available for setting the air temperature of the hot and cold air decks in a dual duct system.
1. Fixed air temperature for both decks.
2. Hot deck temperature set equal to the warmest air required by any of the zones and the cold deck temperature set equal to the coldest air temperature required by any zone.
3. Fixed hot deck temperature with the cold deck temperature set equal to the coldest air temperature required by any zone or vice versa.

Option 1 overheats the air in the hot deck and overcools it in the cold deck, consuming extra energy, and Option 3 provides only a partial improvement. Option 2 is the most economical and should be installed in dual

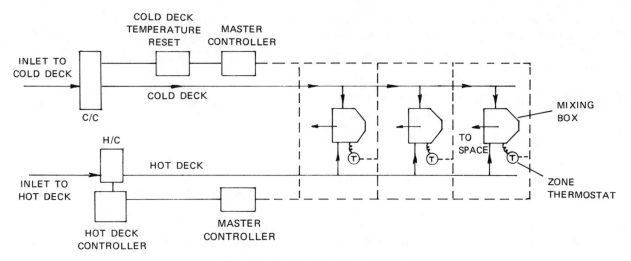

FIG. 10-1. Dual duct air system.

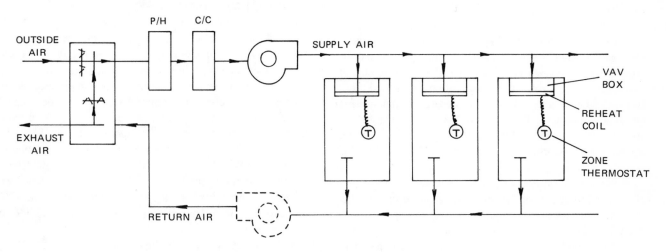

FIG. 10-2. Pure modulating VAV system.

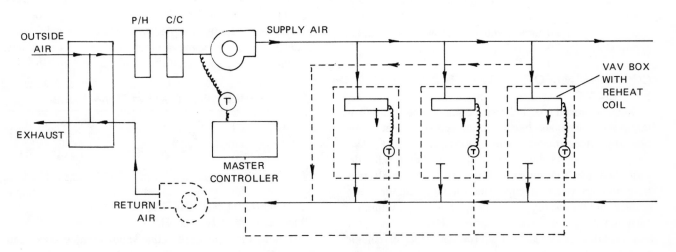

FIG. 10-3. Bypass VAV system.

duct systems to reduce energy consumption. Several control manufacturers market pre-engineered controls (both pneumatic and electronic) for this purpose. A sample control layout is shown in Figure 10-1.

CONSTANT VOLUME SYSTEMS WITH REHEAT

Two options are available for setting the temperature of the supply air entering the terminal boxes of constant volume systems with reheat.

1. Fixed supply air temperature.
2. Varied supply air temperature, set equal to the coolest air required by any zone.

Option 1 wastes energy in overcooling at the cooling coil and overheating at the reheat coil, while Option 2 is highly desirable for energy conservation. The controls are similar to those for the dual duct system.

VARIABLE VOLUME AIR SYSTEMS

There are two types of VAV systems:

1. Pure Modulating VAV System   In a pure modulating VAV system, Figure 10-2, the temperature of the air leaving the supply fan is held at a constant temperature of 53° to 55°F in most cases. Air volume is varied as a function of demand and zones requiring heating receive air at a minimum setting of the VAV box. The reheat coils heat this air to the temperature required by the zone. There is no opportunity to set the temperature for this system.

2. Bypass VAV System   In a bypass system, Figure 10-3, the supply air fan handles a constant volume of air. A part of the supply air from each zone is bypassed into the return in response to the load demand. This system does not provide any savings in fan energy, compared to the constant volume systems. However, it is possible to save energy by raising the supply temperature and decreasing the bypass air as the cooling load decreases.

System 1 is used more frequently because of its higher potential for energy conservation.

COORDINATION OF PERIMETER AND INTERIOR SYSTEMS

Many nonresidential buildings require winter cooling in the interior zones and heating in the perimeter zones, Figure 10-4. A perimeter system, especially a radiator system, can be overheating most of the time because the design load rarely occurs. The interior system will then have to overcool to compensate for the recurrent overheating and the two systems will waste energy by fighting each other.

To minimize wasted energy, perimeter systems should be controlled by the effective outside wall temperature. Heating and cooling control systems must be set to keep

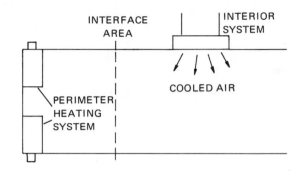

FIG. 10-4. Perimeter and interior systems.

this temperature high enough to prevent discomfort and low enough to minimize mixing of peripheral and interior zone air.

SPACE TEMPERATURE AND HUMIDITY CONTROL

As discussed in Chapter 3, a temperature setback in winter saves heating energy, and a temperature increase in summer saves cooling energy. There is, however, a limit to which the indoor temperature can be reduced in winter or increased in summer based on the occupants' activity, health, type of clothing, etc. The extensive work at Kansas State University[10-2], sponsored by ASHRAE, has produced the comfort zone ranges shown in Figure 10-5 and Table 10-1. These studies show a high dependence on temperature but only a very slight dependence on relative humidity. A relative humidity change of 13% is equivalent to a dry bulb temperature change of 1°F at comfort conditions.

It is recommended that temperatures be set at the upper and lower bounds of the comfort zone for the cooling and heating seasons, respectively. The relative humidity should be specified over a wide band of, say, 20 to 60%. Mechanical systems designed to provide indoor humidity control within a narrow range waste a considerable amount of energy in humidification and dehumidification.

TABLE 10-1. Comfort Zone Limits

Cool	68°F, 25% RH to 65°F, 60% RH
Slightly Cool	74°F, 20% RH to 70.5°F, 60% RH
Comfortable	80°F, 20% RH to 77°F, 60% RH
Slightly Warm	86°, 20% RH to 84.6°F, 60% RH

EFFICIENCY OF CENTRAL CHILLED WATER PLANTS

Chilled water plants use the vapor compression cycle for cooling. As shown in Figure 10-6, the refrigerant in the evaporative coil extracts the heat of evaporation from

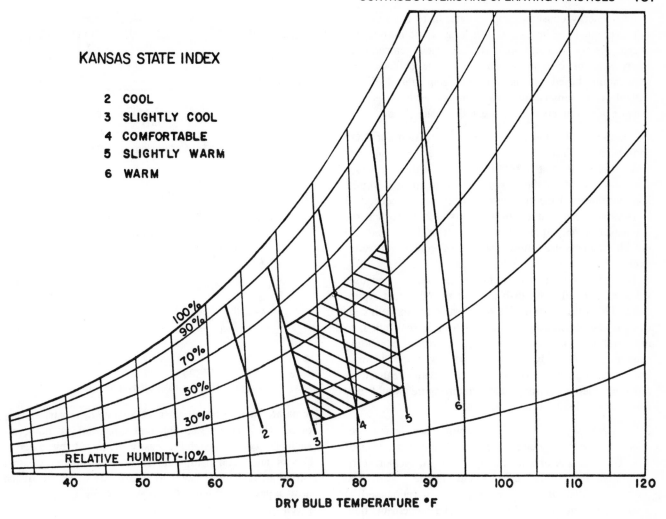

KANSAS STATE INDEX

2 COOL
3 SLIGHTLY COOL
4 COMFORTABLE
5 SLIGHTLY WARM
6 WARM

FIG. 10-5. Comfort levels shown on a psychrometric chart.[10-3]

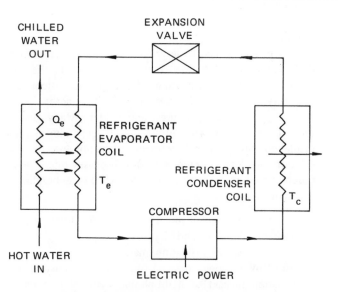

FIG. 10-6. Schematic of a chilled water plant.

water or space being cooled. The coefficient of performance (the ratio of cooling produced per unit of energy supplied to the compressor) for an ideal vapor compression cycle is given by:

$$COP = \frac{T_e}{T_c - T_e} \qquad (10\text{-}1)$$

where

T_e = Temperature of the evaporator coil
T_c = Temperature of the refrigerant condenser coil

From equation 10-1 we see that the higher the temperature, T_e, the higher is the COP and the lower will be the input energy needed to produce a ton of cooling.

The efficiency or COP of central chilled water plants can be improved in several ways:

1. *Increase the Chilled Water Temperature*   As cooling demand decreases, the chilled water temperature should be increased, which will lead to a higher COP and

a lower energy consumption. To date, there is no universally accepted method of establishing a relationship between system demand and chilled water temperature. Many systems can successfully reset the chilled water temperature based on the outdoor temperature. Higher chilled water temperatures save energy as long as they provide air cool enough to meet the space load.

2. *Condenser Temperature*   The expression for COP shows that the lower the T_c, the condenser temperature at which the refrigerant rejects energy, the higher the COP. The power used by the refrigeration equipment is significantly reduced if the condenser temperature is reduced. It is recommended that controls be installed to reduce the condenser temperature when the system loads and the outside temperature allow the use of cooler condenser water.

The extra power needed to obtain cooler condenser water is generally less than the cost of refrigerant equipment. This rule is applicable for most of the modern centrifugal and absorption chillers and, to a lesser degree, to refrigeration machinery using direct expanding evaporators. A study by the Trane Company[10-1] for a large central plant showed that the net savings realized by reducing the condensing water temperature would buy new chillers in 20 years.

3. *Chiller Staging*   Chiller staging uses less power than equipment to produce warmer chilled water (higher T_e). The series arrangement, Figure 10-7a, is more efficient than the parallel arrangement, Figure 10-7b. The upstream chiller of the series arrangement (numbered 1) produces warmer water. In the parallel arrangement, both chillers are producing chilled water at 42°F. The Trane Company study shows that a series arrangement produced a 7 to 8% energy savings for a plant having a total capacity of 550 to 600 tons.

DISTRIBUTION SYSTEM LOSSES

Many sprawling industrial complexes, shopping centers, campus-type housing, defense bases, etc. have central plants for generating chilled water and for boilers generating steam or hot water. To carry heat and/or chilled water to outlying buildings from the central power plant, these complexes use heat distribution systems of two types:

- Aboveground systems
- Underground systems

Aboveground Systems

Some industrial complexes use aboveground heat and/or chilled water distribution systems, especially in areas having a high groundwater table. The aboveground sys-

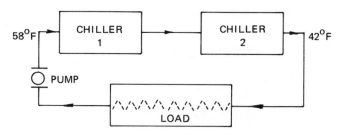

FIG. 10-7a. Chiller staging in series.

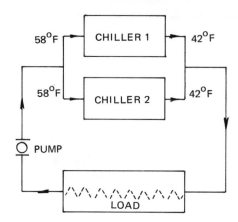

FIG. 10-7b. Chiller staging in parallel.

tems are easy to install, inspect, maintain, repair, and replace. They have certain inherent shortcomings, however, such as being unsightly, easily damaged by external shocks and vandalism, subject to more severe weather variation, and more likely to freeze due to failure of the heat supply. The rate of energy loss for a pipe with a given amount of insulation in an aboveground system is more than for a similar pipe and insulation in an underground system.

Underground Systems

In both above- and underground systems, pipes or conduits are insulated to reduce the heat losses to the ambient air or other surroundings. Underground systems are generally more cost effective in reducing energy losses because ground temperatures are milder. Many underground systems fail when ground water seeps into thermal insulation through poorly constructed and/or installed insulation envelopes. Reference 10-3 discusses this type of failure based on a survey of U.S. Air Force buildings. Problems included:

- Use of pre-fabricated metal conduits with little, if any, space for drainage and drying.
- Use of loose-fill insulation material.
- Use of hygroscopic material (that absorbs moisture) for insulation instead of hydrophobic material that repels moisture.

A careful evaluation of the existing underground distribution systems should be made to determine their thermal distribution efficiency. Corrective action should be undertaken to increase the effectiveness of insulation after evaluating its cost effectiveness. Typical underground distribution systems are:

1. The Air Space Conduit System   One of the most commonly used underground systems, consists of a steel outer casing surrounding one or more interior pipes that are insulated with a pre-formed pipe insulation, Figure 10-8. The pipe support is usually provided by a thermal insulating material. Various materials such as coal tar enamel, glass-fiber-reinforced asphalt compound, and glass-fiber-reinforced epoxies are used to protect the outside casing from corrosion. The system is drainable, driable and testable.

2. Sealed Non-air Space Conduit System   This system is basically a single carrier pipe enclosed in pipe insulation having a tight outer casing, Figure 10-9. The most

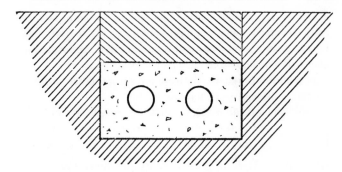

FIG. 10-10. Poured-in-place insulation envelope.[10-3]

commonly used insulation is urethane foam, provided the foam is not exposed to temperatures higher than 200°F. Calcium silicate and cellular glass insulation are used for high temperature systems.

Many of these systems are not air pressure testable, drainable nor driable. Although it is essential that insulation be kept dry under all conditions, this is rather difficult to achieve in practice. These systems are commonly used for chilled water or low temperature, hot water distribution systems.

3. Poured-in-Place Insulation Envelope System   This system consists of an envelope of loose-fill insulation material poured around the carrier pipe in a trench, Figure 10-10. Powdered hydrocarbon material and a powdered chalk material that has been chemically treated (with oleic acid) are commonly used as loose-fill materials. This system is relatively easy to install and repair

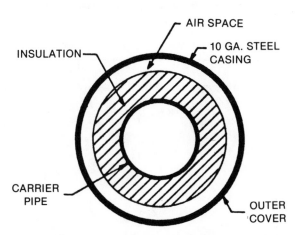

FIG. 10-8. Air-space conduit system.[10-3]

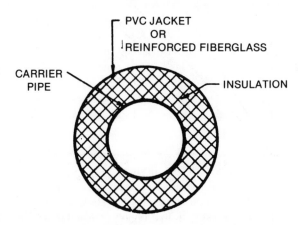

FIG. I0-9. Sealed non-air space conduit system.[10-3]

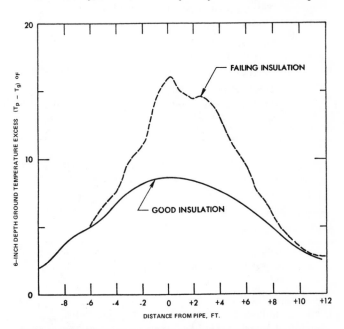

FIG. 10-11. Temperature profile with good insulation and with failing insulation.[10-3]

and is relatively inexpensive, but should not be used with chilled water because the permeability of the insulation is high. Water vapors can easily pass through the insulation and will condense on the chilled water pipe.

In addition to the above, there are several other arrangements mentioned in Reference 10-3.

The manufacturer-recommended preventive maintenance and inspection procedures should be instituted to keep the pipe insulation dry. This will assure satisfactory performance of the underground systems.

Insulation failure in underground heat distribution systems increases heat losses, which elevate the ground surface temperature above the pipe. Failure is sometimes indicated by dead brown grass or melted snow. One inspection method monitors the temperature 6 in. below the ground surface to avoid the rapid surface fluctuations caused by wind and solar heating. Figure 10-11 shows a typical temperature profile with good insulation and with failing insulation.

Heat Loss from Underground Distribution Systems

Two of the factors, C and k_p, must be calculated before the heat loss from an underground pipe can be estimated.

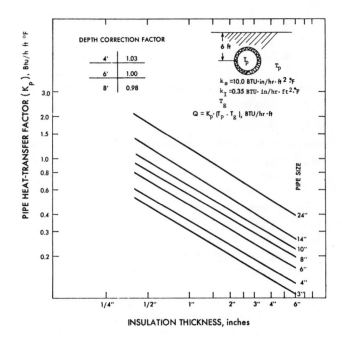

FIG. 10-12. Heat transfer factors (K_p) for insulated underground pipes.[10-3]

$$\frac{1}{C} = \frac{6}{\pi}\left\{\frac{1}{k_w}\cdot\ln\left(\frac{D}{D-2t_w}\right) + \frac{1}{k_I}\cdot\ln\left(\frac{D+2\cdot t_I}{D}\right)\right\} \qquad (10\text{-}2)$$

where

k_w = Pipe thermal conductivity, Btuh·in·ft²
$\ln$ = Natural logarithm
D = Pipe outside diameter, in.
t_w = Pipe thickness, in.
k_I = Thermal conductivity of insulation, Btuh·in·°F
t_I = Thickness of insulation, in.
C = Thermal conductivity of the pipe and insulation, Btuh·ft²·°F

MATERIAL	k_w VALUES Btuh·in·ft²·°F
Steel	360
Plastic	2 to 3
Concrete	12
Cement asbestos	16

$$\frac{1}{k_p} = \frac{1}{C} + \frac{6}{\pi k_s}\times\ln\left\{\frac{y}{r} + \sqrt{\left(\frac{y}{r}\right)^2 - 1}\right\} \qquad (10\text{-}3)$$

when $\frac{y}{r} \gg 1$

$$\frac{1}{k_p} = \frac{1}{C} + \frac{6}{\pi k_s}\times\ln\left(\frac{2y}{r}\right)$$

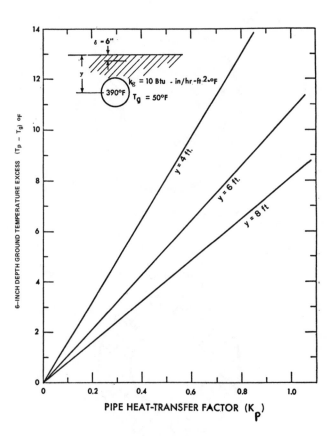

FIG. 10-13. Calculated values of the ground temperature excess.[10-3]

where

k_s = Thermal conductivity of undisturbed earth, Btuh·in·ft^2·°F

y = Distance from the ground surface to the center of the pipe, in.

r = Radius of the conduit, in.

k_p = Thermal conductance of pipe, insulation, conduit and earth, Btuh·ft·°F

Knowing the value of k_p, the heat loss per foot of conduit can now be estimated by

$$Q = k_p(T_p - T_g) \qquad (10\text{-}4)$$

where

T_p = Temperature of the fluid in the pipe, °F

T_g = Undisturbed earth temperature, °F, taken at a distance, δ, which is usually 6″ below the surface and at a point 15 ft or more from the pipe (Figure 10-13).

The heat loss is related to the excess ground temperature by the following equation:

$$Q = 2\pi k_g \cdot \frac{T_p - T_g}{\ln \frac{y+\delta}{y-\delta}} \qquad (10\text{-}5)$$

Figure 10-12 shows pipe heat transfer factors for an underground pipe having an insulation thickness varying from ½″ to 6″ (with $k_I = 0.35$ Btuh·in·ft^2·°F) for pipes that are 3″ to 24″ in diameter.

Figure 10-13 shows typical values of the excess ground temperature $(T_p - T_g)$ measured at a 6″ ground depth. If the observed temperature at the 6″ depth is significant greater than the value shown in Figure 10-14, it indicates a probable failure of the underground insulation.

Heat Loss from Aboveground Pipe

The rate of heat flow (due to conduction, convection and radiation) through the surface of a pipe covered with a layer of insulation at the outer surface may be calculated as follows:

$$Q = \frac{T_p - T_o}{R_t} \qquad (10\text{-}6)$$

where

T_p = Inside surface temperature of pipe, approximated as pipe fluid temperature (usually), °F

T_o = Ambient air temperature, °F

R_t = Total thermal resistance = $R_a + R_c + R_I$

R_a = Surface thermal resistance of pipe

TABLE 10-2. Pipe Thermal Resistance Factor (F), in/ft

Nominal Pipe Size	OD Inches	INSULATION THICKNESS, INCHES							Air Space Resistance
		1/2	1	1 1/2	2	2 1/2	3	3 1/2	
1/2	.840	1.498	2.327	2.903	3.345	3.703	4.005	4.266	1.52
3/4	1.050	1.278	2.037	2.578	3.000	3.345	3.637	3.890	1.21
1	1.315	1.080	1.766	2.269	2.667	2.997	3.277	3.522	0.97
1 1/4	1.660	.901	1.510	1.971	2.343	2.653	2.921	3.155	0.77
1 1/2	1.900	.808	1.373	1.809	2.164	2.463	2.722	2.949	0.67
2	2.375	.671	1.167	1.560	1.886	2.164	2.407	2.622	0.54
2 1/2	2.875	.570	1.009	1.365	1.665	1.924	2.153	2.357	0.44
3	3.500	.480	.863	1.182	1.456	1.695	1.907	2.098	0.36
3 1/2	4.000	.426	.774	1.069	1.324	1.549	1.750	1.932	0.32
4	4.500	.383	.702	.976	1.215	1.427	1.618	1.792	0.28
5	5.563	.316	.587	.824	1.035	1.225	1.397	1.556	0.23
6	6.625	.268	.504	.713	.902	1.074	1.232	1.377	0.19
8	8.625	.210	.398	.570	.728	.873	1.009	1.135	0.15
10	10.750	.170	.326	.470	.604	.729	.847	.958	0.12
12	12.750	.144	.278	.404	.521	.632	.737	.836	0.10
14	14.000	.132	.255	.371	.480	.583	.681	.774	0.09
16	16.000	.166	.225	.328	.426	.519	.608	.693	0.08
18	18.000	.103	.201	.294	.383	.468	.549	.627	0.07

Typical values of "k" @ <u>200° F</u>

Glass fiber	0.3
Asbestos	0.4
Magnesium	0.45
Calcium silicate	0.37
Cellular glass	0.48

$$R_I = F/k_I$$

where F = pipe thermal resistance factor, in/ft

k_I = thermal conductivity for pipe insulation, Btu·in/h·ft^2·°F

R_I = Thermal resistance of pipe insulation

R_c = Conduit air space resistance

The surface resistance, R_a, of the pipe is related to the outside air film coefficient, h_a, as follows:

$$R_a = \frac{\pi}{h_o} = \frac{12}{\pi \times D_o \times h_a} \qquad (10\text{-}7)$$

The overall resistance of the pipe insulation, as shown in Table 10-2, varies with the pipe thermal resistance factor, F, and the thermal conductivity, k_I, of the insulation.

$$R_I = \frac{F}{k_I}$$

In some cases, the pipes are enclosed in an air conduit, Figure 10-8, which increases the total resistance, R_t, by the additional resistance, R_c, offered by the air space. Values for R_c are also found in Table 10-2.

Example 10-1

Calculate the heat loss from a nominal 6″ pipe with 2″ of calcium silicate insulation ($k_I = 0.37$ Btuh·in·ft²·°F) in a 12″ diameter steel conduit. The pipe is exposed to a 10 mph wind and is carrying saturated steam at a pressure of 160 psig (370°F). The ambient temperature is 40°F.

Solution

The first step is to calculate the total resistance using Figure 10-14 and Table 10-2.

R_a, surface air resistance $= \dfrac{12}{\pi D_o h_a} = \dfrac{12}{\pi \times (12)(7)}$

$\qquad = .045$

R_c, conduit air space resistance (Table 10-2) $= 0.19$

R_I, insulation resistance (Table 10-2) $= \dfrac{F}{k_I} = \dfrac{0.902}{0.37}$

$\qquad = 2.44$

Total Resistance, R_t $= .045 + .19 + 2.44$

$\qquad = 2.675$

Rate of heat loss $= \dfrac{T_p - T_o}{R_t} = \dfrac{370 - 40}{2.675}$

$\qquad = 213.36$ Btuh·ft

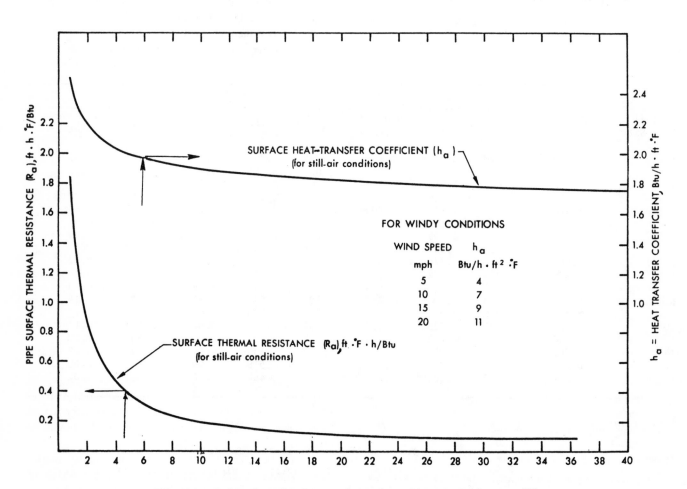

FIG. 10-14. Surface thermal resistance of aboveground heat-distribution pipes.[10-3]

TABLE 10-3. Minimum Recommended Insulation for Various Piping Systems[10-3]

Type of System	Temperature Range, °F	Pipe Size, Inches		
		Up to 2	2½ to 4	Over 4
		Insulation, Inches		
Cooling Systems				
Chilled Water	40–60	1	1	1
Refrigerated Brine	below 32	1½	1½	1½
Heating System				
Hot Water	up to 200	1	1½	2
Hot Water	over 200	1½	2	2½
Condensate	190–220	1½	2	2½
Low-pressure Steam	up to 250	1½	2	2½
Medium-pressure Steam	251–305	2	3	3½
High-pressure Steam	306–400	3	4	4½

Thus we see that heat losses from heat distribution pipes can be reduced by increasing the amount of insulation. The minimum recommended insulation for various systems is given in Table 10-3. Be sure that the various valves, elbows and other joints forming part of the heat distribution system are similarly insulated.

REDUCING THE ENERGY CONSUMPTION OF FANS AND PUMPS

A fan is a weak compressor. It raises the pressure of air to overcome the resistance of the air distribution system. The horsepower required by a fan is given by Equation 4-6.

$$\text{BHP (fan)} = \frac{\text{CFM} \times \Delta P}{6346 \times \eta_t}$$

where

CFM = Air flow in cubic feet per minute

ΔP = Total pressure of fan (inches of water)

$= P_2 - P_1$

η_t = Fan efficiency

Thus, the energy required by a fan is proportional to the air flow rate and the total fan pressure. The fan pressure required to overcome the resistance of any system depends upon the quantity of air circulated and the characteristics of various components such as:

- Size of ducts
- Layout of ductwork, register, grills, diffusers, etc.
- Heating and cooling coils
- Filters
- Intake and discharge openings

Ductwork and Fan Energy

The size of a duct should be as large as is practical, consistent with minimum velocity requirement, availabil-

ity of space, etc. With a larger duct, less pressure is required to overcome inertia and the frictional resistance to flow. Less pressure reduces the velocity and, with it, the energy required.

The layout of the ductwork should minimize the number of bends, abrupt changes in cross-sections, etc. Using long-taper fittings and turning vanes in square bends helps to reduce the pressure drop. Air flow should also be balanced to ensure the correct volume at each register or grille. This is achieved by adjusting the dampers in the various branches to make their resistance equal to the outlet farthese from the fan, called the index run. When the index outlet is supplied through a long run of duct, it creates an excessive pressure drop for the whole system and leads to higher energy consumption by the fan. In such cases, one should consider replacing the index run with a large size, low resistance duct.

Filter Resistance

A filter's resistance to air flow depends upon the following factors:

- Filter construction
- Type of media
- Flow velocity through the filter media (not the face velocity)
- Dirt load

For a given filter, the resistance to air flow increases as the dirt layer increases. Filters should be changed regularly to minimize this resistance. A manometer across each filter bank will indicate when filters should be changed.

The air filtration efficiency of existing and alternative filter systems should be checked by using the ASHRAE Atmospheric Dust Spot Efficiency Test Method or an equivalent.[10-4] If an alternative filter system shows a higher efficiency, its use should be considered.

Coil Resistance

Resistance to air flow through coils depends upon the number of rows or the depth required for adequate heat transfer. Cooling coils have a small temperature difference between the coil and the air; consequently, they usually have many rows and offer a high resistance to air flow. Heating coils use fewer rows and offer less resistance to air flow because of the higher temperature difference.

Fan energy may be reduced by arranging the heating and cooling coils for parallel flow so that the air flow is resisted by only one set of coils. A series arrangement imposes a combined resistance year around. When using chilled or hot water for cooling or heating, consideration may be given to removing the heating coil and then using the cooling coil to provide either cooling or heating. The reduction in system resistance will reduce fan horsepower consumption.

Another benefit of using the cooling coil for heating is the increased surface area for heat transfer. This provides an opportunity to lower the hot water temperature and to use low grade waste heat or low temperature water from solar heating systems.

Installation Factors

Improper installation can affect fan performance and cause it to deviate from the manufacturer's data.

Inlet Conditions

For optimum performance, there should be no obstruction within one wheel diameter of the fan. Fan performance is reduced by 20% if an inlet is placed within one-third of a wheel diameter of the fan. Large fan sheaves and belt guards produce a similar effect by restricting flow.

Fan efficiency is also reduced by the generation of spin at the fan inlet. Outside and return air ducts should be directed to the center of the fan inlet as shown in Figure 10-15. The installation of an elbow-top close to the fan

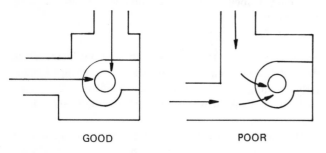

GOOD   POOR

FIG. 10-15. Placing air ducts for fan efficiency.

inlet results in uneven distribution and reduces performance. There should be at least one equivalent duct diameter of straight duct between any inlet elbow and the fan.

Discharge Conditions

A sudden change of diameter or direction at a fan outlet, or the discharge of air into a plenum, leads to a loss of static pressure recovery. This requires an increase in rpm and horsepower consumption and leads to a reduction in fan performance and efficiency. Such arrangements should be avoided as far as possible.

FAN SPEED CONTROL

The horsepower of a centrifugal fan (the common fan for an HVAC System) is related to the speed as follows:

$$BHP = BHP_o \left(\frac{RPM}{RPM_o}\right)^3$$

where
BHP_o = Brake horsepower at a known speed (RPM_o)
BHP = Brake horsepower at a given speed (RPM)
RPM = Revolutions per minute

This shows that the fan horsepower varies with the third power of speed.

As the load on the HVAC system changes, the speed of the fan should be modulated, to produce significant energy savings. Speed modulation can be accomplished in a number of ways, including:

- Multi-speed motors
- Fluid drives
- Mechanical speed reducers
- Solid state devices

The modulation control should be carefully matched to the fan characteristics. These controls are cost effective for buildings having variable loads or those that have not reached their full design load.

In summary, the energy consumption of fans may be reduced by:

- Reducing resistance to air flow
- Reducing heating/cooling loads
- Modulating the fan speed to correspond to the reduced flow requirement in variable volume flow systems
- Matching the phase and frequency with electrical systems.

ENERGY CONSUMED BY PUMPS

The basic laws governing the performance of pumps are similar to those of fans. Thus, a reduction in flow rate or system resistance reduces pump power consumption.

Pumps are used to circulate water (hot water, chilled water, condenser water, etc.) through closed circuits. The flow rate may be adjusted by using various valves. Because most pumps are driven directly by constant speed motors, there is little opportunity to adjust their speed.

There is a high potential for energy savings in reducing the number of operating hours and combating the tendency to overdesign. A complete engineering analysis of possible modifications, such as having a number of small pumps in parallel instead of one large pump, should be considered to determine the cost effectiveness of the modification.

STEAM LEAKS

Because the cost of steam leaks is considerable, the entire system should be inspected and checked for leaks, particularly at the joints, steam traps, and valves. Leaky traps are identified by checking the temperature drop across the steam traps, using a surface pyrometer. No temperature drop indicates steam blow-through. Excessive

temperature drop indicates that the trap is holding back condensate. Faulty traps should be repaired or replaced. Table 10-4 gives an estimate of steam leaks from various sizes of orifices.

HEAT RECOVERY BY FLASHING CONDENSATE

A significant portion of the heat content of the medium and high pressure condensate, which is being vented, can be reclaimed by flashing the condensate into low pressure steam. The quantity of 0 psig steam that can be generated in this way is shown in Table 10-5.

Example 10-2

Calculate the amount of 15 psig steam which is produced by flashing 8000 lb/hour of 100 psig condensate. What are the seasonal savings (5000 hours/season) in fuel oil with an 80% efficient oil-fired furnace (at $5/million Btu) as compared to using water at 80°F?

% of 100 psig condensate flashed to 0 psig steam	= 13.05
% 15 psig condensate flashed to 0 psig steam	= 4.00
Net % of 15 psig steam available = 13.05 − 4.00	= 9.05
Hence the rate of 15 psig steam available	= 8000 lb/hr $\times \frac{9.05}{100}$
	= 724 lb/hour

From the steam tables,

1163.9 = enthalpy of 15 psig (29.7 psia) steam
48.09 = enthalpy of 80°F water
$(1163.9 - 48.09) \times 724 \times 5000 = 4039.2 \times 10^6$ Btu
= seasonal energy savings

Oil consumption $= \frac{4039.2 \times 10^6}{.8} = 5049 \times 10^6$ Btu

Oil cost @$5/$10^6$ Btu = $25,245

RETURNING BOILER CONDENSATE

10% to 30% of the fuel required for steam generation can be saved by returning the condensate to the boiler. In addition to saving energy, this practice also reduces the cost of chemicals for boiler water treatment. Condensate return is becoming a common practice in new plants and should be expanded to existing plants. The economic feasibility of installing return pumps, return pipes and the insulation of condensate return pipes should be studied carefully.

INDUSTRIAL BOILERS

Many industrial boilers use up to 50% more air than the combustion process requires. Heating this excess air, which is exhausted through the chimney, wastes fuel. The air/fuel ratio should be controlled at the optimal level

**TABLE 10-4. Steam Leak Estimate
for 125 psi System** [10-3]

size of orifice	125 psi steam wasted/hour	cost per day @ $5/1000 lb	cost per month
1/32 in.	4.10	$ 0.492	$ 15.00
1/16 in.	15.00	$ 1.80	$ 54.00
1/8 in.	30.00	$ 7.20	$ 216.00
1/4 in.	237.00	$ 28.00	$ 885.00
3/4 in.	552.00	$ 66.24	$1987.00
1/2 in.	997.00	$117.24	$3517.00

Note: For steam at 50 psi, the leak waste is about 75% of the figures given; at 20 psi, it is about 50%; and at 5 psi, the loss is 25%.

**TABLE 10-5. Generation of 0 psig Steam by
Flashing High Pressure Condensate.** [10-3]

Inlet condensate pressure, psig	% condensate flashed as 0 psig steam
200	18.5
180	17.5
160	16.25
140	15.50
120	14.25
100	13.05
80	11.50
60	10.0
40	7.78
20	4.30
15	4.0
10	2.0

which is generally 5 to 10 percent excess air. Excess air can be controlled by using a carbon monoxide monitor to analyze the flue gases from a single burner or a spectral flame analyzer with multiple burners.

Fuel for process heat can be reduced by replacing the old burners with new high-efficiency fuel atomizers. A typical old refinery process burner has a fuel gun that looks like a tube equipped with a shower head. Fuel oil is blown out of the gun and into the combustion chamber by a jet of steam that breaks the oil into a fine combustible mist. This burner might deliver 10 million Btu's per hour and require 200 pounds of steam per hour to atomize its fuel. Modern, more efficient fuel atomizers can reduce this requirement by up to 70 pounds of steam per hour.

OPTIMUM USE OF LIGHTING

No matter how well a lighting system is designed, its utilization of energy is primarily determined by the operating and maintenance procedures. All the energy conservation strategies employed in illumination system design can be retrofitted to existing systems if they can be economically justified. The following steps are necessary for a viable lighting operating and maintenance program.

Lighting Energy Information Systems

A lighting energy information system should first be developed. Lighting levels at all work locations should be metered and recorded, and the tasks evaluated to determine the quantity and quality of illumination that will provide the best visual performance, productivity and safety. Recommended lighting levels are obtained from Figure 9-80 of the IES Lighting Handbook and from the ANSI Standards, covering lighting practices, that are listed in *American Standard Practices*. Existing lighting levels that are higher than the recommended levels should be reduced.

Modernizing Lighting Systems

One of the most effective energy saving methods is to modernize an existing lighting system with high efficacy, high intensity discharge light sources. Table 10-6, showing light source characteristics, is a useful guide in selecting light sources. Generally, high pressure sodium or metal halide lamps can replace mercury and fluorescent lamps with cost and energy savings. For example, a study was conducted for an existing industrial plant having the following characteristics:

Area of the plant = 150,000 sq. ft.
Existing lighting system consists of:
    34 – 400W Mercury lamps
    65 – 1000W Incandescent lamps
    47 – 750W Mercury lamps
Annual energy consumption = 1,001,289 kWh

In order to retrofit this plant, at the same illumination levels, the following illumination system, using high pressure sodium lamps, was proposed:

Number of HP sodium lamps: 34 – 250W
                             126 – 400W
First cost of materials and installation = $53,440
Expected energy use = 599,788 kWh
Net annual savings in energy = 401,501 kWh

An economic analysis, Table 10-7, using an energy cost of 3.1¢ per kWh and an escalation rate of 8%, showed that the capital cost for the new system paid itself back in 5 years and 4 months with a return on investment of 15.7%.

Energy-saving Lamps

Several lamp manufacturers have introduced energy-saving lamps to replace existing lamps where the lighting levels are higher than recommended. Energy-saving lamps have been introduced in incandescent, fluorescent, and mercury vapour lines that also improve the power factor of the system. An economic comparison of conventional and energy-saving lamps is given in Table 10-8. Energy-saving lamps generally preserve the light distribution pattern but reduce the lighting levels.

TABLE 10-6. Lighting Source Characteristics

Type	LPW*	Life (hr × 10³	Lumen Maintenance	Color Rendition	Operating Cost
Incandescent	15–25	0.75–12	fair to excellent	excellent	high
Fluorescent	55–85	7.5–24	fair to excellent	good to excellent	average
Mercury	50–60	16–24	very good	poor to excellent	average
Metal Halide	80–100	1.5–15	good	very good	below average
High Pressure Sodium	75–130	20–24	excellent	good gold white	low
Low Pressure Sodium	up to 180	16	excellent	poor	low

*Lumens per Watt

Cleaning and Relamping

A well-planned program of regular cleaning, relamping, and servicing should be instituted to provide the desired levels of illumination. Figure 10-16 shows the effect of maintenance procedures on energy use. With System A, luminaires are cleaned and relamped every three years. With System B, luminaires are cleaned and one third of the lamps are replaced every year, which saves

---

TABLE 10-7. Economic Analysis of Lighting Systems

CAPITAL INVESTMENT =   53440.0

YEAR	NET SAVINGS	DEPRECIATION	INCOME BEFORE TAXES	INCOME AFTER TAXES	CASH FLOW
1	12447.	5344.	7103.	3551.	8895.
2	13442.	5344.	8098.	4049.	9393.
3	14518.	5344.	9174.	4587.	9931.
4	15679.	5344.	10335.	5168.	10512.
5	16933.	5344.	11589.	5795.	11139.
6	18288.	5344.	12944.	6472.	11816.
7	19751.	5344.	14407.	7204.	12548.
8	21331.	5344.	15987.	7994.	13338.
9	23038.	5344.	17694.	8847.	14191.
10	24881.	5344.	19537.	9768.	15112.

THE PAY BACK PERIOD= 5. YEARS 4. MONTHS

RATE OF RETURN ON INVESTMENT = 15.7PERCENT

TABLE 10-8. Economic Comparison of Conventional and Energy Saving Lamps

Lamp		Power Consumption (watts)	Rated Initial Output (lumens)	Rated Average Life (hours)	Power Cost Savings (dollars)
Incandescent,	Industrial Service				
	Conventional	60	670	3500	
	Energy Saving	54	590	3500	0.63
	Conventional	100	1280	3500	
	Energy Saving	90	1090	3500	1.05
	Conventional	150	2150	3500	
	Energy Saving	135	1790	3500	1.59
Fluorescent,	48″ RS/CW				
	Conventional	40	3150	20000+	
	Energy Saving	34	2800	20000	3.60
Fluorescent,	96″ CW				
	Conventional	75	6300	12000	
	Energy Saving	60	5220	12000	5.40
Mercury Vapor,	Clear				
	Conventional	400	21000	24000+	
	Energy Saving	300	14000	16000+	48.00
Mercury Vapor,	Deluxe White				
	Conventional	400	23000	24000+	
	Energy Saving	300	15700	16000+	48.00

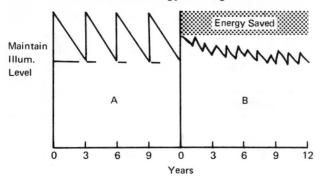

FIG. 10-16. Effect of maintenance on energy use.

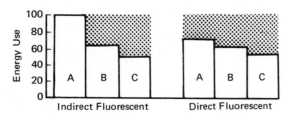

FIG. 10-17. Energy savings through improved reflectances.

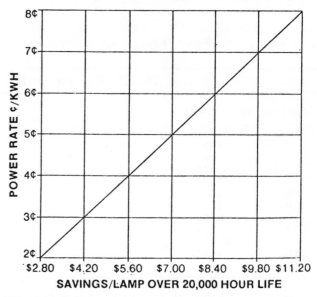

FIG. 10-18. Dollars saved by replacing 40W fluorescents with reduced-wattage fluorescents.

about 15% of the energy used by System A.

Surface Reflectances

Where the room surfaces are at or lower than the recommended reflectance range, they should be upgraded. The IES recommended range for reflectances is given below. Figure 10-17 shows the comparative energy savings, for equal general illumination, in a medium-sized room with reflectances given by three representative systems, A, B, and C.

	Reflectances		
	Ceiling	Walls	Floor
System A	50	30	10
System B	80	40	20
System C	90	60	40

Reduced Wattage Lamps

When lighting levels are above recommended minimum ANSI Standards, it is sometimes desirable to use reduced wattage lamps rather than remove lamps from luminaires. For example, removing both fluorescent lamps from a two-lamp ballast still consumes 6.5 watts of power for a magnetising current. Even though this is a small amount of power, it is a waste of energy. Reduced wattage lamps should be used when relamping. Figure 10-18 shows the dollars saved by substituting a reduced wattage lamp for a 40-watt fluorescent.

Unneeded Lights

Perimeter lighting, adjacent to windows, should be switched off at appropriate times to use daylighting. This can save a significant amount of energy in buildings with large glass areas.

The final step in reducing lighting energy consumption is to encourage personnel to turn off lights as they leave an empty room. Cleaning schedules should also be adjusted to minimize the need for lighting by concentrating the cleaning force in fewer spaces at the same time and by turning off lights in unoccupied areas.

POWER DISTRIBUTION SYSTEMS

Energy loss in an electrical distribution system is the sum of the energy losses in each of the system components. Percent energy loss is the ratio of power consumed by the equipment to the total energy passed through it, the loss generally ending up as dissipated heat. It is clear, from Table 10-9, that a plant that is using a large number of small motors will be wasting an average of 20% of its purchased energy.

To improve the operation of an existing electrical distribution system and to reduce energy losses:

1. Perform a short circuit analysis of the system under normal load conditions. Check every protective device for its proper short circuit rating.
2. Perform a protective device coordination analysis and adjust the settings of every overcurrent and ground fault protective device, so that the total electrical distribution system is selectively coordinated.

TABLE 10-9. Range of Losses in Power System Equipment

Component	% Energy Loss (full load)
Outdoor Circuit Breakers .... (15 to 230 KV)	.002– .015
Generators	.09 – 3.5
Medium Voltage Switchgear ...... (5 & 15 KV)	.005– .02
Current Limiting Reactors .... (600 V to 15 KV)	.09 – .30
Transformers	.55 – 1.90
Load Break Switches	.003– .025
Medium Voltage Starters	.02 – .15
Busway (480 V & Below)	.05 – .5
Low Voltage Switchgear	.13 – .34
Motor Control Centers	.01 – .40
Cable	1.0 – 4.0
Motors	
a. 1– 10 hp............	14.0 –35.0
b. 10– 200 hp..........	6.0 –12.0
c. 200–1500 hp.........	4.0 – 7.0
d. 1500 + up	2.3 – 4.5
Rectifiers (large)	3.0 – 9.0
Static Variable Speed Drives	6.0 –15.0
Capacitors (watts loss/var.)	.5 – 2.0
Lighting (lumens/watt)	3.0 – 9.0

3. Perform a load flow analysis of the system. This will give the voltage drop in the feeders and the power factor of the system at various points. If the voltage drop and power factor are below the values specified by ASHRAE Standards, correct them by using the methods described in Chapter 6.
4. Calculate the load factor of the system. If it is low, investigate the use of a demand controller for the system as described in Chapter 6.

After performing these steps, consideration should be given to the major sources of energy loss: transformers, motors and cables. Cable losses can be reduced by improving the voltage drop and power factor of the system.

Transformers

Transformer losses are composed of core and load losses. Core loss is a fixed quantity depending upon the physical dimensions, type of construction, characteristics of the steel, and other factors. Core loss occurs while a transformer is energized and is independent of transformer loading. Load loss is caused by the ohmic resistance of electrical conductors. It varies as the square of

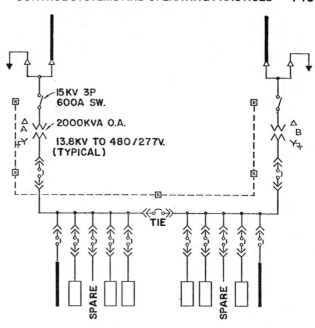

FIG. 10-19. One line diagram of a unit substation.

the transformer current and is a function of transformer loading.

Most of the transformers in industrial and commercial buildings are very lightly loaded (up to 5% of full load) during nonworking hours and weekends. If some of the transformers can be switched off during low demand periods and the loads transfered to other transformers, the core losses of these transformers can be saved. This can be achieved easily in the case of double-ended substations which are commonly used in industrial and commercial buildings, Figure 10-19.

The relationship between total losses and the loading of 2,000 kva substations is shown in Figure 10-20. It is evident that it is more economical, up to 30% of loading, to transfer the load on one 2,000 kva transformer rather than to load two transformers. By transferring the load to one transformer and switching off the other, the no-load losses of one transformer are saved and the power factor of the system is improved.

A study was conducted at an industrial plant having sixteen 2,000 kva double-ended substations and the following load cycle: transformers are loaded up to 90% of full capacity during two working shifts on weekdays and are loaded up to 5% of their capacity for eight hours on weekdays for 24 hours on weekends and holidays. The energy savings are shown in the following analysis:

Transformer rating	= 2000 kva
Total full load loss	= 27.85 kW
No-load loss	= 3.85 kW

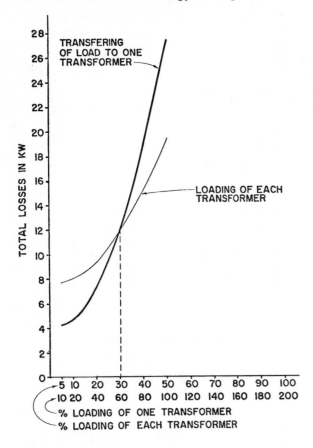

FIG. 10-20. Losses due to loading of two 2000 kva transformers.[10-5]

	TABLE 10-10.				
		Horizontal			
	Syn.	Dripproof		TEFC	
HP	Speed	% Eff.	% Loss	% Eff.	% Loss
½	3600	69.0	31.0	69.0	31.0
	1800	69.0	31.0	69.0	31.0
	1200	70.5	29.5	70.5	29.5
	900	62.5	37.5	62.5	37.5
1	3600	74.5	25.5	74.5	25.5
	1800	75.0	25.0	75.0	25.0
	1200	73.0	27.0	73.0	27.0
	900	70.0	30.0	70.0	30.0
3	3600	80.0	20.0	75.5	24.5
	1800	80.5	19.5	80.5	19.5
	1200	76.0	24.0	76.0	24.0
	900	74.0	26.0	74.0	26.0
5	3600	83.0	17.0	80.5	19.5
	1800	84.0	16.0	84.0	16.0
	1200	79.0	21.0	78.0	22.0
	900	78.0	22.0	78.0	22.0
7½	3600	84.5	15.5	79.0	21.0
	1800	82.0	18.0	81.0	19.0
	1200	83.0	17.0	83.0	17.0
	900	79.0	21.0	79.0	21.0
10	3600	83.0	17.0	82.0	18.0
	1800	82.0	18.0	83.0	17.0
	1200	82.0	18.0	82.0	18.0
	900	81.0	19.0	80.0	20.0
20	3600	87.0	13.0	86.0	14.0
	1800	85.0	15.0	86.0	14.0
	1200	86.0	14.0	85.0	15.0
	900	85.0	15.0	85.0	15.0
30	3600	87.0	13.0	85.0	15.0
	1800	87.0	13.0	85.0	12.0
	1200	87.0	13.0	87.0	13.0
	900	88.0	12.0	90.0	10.0

Losses of 2 transformers at 5%
$$= 2(3.85 + .05^2 \times 24)$$
$$= 7.82 \text{ kW}$$

Loss of one transformer, loaded to 10%, with the other off
$$= (3.85 + .1^2 \times 24) = 4.09 \text{ kW}$$

Saved by switching one transformer
$$= 3.73 \text{ kW}$$

Saved per day $= 3.73 \times 8 = 29.84$ kWh

Saved per holiday $= 3.73 \times 24 = 89.52$ kWh

Saved per year $= 89.52 \times 134 \times 29.84 \times 231$
$$= 18,888.72 \text{ kWh per substation}$$

At a cost of .02/kWh, the savings for 16 substations is
$$= 18,888.72 \times .02 \times 16$$
$$= \$6,044.39 \text{ per year}$$

Cost of installing shutoff switches at each substation
$$= \$1700 \times 16 = \$27,200$$

Using an 8 percent escalation rate for electrical energy, the cost of installing switches to shut off the transformers is repaid in 5 years and 3 months. The rate of return on investment is 14.5%.

Motors

Table 10-10 illustrates the energy loss of ½ through 30 hp motors operating at various speeds. To reduce these losses,

1. Stop the motors when they are not in use. A recent analysis of a multistory building showed that $2,500 a month could be saved by operating the blower motors only when heating and cooling were needed, instead of continuously.

2. Whenever possible, replace existing small motors with the energy-efficient motors described in Chapter 6. Although the initial cost of an energy-efficient motor is higher than that of an ordinary induction motor, the added cost is generally recovered in less than a year. Figure 10-21 shows the payback period for the increased initial cost of energy-efficient motors.
3. Replace any motor that is oversized by 50 percent.

Maintenance

With a proper maintenance program for electrical distribution systems, one should

1. Inspect all connections for overheating.
2. Perform routine load tests.
3. Check conductor insulation for high resistance losses or faults.
4. Check the overall system for undervoltages and unbalanced voltages.
5. Check all equipment to be sure it is not operating in areas where the ambient temperature is higher than the allowed rating.
6. Check the operation of large motors and equipment to

assure that they are being started in an unloaded condition, in order to reduce peak energy consumption.

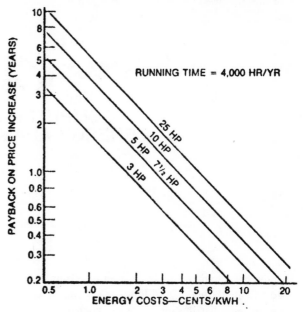

FIG. 10-21. Payback of energy efficient motors.

Chapter 11

COMPUTER PROGRAMS FOR
ENERGY ANALYSIS IN BUILDINGS

In previous chapters we have seen the complexity involved in calculating the building loads and in simulating the energy distribution systems, controls and primary equipment. It is impossible to do the calculations manually without making simplifying assumptions, as in the degree day and bin methods. Modern high speed computers, however, allow greater accuracy in modeling the various processes contributing to energy consumption in a building. Computers have been utilized in developing various energy analysis programs, with the following main components:

- Weather data simulation or input.
- Heating/cooling load calculations for various spaces (zones) in a building, assuming constant space temperature.
- Modification of space loads due to temperature setbacks during nonoccupied hours.
- Energy distribution system simulation (calculation of the energy required for space heating/cooling with a given type of energy distribution system).
- Calculation of primary energy (gas, fuel oil, electric power, steam, etc.) required by the primary equipment to meet a given load.
- Economic analysis.

The results obtained by using the various energy analysis programs will vary for a given building due to the various algorithms (equations) used for the simulation and the compromise between the ability of a program to handle a specific situation and the ability of the user to provide adequate input. A user should know the assumptions governing the equations and related concepts which were used in the development of a program

before deciding to use it. Chen[11-1], Spielvogel[11-2], Ayers[11-3] and others[11-4,11-5] have reviewed the state of the art of computer energy analysis programs.

These programs are available in two types:

- *Public Domain* The source codes for public domain programs are available and can be purchased for a nominal fee for implementation on the buyer's computer facilities. Programs such as NBSLD and NECAP belong to this category.
- *Proprietary Programs* Programs such as TRACE, ECUBE, APEC ESP-1, APEC HCC-III, etc. are proprietary in nature and can be utilized through time sharing facilities, by leased object language codes, or by submitting input data to the program developer.

Brief descriptions of the capabilities of some of these programs are on the following pages, but the list is not complete. Several companies, for example, are developing and using in-house programs for making energy conservation studies. Reference 11-4 should be consulted for additional listings.

A typical input-output flow chart for energy analysis is shown in Figure 11-1. The results obtained by different programs may lead to different values of the energy requirements for a given building. These differences are a result of the number of choices available for simulating the various steps (such as weather data input, heat gain computation, system simulation, etc.) which lead to the calculation of energy requirements. Brief discussions of these choices are given below.

NASA's Energy Cost Analysis Program (Code Name: NECAP)

The NECAP program follows the procedures outlined

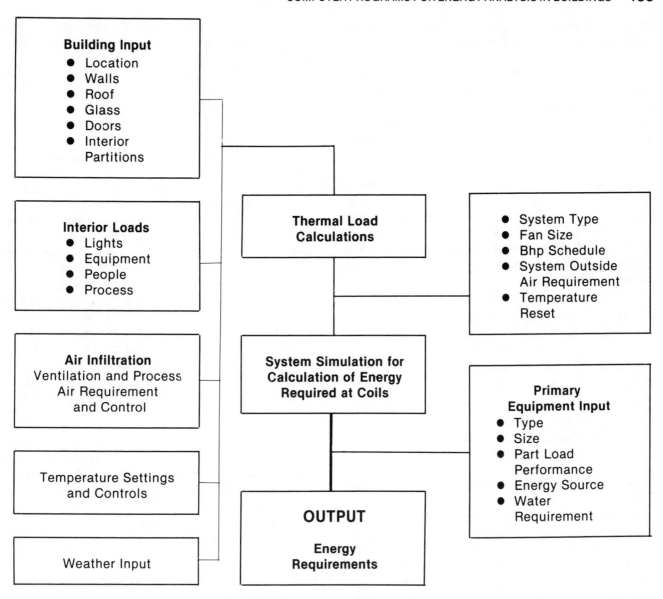

FIG. 11-1. Energy analysis flow chart.

in the ASHRAE booklet *Procedures for Determining Heating and Cooling Loads for Energy Calculation* to estimate the energy requirements for buildings. The program is actually a set of six individual computer programs including (1) a Response Factor program, (2) a Data Verification program, (3) a Thermal Loads Analysis program, (4) a Variable Temperature program, (5) a System and Equipment Simulation program, and (6) an Owning and Operating Cost program. Standard wall construction and schedules can be used to simplify program input. The program is an extension of the Energy Utilization Program developed for the U.S. Postal Service, but incorporates extensive modifications to improve its usability, including completely revised documentation.

This program is available through:

COSMIC
Suite 112
Barrow Hall
University of Georgia
Athens, Georgia 30602
Ph. (404) 542-3265

Alternate Choice Comparison for Energy Systems Selection (Code Name: AXCESS)

AXCESS permits the simultaneous comparison of up to six alternate methods for meeting the energy requirements of a building. The program may be used at various points in the design process. If the structure is still in the conceptual state, the program utilizes routines which approximate the full input information which would normally be available during the later stages of design.

The program calculates hourly zone solar and transmission loads for the year, if not input from another program. These loads, along with the base energy loads, are used to calculate net zone space conditioning requirements. Terminal system operation is simulated and equipment energy consumption calculated in one of three general subroutines.

AXCESS is available through five different means:

(1) Through participating utilities.
(2) Time sharing systems (National C.S.S., Inc.).
(3) Source deck may be purchased.
(4) Object deck may be purchased.
(5) As a service from a participating consulting firm.

For more information, contact:
Edison Electric Institute
1140 Connecticut Ave. N.W.
Washington, D.C. 20036

Trane Air Conditioning Economics (Code Name: TRACE)

The TRACE program calculates peak and hourly zone loads based on coincident hourly climatic data for temperature, solar radiation, wind and humidity during typical days of the year that reflect seasonal variations. The average days are based on the most recent ten years of U.S. Weather Bureau data. The design phase receives input from the load phase, as well as system type information and zone design information, and calculates supply air quantities and temperatures. The system simulation phase utilizes hourly zone loads, calculates return air quantities and temperatures, and accounts for system loads such as reheat loads. The equipment simulation phase takes the hourly output from the system simulation phase and calculates the annual energy consumption based on part-load performance data which is available on tape. The economic phase then does an economic comparison of the various design alternatives based on input consisting of energy consumption data, utility rate structures, and expected installation and maintenance costs.

The program source code is proprietary, but the program may be utilized through local Trane representatives or through time-sharing systems. For more information, contact the local Trane representative or:
Applications Engineering
The Trane Company
3600 Pammel Creek Road
LaCrosse, Wisconsin 54601

Meriwether Energy Systems Analysis Series (Code Name: ESAS)

The Energy Systems Analysis Series is a library of computer programs that determine the annual energy consumption of various types of systems and equipment for a typical year of operation and the relationship between these energy costs and other owning and operating costs. The programs normally used in conducting an energy analysis consist of the Energy Requirements Estimate (ERE) that calculates the hour-by-hour thermal and electrical loads for a building, the Equipment Energy Consumption (EEC) that simulates the operation of the various pieces of equipment, and the Economic Comparison of Systems (ECS) that calculates the total owning and operating costs of each system. A variety of auxiliary programs are available to complement the basic analysis series.

The Energy Systems Analysis Series is available via consulting agreements with R. F. Meriwether and Associates, various time-sharing systems, or through a special lease agreement. For more information, contact:
Ross F. Meriwether and Associates, Inc.
1600 N.E. Loop 410
San Antonio, Texas 78209
Ph. (512) 824-5302

Energy Conservation Utilizing Better Engineering (Code Name: ECUBE)

The ECUBE series of programs (except for ECUBE III) utilizes design point calculations of peak thermal and electrical loads to estimate the hourly, monthly, and annual energy requirements of a building. The energy consumption of various types of systems, which may be used to meet these requirements, is calculated and then the program compares the total owning and operating costs of the systems being considered. The series consists of three basic computer programs: (1) Energy Requirements Program, (2) Equipment Selection and Energy Consumption Program, and (3) Economic Comparison Program.

ECUBE can be utilized via Control Data Corporation's Cybernet time-sharing network or via leasing agreement. For more information contact:

Manager — Energy Systems
American Gas Association
1515 Wilson Boulevard
Arlington, Virginia 22209
Ph. (703) 524-2000

National Bureau of Standards Load Determination Program (Code Name: NBSLD)

NBSLD calculates the hour-by-hour heating and/or cooling load in buildings. It utilizes the thermal response factor technique for calculating transient heat conduction through walls and roofs. The program also includes a routine which can calculate the *floating* temperature of those rooms with limited heating or air-conditioning and natural ventilation, based on the actual net heat loss or gain to the room.

NBSLD, while only a loads program, is generally recognized as the most comprehensive and accurate program for that purpose existing today. It has been linked with the Meriwether ESAS programs and with NECAP to produce two hybrids with complete analysis capability, one of which is called DOE-2.

Source code and documentation may be obtained for a very nominal fee. For more information, contact:

Thermal Engineering Section
Center for Building Technology, IAT
National Bureau of Standards
Washington, D.C. 20234

APEC Heating-Cooling Calculation Program, Version III (Code Name: HCC-III)

The APEC HCC-III program calculates design heating and cooling loads utilizing ASHRAE methodology. The room is the basic level of calculation for which the requisite design data is specified. Heating load calculations consist of conventional transmission loss analyses involving areas, U-factors, and temperature differences. Infiltration is calculated on an input of master factor basis. The maximum glass solar heat gain for the specified winter month and its effect on the heat loss for the room is determined along with lighting and occupant heat gain. Cooling load calculations are made on an hourly basis over a 24-hour period for the selected design day, and radiant load components are time-averaged in accordance with the building mass. Glass loads are separated into transmission and solar components, taking into account the glazing material and shading devices. Hourly room loads and cfm values are accumulated and analyzed psychrometrically, and peak loads on equipment are determined.

This program is available under a fee-for-license arrangement. For more information, contact:

APEC Executive Office
Miami Valley Tower, Suite 2100
Fourth and Ludlow Streets
Dayton, Ohio 45402
Ph. (513) 228-2602

APEC Energy Analysis Program (Code Name: ESP-1)

ESP-1 computes energy requirements by simulating hour-by-hour performance of a building based on the building envelope characteristics, occupant schedules, lighting schedules, equipment and process loads, thermostats and controls, HVAC system type and controls and primary equipment energy consumption.

In addition to summarizing energy consumption, one can monitor energy consumption with up to 20 submeters for various types of equipment.

ESP-1 consists of five independent programs that are used in sequence:

Weather processes a weather tape and adjusts for variations in latitude, longitude and elevation between the weather station and the building site.

Response computes response factors for roof and wall sections as a function of time across the boundary of the wall or roof membrane.

Edit checks for errors and consistency of the user input data before proceeding to the final two subprograms and *Loads* computes the hour-by-hour loads for each building space under study.

Sysim simulates the operation of the actual HVAC systems to be used in the building.

This program is available under a fee-for-license arrangement. For more information, contact:

APEC Executive Office
Miami Valley Tower, Suite 2100
Fourth & Ludlow Streets
Dayton, Ohio 45402

(Code Name: CAL/CON 1)

CAL/CON 1 was developed jointly by the California Energy Commission and ERDA, which is now part of the U.S. Department of Energy. The library includes temperature, wind and cloud data for 60 U.S. locations.

CAL/CON 1 uses a new program language, called Building Design Language, that employs familiar English terminology to input instructions. The user can specify several types of output such as hourly, daily, monthly, and seasonal energy consumption, consumption by zones, and HVAC system performance.

Subprograms are labeled Loads, Systems, Plant, Economics and the Report program produces tables of user selected variables.

For further information, contact:
California Energy Commission
Building & Appliance Standards Office
1111 Howe Ave.
Sacramento, Calif. 95825

Building Loads Analysis and System Thermodynamics Program (Code Name: BLAST)

The BLAST program input uses Building Design Language derived from CAL/CON 1 and has three outputs:

1. Hourly building thermal loads based on the user's input of building characteristics, occupancy schedule and weather data.

2. Hourly consumption by the air handling system to determine the energy requirements of fan coils.

3. Hourly resource requirements of the central plant primary equipment 'such as boilers and chillers.

For further information, contact:
Air Force Civil Engineering Center
Tyndall Air Force Base, Florida 32403

WEATHER DATA

A wide variety of weather data are used in load and energy analysis calculations.

W1 — Any combination of dry and wet bulb temperatures (such as 99% or 97½% for winter and 1% or 2½% for summer design temperature) given in Table 1, Chapter 23 of the ASHRAE Handbook—1977 Fundamentals.

W2 — A set of summer and winter design temperatures plus design hourly temperatures for a typical day for each of the 12 months; to be prepared by user. (Used with APEC HCC-III.)

W3 — Actual weather data with wind effects considered. (Used in NECAP, ECUBE, AXCESS, etc.)

W4 — Twelve model days (each day represents the effective weather profile for one month) generated from yearly hour-by-hour data. (Used in TRACE.)

HEAT GAIN COMPUTATION

The heat gain computation methods used by various programs are another source of difference.

H1 — *TETD (Total Equivalent Temperature Differential Method)*
This is a widely accepted method for calculating heat gain through exterior walls and roofs (Chapter 25, Table 7, ASHRAE Handbook—1977 Fundamentals) as discussed in Chapter 3. It gives reasonably accurate results for design

load computation; however, it can introduce substantial errors for hour-by-hour load calculation using actual weather data.

H2 — *Response Factor Method*
This approach has been adopted for computer programs such as NECAP, NBSLD, etc.

H3 — *Transfer Function Method*
This method is essentially similar to the response factor method, but it assumes a cyclic repetition of the design sol-air temperature for several days.

H4 — *Finite Difference Method*
This is the numerical solution to the Fourier equation for heat conduction. The smaller time steps, which reduce error, will also cause a rapid increase in computation time.

H5 — *Sol-Air Temperature Method*
This method calculates the heat gain through exterior walls and roofs by using the concept of sol-air temperature. Sol-air temperature is the temperature of the outdoor air which, in the absence of all radiation exchanges, would give the same rate of heat entry into the surface as would exist with an actual combination of incident solar radiation, radiant energy exchange with the sky and other outdoor surroundings, and convective heat exchange with outdoor air. The details are discussed on page 25.4 of the ASHRAE Handbook — 1977 Fundamentals.

COOLING LOAD COMPUTATION

The cooling load is not identical with the rate of heat gain. Cooling load is the rate at which heat must be removed from a given space, by HVAC equipment, to maintain a constant room temperature. The exact method involves a lengthy solution of a set of energy equations involving walls, room air, infiltrating air and internal energy sources. The NBSLD program uses this technique for calculating cooling load.

Various approximate methods have been developed to simplify the cooling load calculations. The heat gain is assumed to consist of convective, latent and radiative components. The radiation fraction is not felt until sometime later, since it has a delayed effect, while the other components are gained instantaneously. Most of the computer programs use a set of factors (called the weighting factor method) to identify the instantaneous and delayed fractions of heat gain. The forms and numerical values for these factors vary; some are discussed below.

C1 — *Transfer Function Method*
This method gives a set of precalculated factors

called *Coefficients of Room Transfer Functions* (Table 31, Chapter 25, ASHRAE Handbook—1977 Fundamentals) for different weight structures. The NECAP computer program employs this method with the exception that room transfer functions are actually calculated for a given weight structure. The method closely simulates the time lag phenomenon for the radiative component of heat gain.

C2 — *ASHRAE Algorithms*

The ASHRAE task group on energy requirements for heating and cooling of buildings has compiled two algorithms called RMTMP and HLC for calculating thermal loads, references 11-7 and 11-8. Subroutine RMTMP is a rigorous solution of heat balance simultaneous equations for walls, air, internal sources, etc. The procedures in subroutine HLC are similar to the transfer function method but employ different weighting factors.

C3 — *Trane (TRACE) Method*

This method uses the weighting factors of the NECAP Program for exterior walls and roofs to determine the cooling load. Other weighting factors, relating solar heat gain, lights, etc., are not considered by this method, which may introduce a substantial error, especially if the building has a large window area exposed to the sun.

C4 — *ECUBE Method*

This program assumes that heat is stored only when the hourly load exceeds the coil capacity and completely ignores the heat storage of building components due to the absorption of the radiant portion of the heat gain. Heat gain components are directly used to calculate the total cooling load for the building.

INFILTRATION SIMULATION

The reliability of a mathematical simulation of the air infiltration rate is questionable in most programs. It is difficult to develop an accurate estimate of the air infiltration rate through various cracks, seams, and joints, and infiltration due to the opening and closing of doors and windows with constantly changing conditions of outside temperature and wind. Most of the load and energy analysis programs estimate infiltration by the air change method and assign a single constant value of so many air changes per hour.

SYSTEM SIMULATION

The hourly heating and cooling outputs of the boilers, chillers and other HVAC equipment differ markedly

from the hourly space load calculated in the thermal load analysis part of the program. The purpose of a system simulation program is to translate hourly thermal loads into the hourly thermal requirements imposed upon the heating and cooling plants. These hourly thermal requirements are converted into energy requirements based on part-load characteristics of the heating and cooling plant equipment. The various programs differ in their approach to modeling these steps. The state of the art is discussed below.

S1 — *Energy Distribution Systems Simulation*

The equations required for simulating the various types of energy distribution systems have already been discussed in Chapter 4. Programs such as NECAP and APEC ESP-1 use these types of equations and the procedures outlined in References 11-7 and 11-8, enabling one to estimate the energy required at heating/cooling coils.

S2 — *Primary Equipment Simulation*

The mechanical equipment (boilers, chillers, etc.) operates under part-load conditions during most of the operating period. Many energy programs tend to use overly simplified mathematical models to represent the energy efficiency data for part-load conditions. One such example is the straight line approximation for refrigeration equipment (curve A) in Figure 11-2.

Chen[11-1] has shown that this assumption is inadequate in estimating part-load consumption when condensing water is overcooled or controlled at a constant temperature (curves B and C, Figure 11-2). The NECAP programs[11-6] use an empirical equation supplied by Carrier based on the performance of a large number of chiller

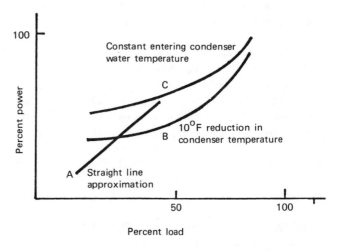

FIG. 11-2. Part load power consumption of a centrifugal chiller.[11-1]

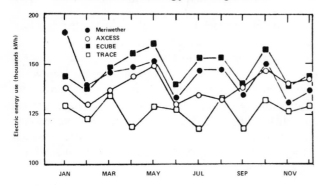

FIG. 11-3. Monthly energy use calculated by various computer programs.[11-3]

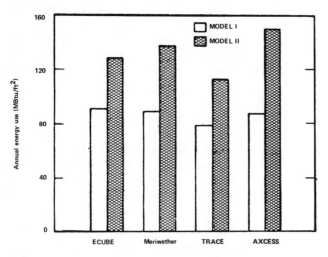

FIG. 11-4. Annual energy use calculated by four computer programs.

machines. The TRACE program uses the concept of a modifier (M), and temperature deviations are introduced to compensate for errors in straight line approximation. The part-load data supplied by the manufacturer should be used, and the energy analysis program structured to accept part-load data for a given project.

CHOICES IN SIMULATION

The above discussions show that computer programs offer many choices for simulating the energy requirements of a building, depending upon the approaches taken for weather simulation, load computation and system simulation. The energy requirements for a given building, as calculated by each computer program, varies because of the approaches used. The user should know the approach taken by a program before deciding to use it.

COMPARISON OF PROGRAM RESULTS

Spielvogel, Ayers, Chen, and others have compared

energy usage calculated by various programs. Ayers' results, using ECUBE, TRACE, AXCESS and MERIWETHER to study an existing 4-story and 12-story office building, are shown in Figures 11-3 and 11-4. The results of simulating the energy requirements for a 20-story office building located in Washington D.C., as presented at the APEC symposium[11-2], are shown in Figures 11-5 and -6.

Variations in the output of each program are expected and attributed to the variety of approaches used for the calculations, but the trends are similar. Once a program is selected for simulating an existing or new building, the same program should be used for evaluating the

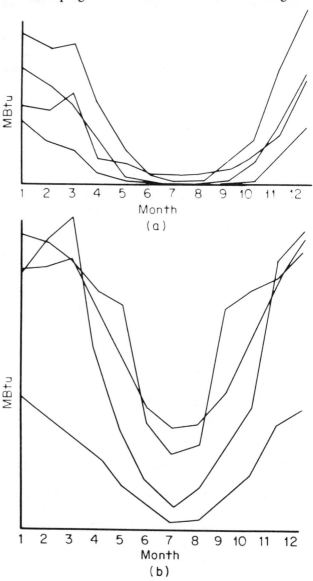

FIG. 11-5. Comparison of heating demand (a) and consumption (b) as calculated by various computer programs.[11-2]

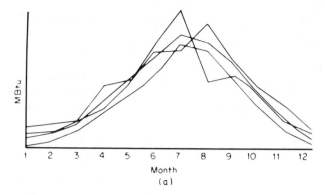

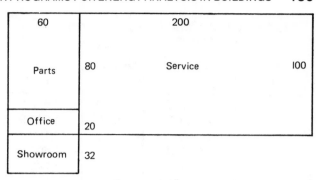

FIG. 11-7. Layout of a typical auto dealership.

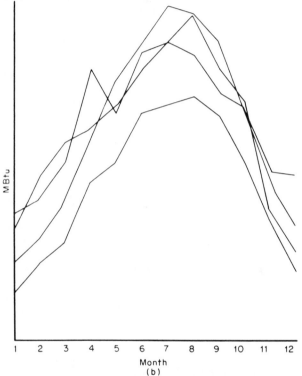

FIG. 11-6. Comparison of cooling demand (a) and consumption (b) as calculated by various computer programs.[11-2]

energy saving alternatives for that building. This is illustrated by using the in-house program called ERA (Energy Requirement Analysis) of Smith, Hinchman & Grylls Associates, Inc., Detroit.

Example 11-1

The energy usage of an auto dealership/service facility is to be analyzed. The typical layout, Figure 11-7, includes the showroom, office, parts storage and service area. The floor area is 27,000 ft² and there is 300 ft² of ¼ in. single-pane glass. The walls are 8 in. concrete block, except the office area which is constructed of 4 in. face brick, 3½ in. insulation and dry wall on the in-

side. The service area has 20 garage doors measuring $12' \times 12'$, $U = 7.0$, that are opened an average of six times a day. The lighting levels are 2 watts/ft² in the parts and service areas and 4 watts/ft² in the office and showroom. The roof is built up on a metal deck with ¾ in. rigid insulation. The office and showroom have acoustic tile ceilings.

Gas-fired heaters are used to heat the service and parts areas, while the office and showroom are heated and cooled by rooftop packaged units. The annual energy consumption in the preceeding 12 months is estimated as

18,900 therms of gas

414,500 kWh of electricity

The following energy-conserving alternatives were considered:

1. Increase the roof insulation by 3 inches.
2. Increase the roof and wall insulation by 3 inches.
3. Insulate to reduce the U of the service area doors from 7.0 to 1.0.
4. Reduce the glass in the showroom area by one-half.
5. Reduce the number of doors to the service area.
6. Use electric unit heaters in the service and parts areas, and three water-source heat pumps in the office.
7. Use two air-source heat pumps in the office and electric unit heaters in the service and parts areas.
8. Use air-source heat pumps for the office and service areas, and electric unit heaters for the parts areas.
9. Increase the roof insulation and decrease the glass area.

The energy consumption, calculated by the computer program ERA, is shown in Table 11-1. The economic analysis of the alternatives is shown in Table 11-2.

This example clearly illustrates the potential of computer programs to evaluate various energy saving alternatives. The importance of selecting the right program cannot be overemphasized.

TABLE 11-1. Dealership Energy Consumption

Alternative No.	Electric kWh	Average Demand kW	Gas Therms
Base*	414,513.6	82.9	18,877.0
1	414,057.0	82.9	14,541.9
2	413,500.4	82.8	10,150.7
3	414,034.8	82.9	14,609.3
4	413,199.1	81.3	17,743.5
5	482,679.9	96.1	7,874.2
6	1,172,393.9	786.2	0
7	913,944.2	442.5	0
8	786,258.2	257.8	0
9	412,817.5	81.3	13,364.0

* Base refers to the existing building.

TABLE 11-2. Economic Analysis of Alternatives*

Alternative No.	Differential Capital Cost	Rate of Return %	Pay Back Years	Comments
1	$ 9,480	8.10	15.75	Better than base
2	21,076	7.3	16.58	Better than base
3	36,000	Negative	—	Does not pay back
4	−70	Instant	0	Better than base
5	102,705	Negative	—	Does not pay back
6	20,250	Negative	—	Does not pay back
7	21,400	Negative	—	Does not pay back
8	44,000	Negative	—	Does not pay back
9	9,410	11.60	12.42	Better than base

* Finance charge for the base system is 10.4%.

Chapter 12

HEATING WITH SOLAR ENERGY

The solar energy impinging on the earth's atmosphere is several thousand times greater than the total amount of all the energy that we use. However, its practical and efficient collection is a difficult problem. For example: the northern U.S. requires large amounts of energy for space heating in winter, a season when climatic conditions minimize the availability of solar energy.

The high cost of solar collectors and low availability of solar energy in winter makes the solar system uneconomic when compared with the present prices of fossil fuels. This situation may not prevail for long, due to the uncontrolled rise of oil, gas and coal prices, and our most abundant renewable resource may become the most popular and lowest-cost form of energy.

SOLAR RADIATION

Solar radiation is normally measured in Langleys. One Langley is equal to one calorie of solar radiation per square centimeter (3.69 Btu/ft^2) and is reported in terms of Langleys received on a horizontal surface at ground level.

The intensity of radiation varies with geographical location, season, time of day, cloud cover and atmospheric contaminants. The solar constant (429.5 Btuh/ft^2 or 1353 W/m^2) is the rate at which energy from the sun is received by a surface, of unit area, which is normal to the sun's rays and at the mean distance of the earth from the sun and above the earth's atmosphere.

As solar energy penetrates the atmosphere, the ultraviolet radiation is absorbed by ozone in the upper atmosphere. A part of the solar wave radiation is diffused and absorbed by dust, gas molecules and water vapor before reaching the surface of the earth.

The total radiation available to a collector on earth is called solar insolation and is composed of:
- Direct (beam) solar radiation
- Diffuse solar radiation
- Reflected solar radiation

The solar energy impinging on a collector depends upon the sun's position in the sky which is expressed in terms of the solar altitude, β, above a horizontal plane and the solar azimuth, γ, measured from the south in a horizontal plane, Figure 12-1a. These angles, in turn, depend on the local latitude, L, and the solar declination, D, which is a function of the day of the year and the hour angle, W. W is equal to the number of minutes of time from local solar noon.

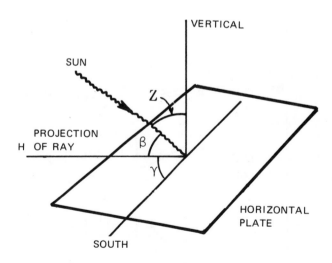

FIG. 12-1a. Solar angles.

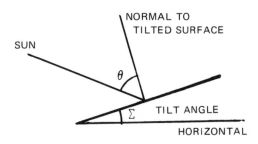

FIG. 12-1b. Collector tilt angle.

Figure 12-1b shows a surface tilted at an angle Σ from the horizontal. The angle of incidence, θ, is defined as the angle between the incoming solar rays and a line normal to that surface. For a horizontal surface, the incident angle is Z and, for a vertical surface, the incident angle is β.

The *ASHRAE Handbook, 1977 Fundamentals* gives the equations relating the angles β and θ to L, D, and W. If the hourly insolation on the horizontal surface, I_h, the solar altitude, β, and the solar azimuth, γ, are known, the hourly insolation on any tilted surface, I_t, may be estimated by the following equation:

$$I_t = 0.9 \, I_h \frac{\cos \theta}{\cos \beta} + 0.1 \, I_h \qquad (12\text{-}1)$$

where 0.9 and 0.1 equal the average hourly direct and diffuse radiation factors. On a clear day, 85-90% of the radiation is direct (beam) solar radiation.

The amount of solar radiation on a given area can be greatly increased[12-1] by tilting the receiver surface southward from the horizontal, Figure 12-1b. The comparative effect of a tilt angle is seen in Figure 12-2 which shows the change in available clear sky solar daily insolation

(Btu/ft²·day) for various tilt angles at 40° N.

TILT ANGLE SELECTION

1. *Heating Only* A tilt angle of latitude plus 10 degrees tends to maximize collection during the winter months. There are other proposals; latitude plus 18 degrees, for example. The difference between latitude plus 10 degrees and plus 18 degrees is small, but may be evaluated by interpolation.

2. *Heating and Cooling* A tilt angle approximately equal to latitude is generally used since a tilt angle of about this order favors year-round heat harvest. There are suggestions that latitude plus 3 degrees will be more favorable. Again, there does not appear to be any significant difference, in terms of collection, between the two tilt angles. The solar system designer can verify differences by interpolation.

ASHRAE Transactions, Volume 80, Part II, 1974 and *ASHRAE Handbook, 1977 Fundamentals* give tables of clear sky radiation at various latitudes, times of day, and for horizontal and vertical surfaces, and should be consulted for details.

SOLAR ENERGY COLLECTORS

Solar energy collectors are classified as flat plate collectors and focusing collectors. Flat plate collectors are less costly, collect heat from diffuse as well as direct radiation, and can operate on bright cloudy days. Focusing collectors can utilize only direct radiation, but produce higher working temperatures.

Flat plate collectors traditionally are made of copper, aluminum or steel to provide good heat transfer. The surfaces are usually painted with a flat black paint or a

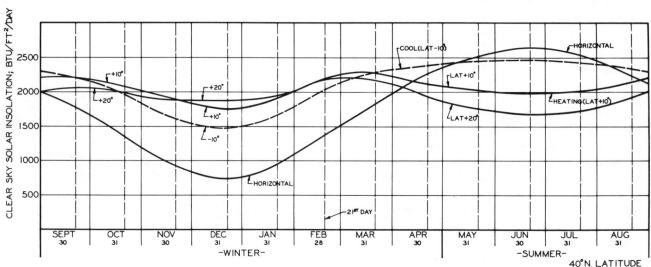

FIG. 12-2. Effect of tilt angle on insolation.[12-5]

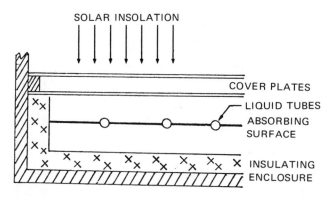

FIG. 12-3. Cross section of a typical flat plate collector.

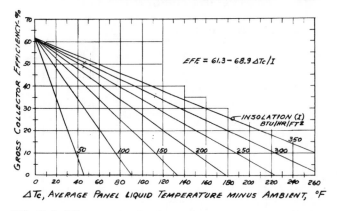

FIG. 12-4. Efficiency of a flat plate collector under varying conditions.[12-5]

selective coating to produce high absorptivity of long wave, infrared, low temperature radiation. Depending upon the service and operating ambient temperatures, the collector has one to three transparent cover plates, usually of glass but are sometimes of transparent plastics such as tedlar or mylar. A typical cross section of a flat plate collector, using liquid as a working medium, is shown in Figure 12-3.

The classical equation for estimating the useful heat gain, Q_u, from a collector is as follows:

$$Q_u/A_c = F_r[S_t\tau\alpha - U_L(T_{fi} - T_a)] \qquad (12\text{-}2)$$

where

A_c = Area of collector

F_r = Heat removal factor ($0.75 - 0.85$ for most collectors)

S_t = Insolation as seen by a tilted surface

τ = Fraction of energy transmitted by cover plates

α = Fraction of energy absorbed by the collector plate

U_L = Overall loss coefficient

T_{fi} = Fluid inlet temperature to collector

T_a = Ambient air temperature

The efficiency of a collector is defined as:

$$\eta = \frac{Q_u}{S_t A_c} \qquad (12\text{-}3)$$

The energy collection capability or efficiency of a specific flat plate collector depends on:

1. Solar insolation (radiation) impinging on the transparent cover, measured on the same plane as the cover.

2. The difference between the average temperature of the heated fluid flowing through the panel and the surrounding or ambient air temperature. For brevity, this difference will be referred to as ΔT_c.

Efficiency rating charts for solar collectors are generally plotted as ΔT_c versus efficiency and for various solar insolation values. A typical plot, Figure 12-4, describes the performance of a flat plate collector having no selective coating. System designers should apply the manufacturer's specific catalogued data for the collectors used in their system. Figure 12-4 shows that, for a given insolation level, the lower the ΔT_c, the higher the collection efficiency. Also, for a given ΔT_c, as the incident

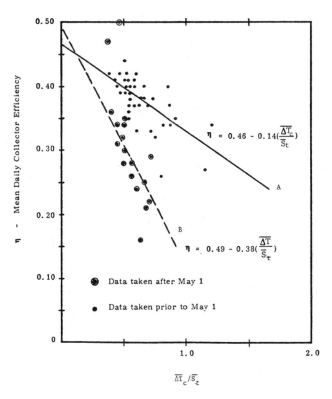

FIG. 12-5. Actual daily collector efficiency.[12-6]

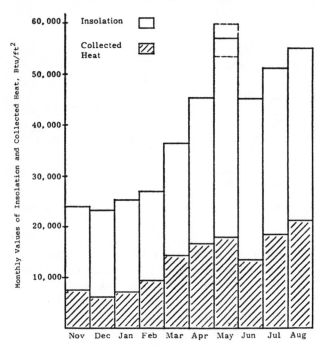

FIG. 12-6. Monthly yield from available insolation.[12-6]

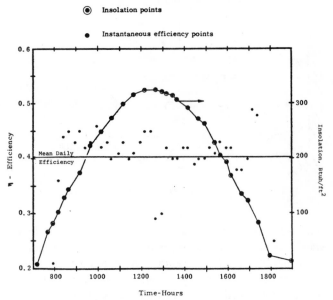

FIG. 12-7. Hourly insolation and collector efficiency, April 29.[12-6]

solar radiation increases, the collector efficiency increases. Thus, a given collector will give optimum output on a clear day at the lowest possible temperature difference between the collector plate and the ambient air. This is confirmed, Figure 12-5, by the operational data of Smith, Hinchman & Grylls for an Owens-Illinois vacuum tube collector. The corresponding monthly insolation data, collected solar heat, and variation in hourly efficiency for a specific day are shown in Figures 12-6 and 12-7.

APPLICATIONS OF SOLAR ENERGY

The U.S. Department of Energy has organized solar energy research and development into five main program areas:

- Bio-conversion to fuels
- Ocean thermal conversion
- Wind energy conversion
- Photovoltaic conversion
- Solar heating and cooling of buildings, including agricultural applications.

The major concern in each area is to develop systems that are economically competitive with alternative energy

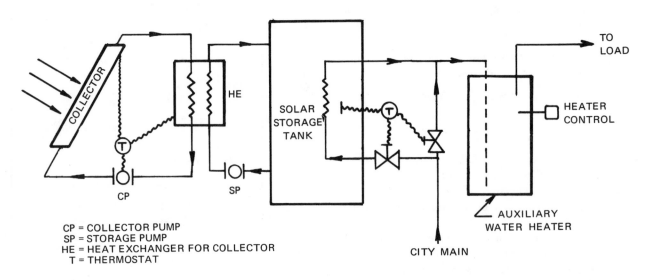

CP = COLLECTOR PUMP
SP = STORAGE PUMP
HE = HEAT EXCHANGER FOR COLLECTOR
T = THERMOSTAT

FIG. 12-8. Schematic of a typical solar domestic water heating system.

sources. The technology for solar heating is in an advanced stage, demonstrated by the construction of a number of buildings with solar heating and cooling systems in various parts of the country.

DOMESTIC WATER HEATING

Heating household water by solar energy is common in Japan, Israel, Australia, and in some parts of the USA, such as southern Florida. In general, solar water heaters utilize a flat plate collector in conjunction with a water storage unit, Figure 12-8.

The energy from the collector is picked up by a circulating water/glycol mixture and transferred to the water in the solar storage tank through a heat exchanger (HE). If freezing temperatures are not a problem, the glycol and the heat exchanger are unnecessary and water from the storage tank is circulated through the collector plate when solar radiation is available. The water from city mains can flow directly to the water heater, when its temperature exceeds that in the solar storage tank, or be preheated in the solar storage tank when energy is available. If the temperature in the solar storage tank is insufficient, heat is supplied by the water heater to raise the water temperature to the required level.

SPACE HEATING

Since the temperature required for heating a building is low, flat plate collectors are adequate in most cases. Heat is transferred from the collector to a heat storage unit by a stream of circulating liquid or air.

A typical layout for a liquid-based solar heating system is shown in Figure 12-9. The heat from the solar storage tank is transferred to the building supply air through a heating coil. If the solar tank cannot provide enough energy, the auxiliary heat source will be turned on.

Domestic hot water can be integrated with the solar space heating system in most cases. Figure 12-10 shows a layout for combined space heating and hot water supply system.

Air-based space heating systems circulate air to pick up the heat energy from the collector and store excess energy in a rock pebble bed. A typical layout for air-type space heating systems is shown in Figure 12-11. Löf, Karakari and Smith[12-2] made a space heating performance comparison between an air solar system and a liquid solar system, under nearly equal conditions, at Colorado State University. They found that both systems had similar performance, but that liquid systems are more compact and advantageous where space is valuable. Liquid systems are required if absorption air conditioning is used.

Air heating systems offer the advantage of simplicity. Corrosion is not a problem, although leaks impair performance. An air system should have more appeal for residential space heating where low maintenance is a signif-

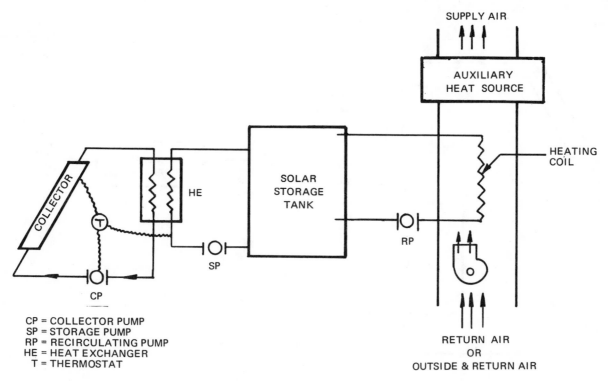

CP = COLLECTOR PUMP
SP = STORAGE PUMP
RP = RECIRCULATING PUMP
HE = HEAT EXCHANGER
 T = THERMOSTAT

FIG. 12-9. Schematic of a typical liquid-based solar heating system.

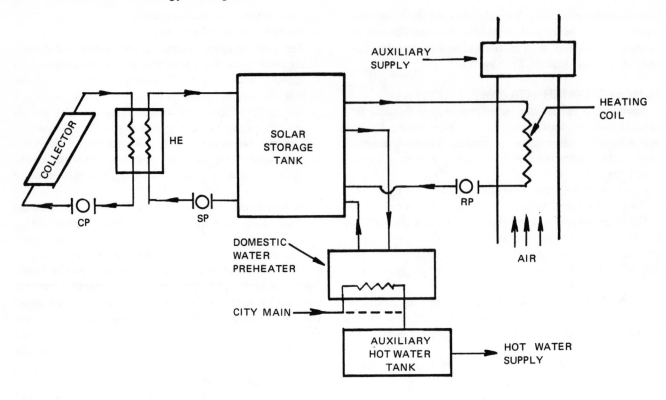

FIG. 12-10. Schematic of a typical combined solar space heating and hot water system.

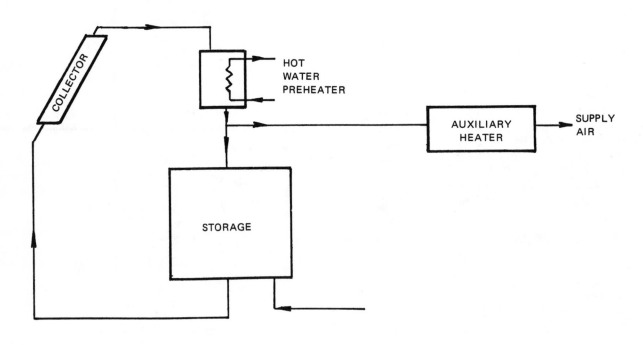

FIG. 12-11. Schematic of a typical air-based space heating system.

icant consideration. Liquid solar heating systems are favored in large commercial buildings when maintenance is available and space for heat storage and equipment conduits is of high value.

DESIGN FACTORS FOR SOLAR SYSTEMS

Based on their detailed computer analysis, Beckman and coauthors[12-3] recommend the following factors for the design of solar space and water heating systems using flat plate collectors.

COLLECTOR FLOW RATE

50/50 ethylene		
glycol/water	0.022 gpm/ft^2	0.015 L/s·m^2
Air	1 to 4 cfm/ft^2	5 to 20 L/s·m^2

STORAGE CAPACITY

Water	1.25 to 2 gal/ft^2	50 to 100 L/m^2
Rock bed	0.5 to 1.15 ft^3·ft^2	0.15 to 0.35 m^3·m^2
Weight of pebbles	50 to 130 lb/ft^2	
Pebble size	0.5 to 1.5 in.	1 to 3 cm

DOMESTIC WATER PREHEAT TANK

1.5 to 2 times the capacity of a conventional water heater

ft^2 and m^2 refer to the area of the collector

ESTIMATING PERFORMANCE

Solar energy is a transient source of energy. Because the intensity of available insolation varies from hour to hour at each location, it is difficult to develop simplified techniques for predicting the performance of a system. The present methods available for calculating the performance of solar systems are classified as:

- Approximate Method
- f-Chart Method
- Detailed Computer Method

Approximate Method

The approximate method is perhaps the easiest way to determine the performance to be expected from a solar energy system and is recommended when only an estimate is needed. The procedure is as follows:

1. *Estimate the Energy Collected*
   Estimate the amount of solar insolation incident on a collector of specified area, tilt and orientation, at a given place, by using Equations 12-1 and 12-2 and the average weather and solar data for a given area on a monthly basis. Use the collector manufacturer's efficiency curve, similar to Figures 12-4 and 12-5, to estimate the average monthly efficiency for the collector system and calculate the amount of energy that can be collected over a given period.

2. *Estimate the Building Load*
   The building space heating, space cooling and domestic water heating requirements are calculated using the techniques outlined in Chapter 3.

3. *Auxiliary Energy Required*
   The difference between Step 2 and Step 1 is the amount of auxiliary energy that must be supplied by conventional systems. The calculations are generally on a monthly basis and the results summed for the desired season of heating and cooling. The error in using this method can be large due to the transient nature of solar energy. This method is useful in determining whether or not a solar energy system is feasible.

f-Chart Method

The f-chart method has been developed by Beckman, Klein and Duffie,[12-4] based on detailed computer analyses using over 300 simulations. Two sets of charts are given in Figures 12-12 and 12-13. These charts give f_L factors for the fraction of energy supplied for heating by a solar system to liquid-based and air-based solar systems, respectively, having layouts similar to Figures 12-9, 12-10 and 12-11.

The value F'_r must be calculated before the values for the X and Y axis can be computed.

$$F'_r = F_r \left[1 + \left(\frac{F_r U_L A_c}{(\dot{m}C_p)_c} \right) \left(\frac{(\dot{m}C_p)_c}{\epsilon_c (\dot{m}C_p)_{min}} - 1 \right) \right]^{-1} \quad (12-4)$$

where

A_c = Area of collector. ft^2

F_r = Solar collector heat removal factor (already discussed)

$\dot{m}_c$ = Mass rate of flow of the fluid through the solar collector (lbs/hour)

C_P = Specific heat of fluid flowing through the solar collector

ϵ_c = Effectiveness of heat exchanger in the collector-storage flow loop

$(\dot{m}C_p)_{min}$ = The smallest value of the two $\dot{m}C_p$ products for the two fluid streams passing through the heat exchanger between the collector and the solar storage

If no heat exchanger is used between the collector and the solar storage, $F'_r = F_r$.

Once F'_r has been determined, the values for the X and Y axis can be calculated.

$$X = \frac{A_c F'_r U_L (T_{ref} - T_a) D_m \times 24}{L} \quad (12-5)$$

$$Y = \frac{A_c F'_r (\overline{\tau \alpha}) S_t \times D_m}{L} \quad (12-6)$$

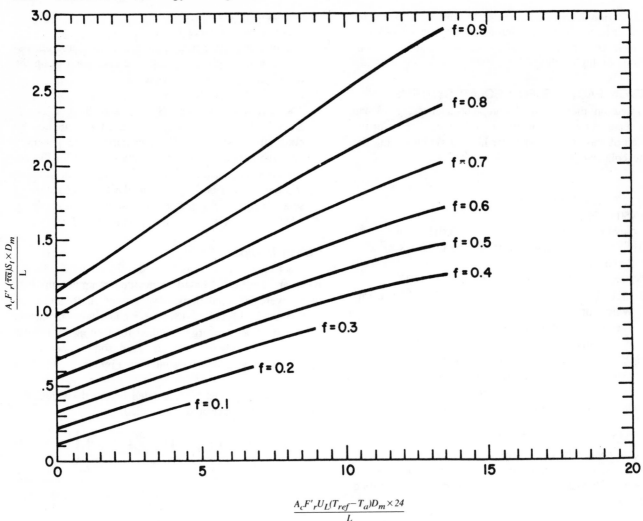

FIG. 12-12. f-Chart for liquid solar heating systems.[12-3]

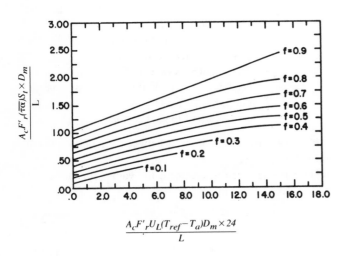

FIG. 12-13. f-Chart for air solar heating systems.[12-3]

where

$\overline{\tau\alpha}$ = The average value of the transmittance-absorptance factor. For tilted collectors, it is taken as 90–95 percent of the value determined for normal incidence.

S_t = Monthly value of the solar radiation incident on the collector array (Btu/month·ft^2)

L = Monthly space heating and hot water load

T_{ref} = 212°F

T_a = Monthly average daytime ambient temperature

D_m = Number of days in the month

U_L = Overall heat transfer loss coefficient for the solar collector (Btuh·ft^2·°F)

In addition to the f-chart, Figure 12-12, Beckman et al. give the following equation for calculating the f-factor for liquid based systems:

$$f = 1.029Y - 0.065X - 0.245Y^2 + 0.0018X^2 + 0.0215Y^3$$
$$\text{for } 0<Y<3 \text{ and } 0<X<18 \qquad (12\text{-}7)$$

Here is a comparison of the values obtained by using the f-chart and the equation:

X	Y	f-value (chart)	f-value (equation)
2.0	0.22	0.1	0.092
5.0	0.5	0.2	0.176
9.7	3.0	*	0.998
10.0	2.1	0.8	0.809

Values falling in the indeterminant area above the f=0.9 line, Figure 12-12, should be calculated using the equation. Any value greater than one should be taken as one since values greater than unity indicate available but unusable energy.

A comparable equation, for calculating the f-factor for solar air systems, is

$$f = 1.04Y - 0.065X - 0.159Y^2 + 0.00187X^2 - 0.0095Y^3$$
$$\text{for } 0<Y<3 \text{ and } 0<X<18 \qquad (12\text{-}8)$$

Comparing the values obtaining by using the f-chart, Figure 12-13, and the equation,

X	Y	f-value (chart)	f-value (equation)
3.0	0.3	0.1	0.12
8.0	0.75	0.3	0.29
8.0	1.55	0.8	0.794
12.0	2.5	*	0.95

Again, no calculated value of f can be taken as greater than one.

The fraction of yearly load that is carried by the solar system is calculated by summarizing the monthly fractions as follows:

$$f_{yearly} = \frac{\Sigma f_{mi} \times L_i}{\Sigma L_i} \qquad (12\text{-}9)$$

where

f_{mi} = Fraction of load carried by solar system for the *i*th month

L_i = Building space heating and hot water load for the *i*th month

Correction Factors For f-Charts

In developing the f-charts, Figures 12-12 and 12-13, the following assumptions were made regarding the design of solar heating systems.

a. *Water-Based Solar Heating System*

The size of the thermal storage unit, for both space heating and domestic hot water, was assumed to be 1.85 gallons/ft² of solar collector. This is equivalent to 15 Btu of energy stored per °F temperature rise in the thermal storage unit per ft² of collector. The correction factor for other storage units is shown in Figure 12-14. The undersized storage unit significantly affects the performance of a solar heating system. On the other hand, oversized storage does not improve the performance to any significant degree.

b. *Heat Exchanger*

The following equation was assumed to hold good for the sizing of a heat exchanger between the building space and the solar thermal storage unit:

$$\frac{\epsilon_L (\dot{m}C_p)_{min}}{U_o A_o} = 2.0 \qquad (12\text{-}10)$$

where

ϵ_L = Effectiveness of the water-air load heat exchanger

$(\dot{m}C_p)_{min}$ = Smallest value of the heat capacitance rate for two fluid flow rates passing through the heat exchanger of the solar storage tank

$U_o A_o$ = Product of the overall transmission loss factor and the building envelope area

A correction factor shown in Figure 12-15 is used for other sizes of heat capacitance rates.

c. *Flow Rate, Air-Type Solar Collector*

In developing an f-chart for the air-heating system, the following assumption was made for the flow rate:

$$\left(\frac{\dot{m}C_p}{F_r A_c}\right)_{collector} = 2.87 \text{ Btu/hr·ft}^2\text{·°F} \qquad (12\text{-}11)$$

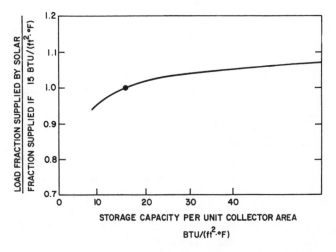

FIG. 12-14. Correction factor for liquid solar systems having storage capacities other than 15 Btu/ft²·°F.[12-3]

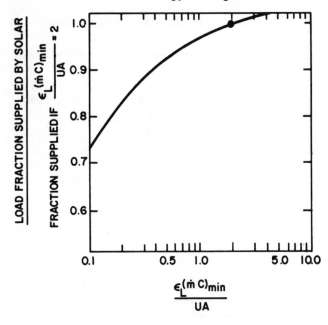

FIG. 12-15. Correction factor for liquid solar heat exchangers.[12-3]

For other values of this expression (1.47 to 5.88 Btu/hr·ft²·°F), the x-axis parameter of the f-chart should be multiplied by

$$\left(\frac{\dot{m}C_p}{F_rA_c} \times \frac{1}{2.87}\right)^{0.28} \qquad (12\text{-}12)$$

before using the chart.

d. *Capacity of Rock Storage*

The capacity of the rock storage was assumed to be 19.6 Btu/°F·ft² of collector area. The performance for other storage capacities (between 9.8 and 78.3 Btu/°F·ft²) is determined by multiplying the x-axis parameter of the f-chart by

$$\left(\frac{V\rho_aC_r}{F_rA_c} \times \frac{1}{19.6}\right)^{-0.3} \qquad (12\text{-}13)$$

when

V = Volume of rock storage, ft³

ρ_a = Apparent density of storage medium (lb/ft³)

C_r = Specific heat of storage medium

Economic Analysis

The cost effectiveness of solar heating/cooling systems is determined by the economic analysis discussed in Chapter 7. The basic formula used in computing the annual dollar savings over the lifetime of the system equates the annual savings in dollars due to the solar system with the cost of the energy provided by the solar system minus annual payments incurred due to the installation of the solar system.

Annual savings $= C_f \times f \times L - [(C_cA_c + C_{st} + C_e)i + C_fQ_s]$

$$(12\text{-}14)$$

where

C_f = unit cost of conventional fuel

f = fraction of energy provided by solar system for heating/cooling

L = building space heating/cooling load

C_c = collection cost per unit area

C_{st} = cost of thermal storage

C_e = cost of all other auxiliary equipment pumps, controls, heat exchangers, etc. for the solar system

i = fraction of total capital investment to be paid annually for a specified mortgage period and interest rate

Q_s = power consumed for solar system operation

Example 12-1

A small shop located in the Detroit Area (42 degrees N latitude) has a floor area of 6,000 square feet. The heating requirements are estimated to be 42,000 Btu/degree day. Consider a liquid type solar collector, having $U_L = 0.79$ Btuh·ft²·°F and a heat removal factor $F'_r = 0.89$, for use as a supplemental source of energy. The collector is tilted at 55 degrees from the horizontal.

The average monthly solar radiation as seen by the collector may be estimated by using Table 12-1 and Figure 12-16.

Estimate the fraction of total heating energy supplied by the collector having an area of 1,000, 1,500 and 2,500 ft².

Solution 12-1

First, estimate the average daily radiation seen by the tilted collector on a monthly basis, $\bar{S}_t$. This process is summarized in Table 12-2.

Next, estimate the f-factor (fraction of the load supplied by solar energy) for the various months. Table 12-4 shows the calculated monthly building loads in column 3. Factors X and Y are based on 1000 ft² of collector and are shown in columns 6 and 7. These factors are needed to determine the f-factors (from Figure 12-12) shown in column 8.

The heating energy provided by solar energy is shown in column 9. The auxiliary energy required from conventional sources is shown in column 10. The calculations are for the months of October through May. For the other months, June through September, the values of X and Y are such that f is greater than 1, indicating that the solar contribution exceeds the load.

Similarly, the solar contribution of the 1,500 ft² and the 2,500 ft² collectors may be determined by recalculat-

TABLE 12-1. Average Daily Terrestrial Solar Energy Received on a Horizontal Surface (ly/day)

LOCATION	JAN	FEB	MAR	APR	MAY	JUN	JUL	AUG	SEP	OCT	NOV	DEC	ANNUAL
Chicago, Il	96	147	227	331	424	458	473	403	313	207	120	76	273
Lemont, Il	171	232	326	390	497	553	527	486	384	265	157	131	343
Moline, Il	159	220	317	402	493	558	565	498	407	290	176	134	352
Indianapolis, In	147	214	312	393	491	547	542	486	405	293	176	130	345
Ames, Ia	174	253	326	403	480	541	436	460	367	274	187	143	345
Dodge City, Ks	255	316	418	528	568	650	642	592	493	380	285	234	447
Kansas City, Ks	182	251	342	441	522	589	579	525	426	327	215	164	380
Manhattan, Ks	192	264	345	433	527	551	531	526	410	292	227	156	371
Topeka, Ks	192	249	337	430	505	554	552	512	424	320	214	165	371
Lexington, Ky	172	263	357	480	581	628	617	563	494	357	245	175	411
Louisville, Ky	164	231	325	420	515	560	550	498	408	303	190	150	360
Lake Charles, La	239	304	396	483	554	582	521	506	448	402	296	232	414
New Orleans, La	237	296	393	479	539	549	502	491	418	389	269	220	399
Shreveport, La	232	292	384	446	558	557	578	528	414	354	254	205	400
Blue Hill, Ma	153	228	319	389	469	510	502	449	354	266	162	135	328
Boston, Ma	139	198	293	364	472	499	496	425	341	238	145	119	311
Cambridge, Ma	153	235	323	400	420	476	482	464	367	253	164	124	322
East Wareham, Ma	140	218	305	385	452	508	495	436	365	258	163	140	322
Lynn, Ma	118	209	300	394	454	549	528	432	341	241	135	107	317
Annapolis, Md	175	243	340	419	488	557	542	469	383	294	189	155	355
Silver Hill, Md	182	244	340	438	513	555	516	459	397	295	202	163	359
Caribou, Me	133	231	364	400	476	470	508	448	336	212	111	107	316
Portland, Me	157	237	359	406	513	541	561	482	383	273	157	138	351
East Lansing, Mi	121	210	309	359	483	547	540	466	373	255	136	108	311
Sault Ste Marie, Mi	130	225	356	416	523	557	573	472	322	216	105	96	333
St. Cloud, Mn	170	251	366	423	499	541	555	491	360	241	146	123	348
Columbia, Mo	173	251	340	434	530	574	574	522	453	322	225	158	380
Glasgow, Mt	154	258	385	466	568	605	645	531	410	267	154	116	388
Great Falls, Mt	140	232	366	434	528	583	639	532	407	264	154	112	366
Summit, Mt	122	162	268	414	462	493	560	510	354	216	102	76	312
Lincoln, Ne	188	259	350	416	494	544	568	484	396	296	199	159	363
North Omaha, Ne	193	229	365	463	516	546	568	519	410	298	204	170	379
North Platte, Ne	200	266	358	475	523	599	598	540	432	322	220	178	393
Bismarck, N.D.	157	250	356	447	550	590	617	516	390	272	161	124	369
Cape Hatteras, NC	244	317	432	571	635	645	629	557	472	361	284	216	447

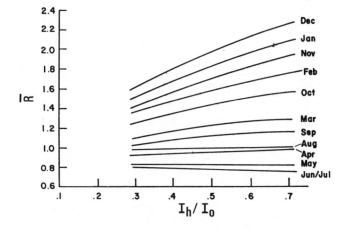

FIG. 12-16. Ratio of solar radiation on a tilted collector to that on a horizontal surface at 40°N Latitude, 55° tilt angle.[12-7]

ing the X and Y factors, as shown in Table 12-5.

Economic Analysis

The economic analysis is based on the following assumptions:

Cost of the solar system, including
storage, controls, pumps, etc. = $20-40/ft^2 of collector
Interest rate = 9%
Term of mortgage = 20 years
Annual cost of system, i = .1095
Cost of natural gas = $3.00/million Btu
adjusted for 80% furnace
efficiency = $3.75/million Btu
Power consumed by system
auxiliaries = 750 kWh
Electric cost = $0.06 kWh
= $17.58/million Btu

TABLE 12-2. Average Daily Radiation Seen by a Tilted Collector on a Monthy Basis

	I_h	I_o	I_h/I_o	$\bar{R}$	$\bar{S}_t$	S_t	T_a
January	121	323	.375	1.65	200	738	27
February	210	451	.465	1.55	326	1203	27
March	309	617	.501	1.22	378	1395	35
April	359	802	.447	0.97	348	1284	48
May	483	931	.518	0.82	396	1461	59
June	547	987	.554	0.78	427	1576	70
July	540	960	.562	0.78	421	1553	74
August	466	851	.548	1.00	466	1720	73
September	373	682	.547	1.15	429	1583	65
October	255	499	.511	1.44	367	1354	54
November	136	354	.384	1.53	208	768	40
December	108	289	.373	1.75	189	697	29

I_h = Average daily terrestrial radiation received on a horizontal surface, Langleys/day (from Table 12-1)

I_o = Average daily extraterrestrial radiation in Langleys (Table 12-3)

$\bar{R}$ = The ratio of average daily terrestrial solar radiation incident on a tilted surface to that on a horizontal surface (tilt angle used here is 55°, Figure 12-16)

$\bar{S}_t$ = Average daily solar radiation in Langleys for the tilted surface = $I_h\bar{R}$

S_t = Average daily solar radiation in Btu/ft² for the tilted surface = $\bar{S}_t \times 3.69$ Btu

T_a = Monthly average daytime temperature, °F.

To estimate the annual savings or penalty (S/P), use a variation of Equation 12-14. Where the fuel is gas

$$S/P = C_f \times fL - (i \times A_cC_c - C_fQ_s)$$
$$= 3.75 \times f \times 256.33 - (.1095 \times A_cC_c + .06 \times 750)$$
$$= 961.23 \times f - (.1095 \times A_cC_c + 45)$$

Where the fuel is electricity

$$S/P = 17.58 \times f \times 256.33 - (.1095 \times A_cC_c + 45)$$
$$= 4506.28 \times f - (.1095 \times A_cC_c + 45)$$

If electric rates should increase 25%, to 0.075/kWh, or gas rates increase six-fold, a solar heating system priced

TABLE 12-3. Average Daily Extraterrestrial Solar Energy Received on a Horizontal Surface (Langley/Day)

°N	Jan	Feb	Mar	Apr	May	Jun	Jul	Aug	Sep	Oct	Nov	Dec
20	636	725	819	896	929	935	929	904	843	750	657	609
25	571	671	785	806	940	956	946	904	818	702	585	541
30	502	613	744	870	945	972	957	896	786	650	529	471
35	432	551	697	848	944	982	962	883	749	593	461	399
40	359	485	646	819	937	987	962	864	706	532	390	325
45	287	417	589	785	925	987	957	839	657	467	319	253
50	215	346	529	745	908	983	947	808	604	399	247	182
55	145	274	464	701	887	976	934	774	546	329	177	114
60	81	203	396	652	864	969	920	735	484	258	110	54

TABLE 12-4. $A_c = 1000$ ft²

1	2	3	4	5	6	7	8	9	10
M	DD	L	D_m	TD	X	Y	f	f×L	L−fL
January	1181	49.60	31	185	1.95	0.32	0.18	8.93	40.67
February	1058	44.44	28	185	1.97	0.53	0.36	16.00	28.44
March	936	39.31	31	177	2.36	0.77	0.51	20.05	19.26
April	522	21.92	30	164	3.79	1.24	0.73	16.00	5.92
May	220	9.24	31	153	8.66	3.45	1.00	9.24	----
October	360	15.12	31	158	5.47	1.95	0.93	14.06	1.06
November	738	31.00	30	172	2.81	0.52	0.30	9.30	21.70
December	1088	45.70	31	183	2.09	0.33	0.19	8.68	37.02

Annual fraction supplied by solar = $\dfrac{\Sigma \text{ column } 9}{\Sigma \text{ column } 3} = \dfrac{102.26}{256.33} = 39.89\%$

Col. 1. M = Month
2. DD = Degree days
3. L = Building load, Btu/month × 10
4. D_m = Number of days in the month
5. TD = $212 - T_a$
6. $X = \dfrac{A_cF'_rU_L(212 - T_a) \times 24 \times D_m}{L}$
$= \dfrac{1000 \times 0.89 \times 0.79(212 - T_a) \times 24 \times D_m}{L}$
$= \dfrac{16874(212 - T_a)D_m}{L}$

7. $Y = \dfrac{A_cF'_r(\tau\alpha)S_t \times D_m}{L}$
$= \dfrac{1000 \times 0.89 = 0.79S_t \times D_m}{L}$
$= \dfrac{703.1 \times S_t \times D_m}{L}$

8. f = Fraction of load supplied by solar system
9. f×L = Solar contribution, Btu × 10⁶
10. L−fL = Auxiliary energy required, Btu × 10⁶

TABLE 12-5.

M	\(A_c = 1500\) ft²					\(A_c = 2500\) ft²				
	X	Y	f	fL	L−fL	X	Y	f	fL	L−fL
October	8.21	2.93	1.00	15.12	----	13.68	4.88	1.00	15.12	----
November	4.22	0.78	0.48	14.88	16.12	7.03	1.30	0.61	18.91	12.09
December	3.14	0.50	0.28	12.80	32.90	5.23	0.83	0.41	18.74	26.96
January	2.93	0.48	0.28	13.89	35.71	4.88	0.80	0.40	19.84	29.76
February	2.96	0.80	0.50	22.22	22.22	4.93	1.33	0.73	32.44	12.00
March	3.54	1.16	0.68	26.73	12.58	5.90	1.93	0.91	35.77	3.54
April	5.69	1.86	0.88	19.29	2.63	9.48	3.10	1.00	21.92	----
May	12.99	5.13	1.00	9.24	----	21.65	8.63	1.00	9.24	----
	Solar fraction = $\frac{134.17}{256.33}$ = 52.34%					Solar fraction = $\frac{171.98}{256.33}$ = 67.09%				

FIRST YEAR DOLLAR SAVINGS/PENALTY WITH SOLAR SYSTEM AND GAS

C_c	$A_c = 1000$ ft² f = .3989	$A_c = 1500$ ft² f = .5234	$A_c = 2500$ ft² f = .6709
$20/ft²	P = −1852	P = −2826	P = −4875
$30/ft²	P = −2947	P = −4469	P = −7613
$40/ft²	P = −4042	P = −6112	P = −10350

FIRST YEAR DOLLAR SAVINGS/PENALTY WITH SOLAR SYSTEM AND ELECTRICITY

$20/ft²	P = −437	P = −971	P = −2497
$30/ft²	P = −1532	P = −2614	P = −5234
$40/ft²	P = −2635	P = −4256	P = −7972

at $20/ft² would become feasible in a northern area like Detroit. However, it might be presently economical to use a solar system to provide process hot water, at temperatures of 120° to 160°F, for year-round applications in factories, restaurants and hospitals.

Example 12-2

A restaurant in Detroit, Michigan, requires approximately 860 gallons/day of hot water at 120°F. The average temperature of the supply water is 50°F. Estimate the first year savings/penalty resulting from using a 500 ft² collector, at $20/ft² to provide part of the hot water load. The performance factors of the collector (F_r, $\tau\alpha$, U_L, etc.) may be taken from the previous example. The conventional fuel is either gas, at $3.00/million Btu or electricity, at .06/kWh.

Solution 12-2

The procedure to be followed is similar to that used in the previous problem. Beckman et al. have suggested the following empirical correction factor for the x-axis:

$$C.F. = \frac{11.6 + 0.655(T_w - 32) + 2.145(T_m - 32) - 1.29(T_a - 32)}{100 - (T_a - 32)/1.8}$$

(12-15)

where

T_w = Hot water temperature
T_m = Supply water temperature
T_a = Ambient temperature

Table 12-6 shows the results of the calculations incorporating the following:

$$X = \frac{8437.2(212 - T_a)D_m}{L}$$

$$C.F. = \frac{11.6 + 0.655(120 - 32) + 2.145(50 - 32) - 1.29(T_a - 32)}{100 - (T_a - 32)/1.8}$$

TABLE 12-6. Water heater solar system, $A_c = 500$ ft²

M	L	C.F.	X	Y	f	fL	L−fL
January	15.5	1.11	3.47	0.52	0.27	4.19	11.31
February	14.0	1.11	3.47	0.85	0.51	7.14	6.86
March	15.5	1.05	3.14	0.98	0.61	9.46	6.04
April	15.0	0.96	2.66	0.90	0.58	8.70	6.30
May	15.5	0.86	2.22	1.03	0.69	10.70	4.80
June	15.0	0.75	1.80	1.11	0.76	11.40	3.60
July	15.5	0.70	1.63	1.09	0.76	11.78	3.72
August	15.5	0.71	1.67	1.21	0.82	12.71	2.79
September	15.0	0.80	1.98	1.11	0.75	11.25	3.75
October	15.5	0.90	2.40	0.95	0.63	9.76	5.74
November	15.0	1.02	2.96	0.54	0.31	4.65	10.35
December	15.5	1.10	3.40	0.49	0.25	4.80	10.70
Total	182.5					106.54	75.96

```
SOLAR COLLECTOR SYSTEM EFFICIENCY EVALUATION PROGRAM-WINTER CYCLE
----------------------------------------------------------------


INPUT-MONTH & WATER TEMPERATURE AT PUMP START-UP
?MAR,100

INPUT - NO. OF PANELS IN SERIES & GPM FLOW RATE PER PANEL
?1,.50

INPUT-FR,UL,TA,G (GALS.STORAGE)
?.92,.90,.80,7500

INPUT COLLECTOR AREA,FLOOR AREA,DAYS IN MONTH
?8000,42000,31

INPUT T AMBIENT,DAY LENGTH,NOON SOLAR RADIATION
?36,12,198

INPUT BLDG. HEAT LOSS,DELTA T,DEGREE-DAYS IN MONTH
?865000,80,927

DAILY LOSS FACTOR= 184.759
LBS. STORAGE PER SQ. FT. OF COLLECTOR (M) = 7.80938
COLLECTION SYSTEM PUMP START TIME= 7.42158 A.M.
MAX STORAGE TANK TEMPERATURE= 151.56
BTU/SQ.FT./DAY OUTPUT= 70.3175
MBTU SOLAR HEAT GAIN= 91.5534
MBTU TOTAL HEAT LOSS= 240.556
FLAT PLATE COLLECTOR EFFICIENCY= 0.27068
CIRCUITING CORRECTION FACTOR = 1.
COLLECTOR SYSTEM EFFICIENCY FOR MONTH= 0.38059


INPUT-MONTH & WATER TEMPERATURE AT PUMP START-UP
?MAR,100

INPUT - NO. OF PANELS IN SERIES & GPM FLOW RATE PER PANEL
?6,.50

INPUT-FR,UL,TA,G (GALS.STORAGE)
?.92,.90,.80,7500

INPUT COLLECTOR AREA,FLOOR AREA,DAYS IN MONTH
?8000,42000,31

INPUT T AMBIENT,DAY LENGTH,NOON SOLAR RADIATION
?36,12,198

INPUT BLDG. HEAT LOSS,DELTA T,DEGREE-DAYS IN MONTH
?865000,80,927

DAILY LOSS FACTOR= 184.759
LBS. STORAGE PER SQ. FT. OF COLLECTOR (M) = 7.80938
COLLECTION SYSTEM PUMP START TIME= 7.42158 A.M.
MAX STORAGE TANK TEMPERATURE= 169.096
BTU/SQ.FT./DAY OUTPUT= 59.6517
MBTU SOLAR HEAT GAIN= 77.6665
MBTU TOTAL HEAT LOSS= 240.556
FLAT PLATE COLLECTOR EFFICIENCY= 0.19215
CIRCUITING CORRECTION FACTOR = 0.84832
COLLECTOR SYSTEM EFFICIENCY FOR MONTH= 0.32158
```

FIG. 12-17. Printout from a typical solar system analysis program.[12-6]

$$= \frac{11.6 + 57.64 + 38.61 - 1.29(T_a - 32)}{100 - (T_a - 32)/1.8}$$

$$= \frac{107.85 - 1.29(T_a - 32)}{100 - (T_a - 32)/1.8}$$

Economic Analysis

From the preceeding example, the first year annual saving/penalty when gas is used

$$S/P = 3.75 \times 106.54 - (.1095 \times 500 \times 20 + 45)$$
$$P = -\$740.45$$

When electricity is the source of energy

$$S/P = 17.58 \times 106.54 - (.1095 \times 500 \times 20 + 45)$$
$$S = \$732.97$$

Thus a solar water heating system, at \$20.00/ft^2 is competitive with electric resistance heat in northern areas such as Detroit, Michigan.

Computer Programs

Detailed computer programs, such as TRNSYS which was developed at the University of Wisconsin, are being used to simulate the performance of solar heating/cooling systems. These programs use hour-by-hour weather data, collector and storage characteristics, control modes, heating/cooling system characteristics, etc., as inputs. The fraction of the load supplied by a solar system is the primary output. Figure 12-17 shows the input to and the output from a typical computer program. Results from the computer programs are more accurate than those obtained using the f-chart method.

SOLAR ASSISTED HEAT PUMPS

Figure 12-4 showed that operating with lower mean fluid temperatures in the collector increases the useful gain or fraction of energy provided by the solar heating system. When solar heating systems are combined with conventional systems, the operating temperature is 110° to 140°F for most collectors. The operating temperature of a collector can be reduced to the 60° to 90° range when it is combined with a heat pump. A typical water source heat pump layout is shown in Figure 12-18. This system is comprised of two main circuits: the solar collector circuit and the heat pump circuit. The liquid storage tank receives energy from the solar collector, through the heat exchanger HE, when the collector temperature is greater than the storage temperature.

The heat pump uses the thermal storage tank as a source of heat energy and transfers it to the supply air through the heating coil. If the temperature of the storage tank drops below 60°F, an auxiliary energy source is activated. In this way, 60-90°F water is used for operating the heat pump at a high COP. At the same time, the solar collector operates at a higher efficiency because of the lower average temperature of the fluid.

The collector/rock bed storage can act as a preheater for the air-source heat pump in Figure 12-19. The outside air and the return air are mixed and preheated either by the collector or by the thermal rock storage depending on which is at a temperature greater than that of the mixed air. This heated, mixed air is a source of heat energy for the heat pump, through the evaporative coil.

The heat pump transfers this energy to the supply air

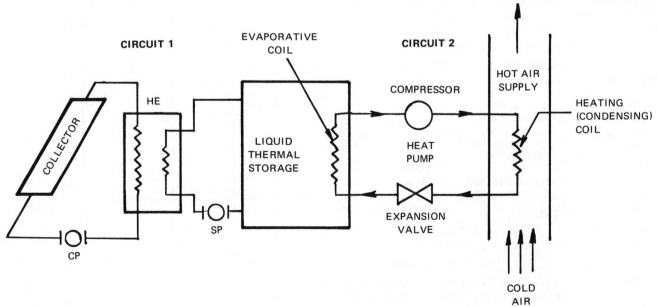

FIG. 12-18. Schematic of a solar system with a water source heat pump.

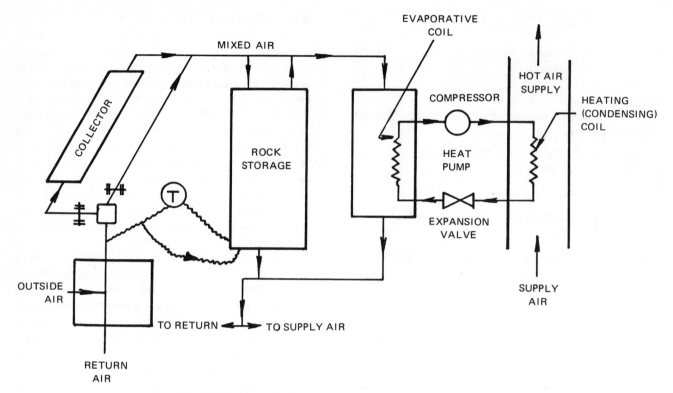

FIG. 12-19. Schematic of an air source solar system with a heat pump.

through the heating coil. The mixed air temperature at the evaporative coil is maintained above 50°F so that the heat pump operates at a high COP. If the mixed air temperature at the evaporative coil exceeds the required supply temperature, the heat pump is deactivated since the solar circuit can meet the space heating load by directly using the heated mixed air.

Here is an example of the improvement in solar system efficiency due to lower operating temperatures or lower ΔT_c ($T_{fluid} - T_{ambient}$).

Example 12-3

The input conditions are for January 21, with the collectors positioned at 40°N latitude and set at a tilt angle of 50 degrees (L + 10°). The outdoor ambient range is from 8°F, at 7:00 a.m., to 28°F, at 5:00 p.m.

A conventional solar system requires a high temperature for the operation of its terminal heating equipment. System 1 will be operated with a starting mean liquid collector temperature of 120°F at 7:00 a.m. and with a final temperature of 150°F at about 5:00 p.m.

Solar System 2, using a heat pump, requires only a minimal water temperature for the operation of its heating equipment. It will have a mean starting temperature of 60° that will rise to a final temperature of 90°F at about 5:00 p.m.

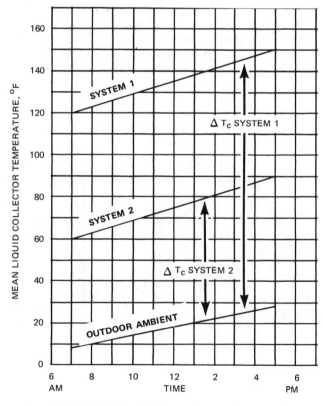

FIG. 12-20. ΔT_c with and without a heat pump.

TABLE 12-7.

Time	ASHRAE* Insolation	System 1			System 2		
		ΔT_c	Collector Efficiency	Collected Insolation	ΔT_c	Collector Efficiency	Collected Insolation
7 a.m.	0	112	0	0	42	0	0
8 a.m.	81	113	0	0	43	25	20
9 a.m.	182	114	18	33	44	45	82
10 a.m.	249	115	29	72	45	49	122
11 a.m.	290	116	34	99	46	50	145
12 noon	303	117	35	106	47	51	155
1 p.m.	290	118	33	96	48	50	145
2 p.m.	249	119	28	70	49	48	120
3 p.m.	182	120	16	29	50	42	76
4 p.m.	81	121	0	0	51	18	15
5 p.m.	0	122	0	0	52	0	0
Totals	1,907 Btu			505 Btu			880 Btu
Mean daily collector efficiency		$\frac{505}{1907} \times 100 = 26.48\%$			$\frac{880}{1907} \times 100 = 46.1\%$		

*January 21 at 40° N Latitude, 50° tilt.

Figure 12-20 illustrates the changing temperature conditions for the two systems and allows the estimation of ΔT_c for each system. The daily operating efficiency of these collector systems can be determined by using the collector performance chart, Figure 12-4, along with the hourly insolation values and hourly ΔT_c.

The results are shown in Table 12-7. Note that the efficiency of solar System 2, with heat pump assist, is 46.1% while the efficiency of System 1 is only 26.48%. Thus, the difference between the outdoor ambient and the mean fluid temperature in the collector is of major importance in solar system design. Increased panel efficiency, established by designing to the lowest possible temperatures in the collector, reduces the amount of collector area. Because solar collectors are a major element in overall cost, system cost is thus reduced. The combination of a heat pump and solar heating system approaches these conditions. The cost effectiveness of this approach has to be carefully evaluated because of the higher cost of the heat pump. Where a heat pump is already in operation, as in a retrofit, this combination may prove attractive.

APPENDIX A

i = 5%

n	$\left(\dfrac{F}{P}\right)$	$\left(\dfrac{P}{F}\right)$	$\left(\dfrac{F}{A}\right)$	$\left(\dfrac{A}{F}\right)$	$\left(\dfrac{P}{A}\right)$	$\left(\dfrac{A}{P}\right)$
1	1.0500	0.9524	1.0000	1.0000	0.9524	1.0500
2	1.1025	0.9070	2.0500	0.4878	1.8594	0.5378
3	1.1576	0.8638	3.1525	0.3172	2.7232	0.3672
4	1.2155	0.8227	4.3101	0.2320	3.5460	0.2820
5	1.2763	0.7835	5.5256	0.1810	4.3295	0.2310
6	1.3401	0.7462	6.8019	0.1470	5.0757	0.1970
7	1.4071	0.7107	8.1420	0.1228	5.7864	0.1728
8	1.4775	0.6768	9.5491	0.1047	6.4632	0.1547
9	1.5513	0.6446	11.0266	0.0907	7.1078	0.1407
10	1.6289	0.6139	12.5779	0.0795	7.7217	0.1295
11	1.7103	0.5847	14.2068	0.0704	8.3064	0.1204
12	1.7959	0.5568	15.9171	0.0628	8.8633	0.1128
13	1.8856	0.5303	17.7130	0.0565	9.3936	0.1065
14	1.9799	0.5051	19.5986	0.0510	9.8986	0.1010
15	2.0789	0.4810	21.5786	0.0463	10.3797	0.0963
16	2.1829	0.4581	23.6575	0.0423	10.8378	0.0923
17	2.2920	0.4363	25.8404	0.0387	11.2741	0.0887
18	2.4066	0.4155	28.1324	0.0355	11.6896	0.0855
19	2.5270	0.3957	30.5390	0.0327	12.0853	0.0827
20	2.6533	0.3769	33.0660	0.0302	12.4622	0.0802
21	2.7860	0.3589	35.7192	0.0280	12.8212	0.0780
22	2.9253	0.3418	38.5052	0.0260	13.1630	0.0760
23	3.0715	0.3256	41.4305	0.0241	13.4886	0.0741
24	3.2251	0.3101	44.5020	0.0225	13.7986	0.0725
25	3.3864	0.2953	47.7271	0.0210	14.0939	0.0710
26	3.5557	0.2812	51.1134	0.0196	14.3752	0.0696
27	3.7335	0.2678	54.6691	0.0183	14.6430	0.0683
28	3.9201	0.2551	58.4026	0.0171	14.8981	0.0671
29	4.1161	0.2429	62.3227	0.0160	15.1411	0.0660
30	4.3219	0.2314	66.4388	0.0151	15.3725	0.0651
31	4.5380	0.2204	70.7608	0.0141	15.5928	0.0641
32	4.7649	0.2099	75.2988	0.0133	15.8027	0.0633
33	5.0032	0.1999	80.0638	0.0125	16.0025	0.0625
34	5.2533	0.1904	85.0669	0.0118	16.1929	0.0618
35	5.5160	0.1813	90.3203	0.0111	16.3742	0.0611
36	5.7918	0.1727	95.8363	0.0104	16.5469	0.0604
37	6.0814	0.1644	101.6281	0.0098	16.7113	0.0598
38	6.3855	0.1566	107.7095	0.0093	16.8679	0.0593
39	6.7048	0.1491	114.0950	0.0088	17.0170	0.0588
40	7.0400	0.1420	120.7998	0.0083	17.1591	0.0583
41	7.3920	0.1353	127.8397	0.0078	17.2944	0.0578
42	7.7616	0.1288	135.2317	0.0074	17.4232	0.0574
43	8.1497	0.1227	142.9933	0.0070	17.5459	0.0570
44	8.5571	0.1169	151.1430	0.0066	17.6628	0.0566
45	8.9850	0.1113	159.7001	0.0063	17.7741	0.0563
46	9.4343	0.1060	168.6851	0.0059	17.8801	0.0559
47	9.9060	0.1009	178.1194	0.0056	17.9810	0.0556
48	10.4013	0.0961	188.0254	0.0053	18.0772	0.0553
49	10.9213	0.0916	198.4266	0.0050	18.1687	0.0550
50	11.4674	0.0872	209.3480	0.0048	18.2559	0.0548
51	12.0408	0.0831	220.8154	0.0045	18.3390	0.0545
52	12.6428	0.0791	232.8561	0.0043	18.4181	0.0543
53	13.2749	0.0753	245.4989	0.0041	18.4934	0.0541
54	13.9387	0.0717	258.7739	0.0039	18.5651	0.0539
55	14.6356	0.0683	272.7126	0.0037	18.6335	0.0537
56	15.3674	0.0651	287.3482	0.0035	18.6985	0.0535
57	16.1358	0.0620	302.7156	0.0033	18.7605	0.0533
58	16.9426	0.0590	318.8514	0.0031	18.8195	0.0531
59	17.7397	0.0562	335.7940	0.0030	18.8758	0.0530
60	18.6792	0.0535	353.5837	0.0028	18.9293	0.0528
61	19.6131	0.0510	372.2628	0.0027	18.9803	0.0527
62	20.5938	0.0486	391.8760	0.0026	19.0288	0.0526
63	21.6235	0.0462	412.4698	0.0024	19.0751	0.0524
64	22.7047	0.0440	434.0933	0.0023	19.1191	0.0523
65	23.8399	0.0419	456.7979	0.0022	19.1611	0.0522
66	25.0319	0.0399	480.6378	0.0021	19.2010	0.0521
67	26.2835	0.0380	505.6697	0.0020	19.2391	0.0520
68	27.5977	0.0362	531.9532	0.0019	19.2753	0.0519

i = 6%

n	$\left(\frac{F}{P}\right)$	$\left(\frac{P}{F}\right)$	$\left(\frac{F}{A}\right)$	$\left(\frac{A}{F}\right)$	$\left(\frac{P}{A}\right)$	$\left(\frac{A}{P}\right)$
1	1.0600	0.9434	1.0000	1.0000	0.9434	1.0600
2	1.1236	0.8900	2.0600	0.4854	1.8334	0.5454
3	1.1910	0.8396	3.1836	0.3141	2.6730	0.3741
4	1.2625	0.7921	4.3746	0.2286	3.4651	0.2886
5	1.3382	0.7473	5.6371	0.1774	4.2124	0.2374
6	1.4185	0.7050	6.9753	0.1434	4.9173	0.2034
7	1.5036	0.6651	8.3938	0.1191	5.5824	0.1791
8	1.5938	0.6274	9.8975	0.1010	6.2098	0.1610
9	1.6895	0.5919	11.4913	0.0870	6.8017	0.1470
10	1.7908	0.5584	13.1808	0.0759	7.3601	0.1359
11	1.8983	0.5268	14.9716	0.0668	7.8869	0.1268
12	2.0122	0.4970	16.8699	0.0593	8.3838	0.1193
13	2.1329	0.4688	18.8821	0.0530	8.8527	0.1130
14	2.2609	0.4423	21.0151	0.0476	9.2950	0.1076
15	2.3966	0.4173	23.2760	0.0430	9.7122	0.1030
16	2.5404	0.3936	25.6725	0.0390	10.1059	0.0990
17	2.6928	0.3714	28.2129	0.0354	10.4773	0.0954
18	2.8543	0.3503	30.9057	0.0324	10.8276	0.0924
19	3.0256	0.3305	33.7600	0.0296	11.1581	0.0896
20	3.2071	0.3118	36.7856	0.0272	11.4699	0.0872
21	3.3996	0.2942	39.9927	0.0250	11.7641	0.0850
22	3.6035	0.2775	43.3923	0.0230	12.0416	0.0830
23	3.8197	0.2618	46.9958	0.0213	12.3034	0.0813
24	4.0489	0.2470	50.8156	0.0197	12.5504	0.0797
25	4.2919	0.2330	54.8645	0.0182	12.7834	0.0782
26	4.5494	0.2198	59.1564	0.0169	13.0032	0.0769
27	4.8223	0.2074	63.7058	0.0157	13.2105	0.0757
28	5.1117	0.1956	68.5281	0.0146	13.4062	0.0746
29	5.4184	0.1846	73.6398	0.0136	13.5907	0.0736
30	5.7435	0.1741	79.0582	0.0126	13.7648	0.0726
31	6.0881	0.1643	84.8017	0.0118	13.9291	0.0718
32	6.4534	0.1550	90.8898	0.0110	14.0840	0.0710
33	6.8406	0.1462	97.3432	0.0103	14.2302	0.0703
34	7.2510	0.1379	104.1838	0.0096	14.3681	0.0696
35	7.6861	0.1301	111.4348	0.0090	14.4982	0.0690
36	8.1473	0.1227	119.1209	0.0084	14.6210	0.0684
37	8.6361	0.1158	127.2681	0.0079	14.7368	0.0679
38	9.1543	0.1092	135.9042	0.0074	14.8460	0.0674
39	9.7035	0.1031	145.0585	0.0069	14.9491	0.0669
40	10.2857	0.0972	154.7620	0.0065	15.0463	0.0665
41	10.9029	0.0917	165.0477	0.0061	15.1380	0.0661
42	11.5570	0.0865	175.9506	0.0057	15.2245	0.0657
43	12.2505	0.0816	187.5076	0.0053	15.3062	0.0653
44	12.9855	0.0770	199.7581	0.0050	15.3832	0.0650
45	13.7646	0.0727	212.7435	0.0047	15.4558	0.0647
46	14.5905	0.0685	226.5082	0.0044	15.5244	0.0644
47	15.4659	0.0647	241.0986	0.0041	15.5890	0.0641
48	16.3939	0.0610	256.5646	0.0039	15.6500	0.0639
49	17.3775	0.0575	272.9584	0.0037	15.7076	0.0637
50	18.4202	0.0543	290.3359	0.0034	15.7619	0.0634
51	19.5254	0.0512	308.7561	0.0032	15.8131	0.0632
52	20.6969	0.0483	328.2815	0.0030	15.8614	0.0630
53	21.9387	0.0456	348.9784	0.0029	15.9070	0.0629
54	23.2550	0.0430	370.9171	0.0027	15.9500	0.0627
55	24.6503	0.0406	394.1721	0.0025	15.9905	0.0625
56	26.1293	0.0383	418.8224	0.0024	16.0288	0.0624
57	27.6971	0.0361	444.9517	0.0022	16.0649	0.0622
58	29.3589	0.0341	472.6489	0.0021	16.0990	0.0621
59	31.1205	0.0321	502.0078	0.0020	16.1311	0.0620
60	32.9877	0.0303	533.1283	0.0019	16.1614	0.0619
61	34.9670	0.0286	566.1160	0.0018	16.1900	0.0618
62	37.0650	0.0270	601.0829	0.0017	16.2170	0.0617
63	39.2889	0.0255	638.1479	0.0016	16.2425	0.0616
64	41.6462	0.0240	677.4368	0.0015	16.2665	0.0615
65	44.1450	0.0227	719.0830	0.0014	16.2891	0.0614
66	46.7937	0.0214	763.2280	0.0013	16.3105	0.0613
67	49.6013	0.0202	810.0216	0.0012	16.3307	0.0612
68	52.5774	0.0190	859.6229	0.0012	16.3497	0.0612

i = 7%

n	$\left(\dfrac{F}{P}\right)$	$\left(\dfrac{P}{F}\right)$	$\left(\dfrac{F}{A}\right)$	$\left(\dfrac{A}{F}\right)$	$\left(\dfrac{P}{A}\right)$	$\left(\dfrac{A}{P}\right)$
1	1.0700	0.9346	1.0000	1.0000	0.9346	1.0700
2	1.1449	0.8734	2.0700	0.4831	1.8080	0.5531
3	1.2250	0.8163	3.2149	0.3111	2.6243	0.3811
4	1.3108	0.7629	4.4399	0.2252	3.3872	0.2952
5	1.4026	0.7130	5.7507	0.1739	4.1002	0.2439
6	1.5007	0.6663	7.1533	0.1398	4.7665	0.2098
7	1.6058	0.6227	8.6540	0.1156	5.3893	0.1856
8	1.7182	0.5820	10.2598	0.0975	5.9713	0.1675
9	1.8385	0.5439	11.9780	0.0835	6.5152	0.1535
10	1.9672	0.5083	13.8164	0.0724	7.0236	0.1424
11	2.1049	0.4751	15.7836	0.0634	7.4987	0.1334
12	2.2522	0.4440	17.8885	0.0559	7.9427	0.1259
13	2.4098	0.4150	20.1406	0.0497	8.3577	0.1197
14	2.5785	0.3878	22.5505	0.0443	8.7455	0.1143
15	2.7590	0.3624	25.1290	0.0398	9.1079	0.1098
16	2.9522	0.3387	27.8881	0.0359	9.4466	0.1059
17	3.1588	0.3166	30.8402	0.0324	9.7632	0.1024
18	3.3799	0.2959	33.9990	0.0294	10.0591	0.0994
19	3.6165	0.2765	37.3790	0.0268	10.3356	0.0968
20	3.8697	0.2584	40.9955	0.0244	10.5940	0.0944
21	4.1406	0.2415	44.8652	0.0223	10.8355	0.0923
22	4.4304	0.2257	49.0057	0.0204	11.0612	0.0904
23	4.7405	0.2109	53.4362	0.0187	11.2722	0.0887
24	5.0724	0.1971	58.1767	0.0172	11.4693	0.0872
25	5.4274	0.1842	63.2491	0.0158	11.6536	0.0858
26	5.8074	0.1722	68.6765	0.0146	11.8258	0.0846
27	6.2139	0.1609	74.4838	0.0134	11.9867	0.0834
28	6.6488	0.1504	80.6977	0.0124	12.1371	0.0824
29	7.1143	0.1406	87.3465	0.0114	12.2777	0.0814
30	7.6123	0.1314	94.4608	0.0106	12.4090	0.0806
31	8.1451	0.1228	102.0731	0.0098	12.5318	0.0798
32	8.7153	0.1147	110.2182	0.0091	12.6466	0.0791
33	9.3253	0.1072	118.9335	0.0084	12.7538	0.0784
34	9.9781	0.1002	128.2588	0.0078	12.8540	0.0778
35	10.6766	0.0937	138.2369	0.0072	12.9477	0.0772
36	11.4239	0.0875	148.9135	0.0067	13.0352	0.0767
37	12.2236	0.0818	160.3374	0.0062	13.1170	0.0762
38	13.0793	0.0765	172.5611	0.0058	13.1935	0.0758
39	13.9948	0.0715	185.6403	0.0054	13.2649	0.0754
40	14.9745	0.0668	199.6352	0.0050	13.3317	0.0750
41	16.0227	0.0624	214.6096	0.0047	13.3941	0.0747
42	17.1443	0.0583	230.6323	0.0043	13.4524	0.0743
43	18.3444	0.0545	247.7766	0.0040	13.5070	0.0740
44	19.6285	0.0509	266.1209	0.0038	13.5579	0.0738
45	21.0025	0.0476	285.7494	0.0035	13.6055	0.0735
46	22.4726	0.0445	306.7519	0.0033	13.6500	0.0733
47	24.0457	0.0416	329.2245	0.0030	13.6916	0.0730
48	25.7289	0.0339	353.2702	0.0028	13.7305	0.0728
49	27.5299	0.0363	378.9991	0.0026	13.7668	0.0726
50	29.4570	0.0339	406.5291	0.0025	13.8007	0.0725
51	31.5190	0.0317	435.9861	0.0023	13.8325	0.0723
52	33.7254	0.0297	467.5051	0.0021	13.8621	0.0721
53	36.0861	0.0277	501.2305	0.0020	13.8898	0.0720
54	38.6122	0.0259	537.3167	0.0019	13.9157	0.0719
55	41.3150	0.0242	575.9288	0.0017	13.9399	0.0717
56	44.2071	0.0226	617.2439	0.0016	13.9626	0.0716
57	47.3016	0.0211	661.4509	0.0015	13.9837	0.0715
58	50.6127	0.0198	708.7525	0.0014	14.0035	0.0714
59	54.1556	0.0185	759.3652	0.0013	14.0219	0.0713
60	57.9465	0.0173	813.5207	0.0012	14.0392	0.0712
61	62.0027	0.0161	871.4672	0.0011	14.0553	0.0711
62	66.3429	0.0151	933.4699	0.0011	14.0704	0.0711
63	70.9869	0.0141	999.8128	0.0010	14.0845	0.0710
64	75.9560	0.0132	1070.7997	0.0009	14.0976	0.0709
65	81.2729	0.0123	1146.7557	0.0009	14.1099	0.0709
66	86.9620	0.0115	1228.0286	0.0008	14.1214	0.0708
67	93.0493	0.0107	1314.9906	0.0008	14.1322	0.0708
68	99.5628	0.0100	1408.0400	0.0007	14.1422	0.0707

i = 8%

n	$\left(\dfrac{F}{P}\right)$	$\left(\dfrac{P}{F}\right)$	$\left(\dfrac{F}{A}\right)$	$\left(\dfrac{A}{F}\right)$	$\left(\dfrac{P}{A}\right)$	$\left(\dfrac{A}{P}\right)$
1	1.0800	0.9259	1.0000	1.0000	0.9259	1.0800
2	1.1664	0.8573	2.0800	0.4808	1.7833	0.5608
3	1.2597	0.7938	3.2464	0.3080	2.5771	0.3880
4	1.3605	0.7350	4.5061	0.2219	3.3121	0.3019
5	1.4693	0.6806	5.8666	0.1705	3.9927	0.2505
6	1.5869	0.6302	7.3359	0.1363	4.6229	0.2163
7	1.7138	0.5835	8.9228	0.1121	5.2064	0.1921
8	1.8509	0.5403	10.6366	0.0940	5.7466	0.1740
9	1.9990	0.5002	12.4876	0.0801	6.2469	0.1601
10	2.1589	0.4632	14.4866	0.0690	6.7101	0.1490
11	2.3316	0.4289	16.6455	0.0601	7.1390	0.1401
12	2.5182	0.3971	18.9771	0.0527	7.5361	0.1327
13	2.7196	0.3677	21.4953	0.0465	7.9038	0.1265
14	2.9372	0.3405	24.2149	0.0413	8.2442	0.1213
15	3.1722	0.3152	27.1521	0.0368	8.5595	0.1168
16	3.4259	0.2919	30.3243	0.0330	8.8514	0.1130
17	3.7000	0.2703	33.7502	0.0296	9.1216	0.1096
18	3.9960	0.2502	37.4502	0.0267	9.3719	0.1067
19	4.3157	0.2317	41.4463	0.0241	9.6036	0.1041
20	4.6610	0.2145	45.7620	0.0219	9.8181	0.1019
21	5.0338	0.1987	50.4229	0.0198	10.0168	0.0998
22	5.4365	0.1839	55.4568	0.0180	10.2007	0.0980
23	5.8715	0.1703	60.8933	0.0164	10.3711	0.0964
24	6.3412	0.1577	66.7648	0.0150	10.5288	0.0950
25	6.8485	0.1460	73.1059	0.0137	10.6748	0.0937
26	7.3964	0.1352	79.9544	0.0125	10.8100	0.0925
27	7.9881	0.1252	87.3508	0.0114	10.9352	0.0914
28	8.6271	0.1159	95.3388	0.0105	11.0511	0.0905
29	9.3173	0.1073	103.9659	0.0096	11.1584	0.0896
30	10.0627	0.0994	113.2832	0.0088	11.2578	0.0888
31	10.8677	0.0920	123.3459	0.0081	11.3498	0.0881
32	11.7371	0.0852	134.2135	0.0075	11.4350	0.0875
33	12.6760	0.0789	145.9506	0.0069	11.5139	0.0869
34	13.6901	0.0730	158.6267	0.0063	11.5869	0.0863
35	14.7853	0.0676	172.3168	0.0058	11.6546	0.0858
36	15.9682	0.0626	187.1021	0.0053	11.7172	0.0853
37	17.2456	0.0580	203.0703	0.0049	11.7752	0.0849
38	18.6253	0.0537	220.3159	0.0045	11.8289	0.0845
39	20.1153	0.0497	238.9412	0.0042	11.8786	0.0842
40	21.7245	0.0460	259.0565	0.0039	11.9246	0.0839
41	23.4625	0.0426	280.7810	0.0036	11.9672	0.0836
42	25.3395	0.0395	304.2435	0.0033	12.0067	0.0833
43	27.3666	0.0365	329.5830	0.0030	12.0432	0.0830
44	29.5560	0.0338	356.9496	0.0028	12.0771	0.0828
45	31.9204	0.0313	386.5056	0.0026	12.1084	0.0826
46	34.4741	0.0290	418.4260	0.0024	12.1374	0.0824
47	37.2320	0.0269	452.9001	0.0022	12.1643	0.0822
48	40.2106	0.0249	490.1321	0.0020	12.1891	0.0820
49	43.4274	0.0230	530.3427	0.0019	12.2122	0.0819
50	46.9016	0.0213	573.7701	0.0017	12.2335	0.0817
51	50.6537	0.0197	620.6717	0.0016	12.2532	0.0816
52	54.7060	0.0183	671.3255	0.0015	12.2715	0.0815
53	59.0825	0.0169	726.0315	0.0014	12.2884	0.0814
54	63.8091	0.0157	785.1140	0.0013	12.3041	0.0813
55	68.9138	0.0145	848.9231	0.0012	12.3186	0.0812
56	74.4270	0.0134	917.8370	0.0011	12.3321	0.0811
57	80.3811	0.0124	992.2639	0.0010	12.3445	0.0810
58	86.8116	0.0115	1072.6450	0.0009	12.3560	0.0809
59	93.7565	0.0107	1159.4566	0.0009	12.3667	0.0809
60	101.2571	0.0099	1253.2132	0.0008	12.3766	0.0808
61	109.3576	0.0091	1354.4702	0.0007	12.3857	0.0807
62	118.1062	0.0085	1463.8278	0.0007	12.3942	0.0807
63	127.5547	0.0078	1581.9341	0.0006	12.4020	0.0806
64	137.7591	0.0073	1709.4888	0.0006	12.4093	0.0806
65	148.7798	0.0067	1847.2479	0.0005	12.4160	0.0805
66	160.6822	0.0062	1996.0277	0.0005	12.4222	0.0805
67	173.5368	0.0058	2156.7099	0.0005	12.4280	0.0805
68	187.4197	0.0053	2330.2467	0.0004	12.4333	0.0804

i = 9%

n	$\left(\frac{F}{P}\right)$	$\left(\frac{P}{F}\right)$	$\left(\frac{F}{A}\right)$	$\left(\frac{A}{F}\right)$	$\left(\frac{P}{A}\right)$	$\left(\frac{A}{P}\right)$
1	1.0900	0.9174	1.0000	1.0000	0.9174	1.0900
2	1.1881	0.8417	2.0900	0.4785	1.7591	0.5685
3	1.2950	0.7722	3.2781	0.3051	2.5313	0.3951
4	1.4116	0.7084	4.5731	0.2187	3.2397	0.3087
5	1.5386	0.6499	5.9847	0.1671	3.8897	0.2571
6	1.6771	0.5963	7.5233	0.1329	4.4859	0.2229
7	1.8280	0.5470	9.2004	0.1087	5.0330	0.1987
8	1.9926	0.5019	11.0285	0.0907	5.5348	0.1807
9	2.1719	0.4604	13.0210	0.0768	5.9952	0.1668
10	2.3674	0.4224	15.1929	0.0658	6.4177	0.1558
11	2.5804	0.3875	17.5603	0.0569	6.8052	0.1469
12	2.8127	0.3555	20.1407	0.0497	7.1607	0.1397
13	3.0658	0.3262	22.9534	0.0436	7.4869	0.1336
14	3.3417	0.2992	26.0192	0.0384	7.7862	0.1284
15	3.6425	0.2745	29.3609	0.0341	8.0607	0.1241
16	3.9703	0.2519	33.0034	0.0303	8.3126	0.1203
17	4.3276	0.2311	36.9737	0.0270	8.5436	0.1170
18	4.7171	0.2120	41.3013	0.0242	8.7556	0.1142
19	5.1417	0.1945	46.0185	0.0217	8.9501	0.1117
20	5.6044	0.1784	51.1601	0.0195	9.1285	0.1095
21	6.1088	0.1637	56.7645	0.0176	9.2922	0.1076
22	6.6586	0.1502	62.8733	0.0159	9.4424	0.1059
23	7.2579	0.1378	69.5319	0.0144	9.5802	0.1044
24	7.9111	0.1264	76.7898	0.0130	9.7066	0.1030
25	8.6231	0.1160	84.7009	0.0118	9.8226	0.1018
26	9.3992	0.1064	93.3240	0.0107	9.9290	0.1007
27	10.2451	0.0976	102.7231	0.0097	10.0266	0.0997
28	11.1671	0.0895	112.9682	0.0089	10.1161	0.0989
29	12.1722	0.0822	124.1354	0.0081	10.1983	0.0981
30	13.2677	0.0754	136.3076	0.0073	10.2737	0.0973
31	14.4618	0.0691	149.5752	0.0067	10.3428	0.0967
32	15.7633	0.0634	164.0370	0.0061	10.4062	0.0961
33	17.1820	0.0582	179.8003	0.0056	10.4644	0.0956
34	18.7284	0.0534	196.9824	0.0051	10.5178	0.0951
35	20.4140	0.0490	215.7108	0.0046	10.5668	0.0946
36	22.2512	0.0449	236.1248	0.0042	10.6118	0.0942
37	24.2538	0.0412	258.3760	0.0039	10.6530	0.0939
38	26.4367	0.0378	282.6298	0.0035	10.6908	0.0935
39	28.8160	0.0347	309.0665	0.0032	10.7255	0.0932
40	31.4094	0.0318	337.8825	0.0030	10.7574	0.0930
41	34.2363	0.0292	369.2919	0.0027	10.7866	0.0927
42	37.3175	0.0268	403.5282	0.0025	10.8134	0.0925
43	40.6761	0.0246	440.8457	0.0023	10.8380	0.0923
44	44.3370	0.0226	481.5218	0.0021	10.8605	0.0921
45	48.3273	0.0207	525.8588	0.0019	10.8812	0.0919
46	52.6767	0.0190	574.1861	0.0017	10.9002	0.0917
47	57.4177	0.0174	626.8629	0.0016	10.9176	0.0916
48	62.5852	0.0160	684.2805	0.0015	10.9336	0.0915
49	68.2179	0.0147	746.8658	0.0013	10.9482	0.0913
50	74.3575	0.0134	815.0837	0.0012	10.9617	0.0912
51	81.0497	0.0123	889.4412	0.0011	10.9740	0.0911
52	88.3442	0.0113	970.4909	0.0010	10.9853	0.0910
53	96.2952	0.0104	1058.8351	0.0009	10.9957	0.0909
54	104.9617	0.0095	1155.1303	0.0009	11.0053	0.0909
55	114.4083	0.0087	1260.0920	0.0008	11.0140	0.0908
56	124.7050	0.0080	1374.5003	0.0007	11.0220	0.0907
57	135.9285	0.0074	1499.2054	0.0007	11.0294	0.0907
58	148.1620	0.0067	1635.1338	0.0006	11.0361	0.0906
59	161.4966	0.0062	1783.2959	0.0006	11.0423	0.0906
60	176.0313	0.0057	1944.7925	0.0005	11.0480	0.0905
61	191.8741	0.0052	2120.8239	0.0005	11.0532	0.0905
62	209.1428	0.0048	2312.6980	0.0004	11.0580	0.0904
63	227.9657	0.0044	2521.8409	0.0004	11.0624	0.0904
64	248.4826	0.0040	2749.8065	0.0004	11.0664	0.0904
65	270.8460	0.0037	2998.2891	0.0003	11.0701	0.0903
66	295.2222	0.0034	3269.1352	0.0003	11.0735	0.0903
67	321.7922	0.0031	3564.3573	0.0003	11.0766	0.0903
68	350.7535	0.0029	3886.1495	0.0003	11.0794	0.0903

i = 10%

n	$\left(\dfrac{F}{P}\right)$	$\left(\dfrac{P}{F}\right)$	$\left(\dfrac{F}{A}\right)$	$\left(\dfrac{A}{F}\right)$	$\left(\dfrac{P}{A}\right)$	$\left(\dfrac{A}{P}\right)$
1	1.1000	0.9091	1.0000	1.0000	0.9091	1.1000
2	1.2100	0.8264	2.1000	0.4762	1.7355	0.5762
3	1.3310	0.7513	3.3100	0.3021	2.4869	0.4021
4	1.4641	0.6830	4.6410	0.2155	3.1699	0.3155
5	1.6105	0.6209	6.1051	0.1638	3.7908	0.2638
6	1.7716	0.5645	7.7156	0.1296	4.3553	0.2296
7	1.9487	0.5132	9.4872	0.1054	4.8684	0.2054
8	2.1436	0.4665	11.4359	0.0874	5.3349	0.1874
9	2.3579	0.4241	13.5795	0.0736	5.7590	0.1736
10	2.5937	0.3855	15.9374	0.0627	6.1446	0.1627
11	2.8531	0.3505	18.5312	0.0540	6.4951	0.1540
12	3.1384	0.3186	21.3843	0.0468	6.8137	0.1468
13	3.4523	0.2897	24.5227	0.0408	7.1034	0.1408
14	3.7975	0.2633	27.9750	0.0357	7.3667	0.1357
15	4.1772	0.2394	31.7725	0.0315	7.6061	0.1315
16	4.5950	0.2176	35.9497	0.0278	7.8237	0.1278
17	5.0545	0.1978	40.5447	0.0247	8.0216	0.1247
18	5.5599	0.1799	45.5992	0.0219	8.2014	0.1219
19	6.1159	0.1635	51.1591	0.0195	8.3649	0.1195
20	6.7275	0.1486	57.2750	0.0175	8.5136	0.1175
21	7.4002	0.1351	64.0025	0.0156	8.6487	0.1156
22	8.1403	0.1228	71.4027	0.0140	8.7715	0.1140
23	8.9543	0.1117	79.5430	0.0126	8.8832	0.1126
24	9.8497	0.1015	88.4973	0.0113	8.9847	0.1113
25	10.8347	0.0923	98.3470	0.0102	9.0770	0.1102
26	11.9182	0.0839	109.1817	0.0092	9.1609	0.1092
27	13.1100	0.0763	121.0999	0.0083	9.2372	0.1083
28	14.4210	0.0693	134.2099	0.0075	9.3066	0.1075
29	15.8631	0.0630	148.6309	0.0067	9.3696	0.1067
30	17.4494	0.0573	164.4940	0.0061	9.4269	0.1061
31	19.1943	0.0521	181.9434	0.0055	9.4790	0.1055
32	21.1138	0.0474	201.1377	0.0050	9.5264	0.1050
33	23.2252	0.0431	222.2515	0.0045	9.5694	0.1045
34	25.5477	0.0391	245.4767	0.0041	9.6086	0.1041
35	28.1024	0.0356	271.0243	0.0037	9.6442	0.1037
36	30.9127	0.0323	299.1267	0.0033	9.6765	0.1033
37	34.0039	0.0294	330.7394	0.0030	9.7059	0.1030
38	37.4043	0.0267	364.0434	0.0027	9.7327	0.1027
39	41.1448	0.0243	401.4477	0.0025	9.7570	0.1025
40	45.2592	0.0221	442.5925	0.0023	9.7791	0.1023
41	49.7852	0.0201	487.8517	0.0020	9.7991	0.1020
42	54.7637	0.0183	537.6369	0.0019	9.8174	0.1019
43	60.2401	0.0166	592.4006	0.0017	9.8340	0.1017
44	66.2641	0.0151	652.6406	0.0015	9.8491	0.1015
45	72.8905	0.0137	718.9047	0.0014	9.8628	0.1014
46	80.1795	0.0125	791.7951	0.0013	9.8753	0.1013
47	88.1975	0.0113	871.9746	0.0011	9.8866	0.1011
48	97.0172	0.0103	960.1721	0.0010	9.8969	0.1010
49	106.7189	0.0094	1057.1893	0.0009	9.9063	0.1009
50	117.3908	0.0085	1163.9082	0.0009	9.9148	0.1009
51	129.1299	0.0077	1281.2990	0.0008	9.9226	0.1008
52	142.0429	0.0070	1410.4289	0.0007	9.9296	0.1007
53	156.2472	0.0064	1552.4718	0.0006	9.9360	0.1006
54	171.8719	0.0058	1708.7190	0.0006	9.9418	0.1006
55	189.0591	0.0053	1880.5909	0.0005	9.9471	0.1005
56	207.9650	0.0048	2069.6499	0.0005	9.9519	0.1005
57	223.7615	0.0044	2277.6149	0.0004	9.9563	0.1004
58	251.6376	0.0040	2506.3764	0.0004	9.9603	0.1004
59	276.3014	0.0036	2758.0140	0.0004	9.9639	0.1004
60	304.4815	0.0033	3034.8154	0.0003	9.9672	0.1003
61	334.9297	0.0030	3339.2969	0.0003	9.9701	0.1003
62	368.4227	0.0027	3674.2266	0.0003	9.9729	0.1003
63	405.2649	0.0025	4042.6493	0.0002	9.9753	0.1002
64	445.7914	0.0022	4447.9142	0.0002	9.9776	0.1002
65	490.3706	0.0020	4893.7056	0.0002	9.9796	0.1002
66	539.4076	0.0019	5384.0760	0.0002	9.9815	0.1002
67	593.3484	0.0017	5923.4836	0.0002	9.9831	0.1002
68	652.6832	0.0015	6516.8320	0.0002	9.9847	0.1002

i = 11%

n	$\left(\frac{F}{P}\right)$	$\left(\frac{P}{F}\right)$	$\left(\frac{F}{A}\right)$	$\left(\frac{A}{F}\right)$	$\left(\frac{P}{A}\right)$	$\left(\frac{A}{P}\right)$
1	1.1100	0.9009	1.0000	1.0000	0.9009	1.1100
2	1.2321	0.8116	2.1100	0.4739	1.7125	0.5839
3	1.3676	0.7312	3.3421	0.2992	2.4437	0.4092
4	1.5181	0.6587	4.7097	0.2123	3.1024	0.3223
5	1.6851	0.5935	6.2278	0.1606	3.6959	0.2706
6	1.8704	0.5346	7.9129	0.1264	4.2305	0.2364
7	2.0762	0.4817	9.7833	0.1022	4.7122	0.2122
8	2.3045	0.4339	11.8594	0.0843	5.1461	0.1943
9	2.5580	0.3909	14.1640	0.0706	5.5370	0.1806
10	2.8394	0.3522	16.7220	0.0598	5.8892	0.1698
11	3.1518	0.3173	19.5614	0.0511	6.2065	0.1611
12	3.4985	0.2858	22.7132	0.0440	6.4924	0.1540
13	3.8833	0.2575	26.2116	0.0382	6.7499	0.1482
14	4.3104	0.2320	30.0949	0.0332	6.9819	0.1432
15	4.7846	0.2090	34.4054	0.0291	7.1909	0.1391
16	5.3109	0.1883	39.1899	0.0255	7.3792	0.1355
17	5.8951	0.1696	44.5008	0.0225	7.5488	0.1325
18	6.5436	0.1528	50.3959	0.0198	7.7016	0.1298
19	7.2633	0.1377	56.9395	0.0176	7.8393	0.1276
20	8.0623	0.1240	64.2028	0.0156	7.9633	0.1256
21	8.9492	0.1117	72.2651	0.0138	8.0751	0.1238
22	9.9336	0.1007	81.2143	0.0123	8.1757	0.1223
23	11.0263	0.0907	91.1479	0.0110	8.2664	0.1210
24	12.2392	0.0817	102.1741	0.0098	8.3481	0.1198
25	13.5855	0.0736	114.4133	0.0087	8.4217	0.1187
26	15.0799	0.0663	127.9988	0.0078	8.4881	0.1178
27	16.7386	0.0597	143.0786	0.0070	8.5478	0.1170
28	18.5799	0.0538	159.8173	0.0063	8.6016	0.1163
29	20.6237	0.0485	178.3972	0.0056	8.6501	0.1156
30	22.8923	0.0437	199.0209	0.0050	8.6938	0.1150
31	25.4104	0.0394	221.9132	0.0045	8.7331	0.1145
32	28.2056	0.0355	247.3236	0.0040	8.7686	0.1140
33	31.3082	0.0319	275.5292	0.0036	8.8005	0.1136
34	34.7521	0.0288	306.8374	0.0033	8.8293	0.1133
35	38.5749	0.0259	341.5895	0.0029	8.8552	0.1129
36	42.8181	0.0234	380.1644	0.0026	8.8786	0.1126
37	47.5281	0.0210	422.9825	0.0024	8.8996	0.1124
38	52.7562	0.0190	470.5106	0.0021	8.9186	0.1121
39	58.5593	0.0171	523.2667	0.0019	8.9357	0.1119
40	65.0009	0.0154	581.8260	0.0017	8.9511	0.1117
41	72.1510	0.0139	646.8269	0.0015	8.9649	0.1115
42	80.0876	0.0125	718.9779	0.0014	8.9774	0.1114
43	88.8972	0.0112	799.0654	0.0013	8.9886	0.1113
44	98.6759	0.0101	887.9626	0.0011	8.9988	0.1111
45	109.5302	0.0091	986.6385	0.0010	9.0079	0.1110
46	121.5786	0.0082	1096.1688	0.0009	9.0161	0.1109
47	134.9522	0.0074	1217.7473	0.0008	9.0235	0.1108
48	149.7969	0.0067	1352.6995	0.0007	9.0302	0.1107
49	166.2746	0.0060	1502.4965	0.0007	9.0362	0.1107
50	184.5648	0.0054	1668.7711	0.0006	9.0417	0.1106
51	204.8670	0.0049	1853.3359	0.0005	9.0465	0.1105
52	227.4023	0.0044	2058.2029	0.0005	9.0509	0.1105
53	252.4166	0.0040	2285.6052	0.0004	9.0549	0.1104
54	280.1824	0.0036	2538.0218	0.0004	9.0585	0.1104
55	311.0025	0.0032	2818.2042	0.0004	9.0617	0.1104
56	345.2127	0.0029	3129.2066	0.0003	9.0646	0.1103
57	383.1861	0.0026	3474.4193	0.0003	9.0672	0.1103
58	425.3366	0.0024	3857.6054	0.0003	9.0695	0.1103
59	472.1236	0.0021	4282.9421	0.0002	9.0717	0.1102
60	524.0572	0.0019	4755.0657	0.0002	9.0736	0.1102
61	581.7035	0.0017	5279.1229	0.0002	9.0753	0.1102
62	645.6909	0.0015	5860.8264	0.0002	9.0768	0.1102
63	716.7169	0.0014	6506.5173	0.0002	9.0782	0.1102
64	795.5558	0.0013	7223.2342	0.0001	9.0795	0.1101
65	883.0669	0.0011	8018.7900	0.0001	9.0806	0.1101
66	980.2043	0.0010	8901.8568	0.0001	9.0816	0.1101
67	1088.0267	0.0009	9882.0612	0.0001	9.0826	0.1101
68	1207.7097	0.0008	10970.0878	0.0001	9.0834	0.1101

i = 12%

n	$\left(\dfrac{F}{P}\right)$	$\left(\dfrac{P}{F}\right)$	$\left(\dfrac{F}{A}\right)$	$\left(\dfrac{A}{F}\right)$	$\left(\dfrac{P}{A}\right)$	$\left(\dfrac{A}{P}\right)$
1	1.1200	0.8929	1.0000	1.0000	0.8929	1.1200
2	1.2544	0.7972	2.1200	0.4717	1.6901	0.5917
3	1.4049	0.7118	3.3744	0.2963	2.4018	0.4163
4	1.5735	0.6355	4.7793	0.2092	3.0373	0.3292
5	1.7623	0.5674	6.3528	0.1574	3.6048	0.2774
6	1.9738	0.5066	8.1152	0.1232	4.1114	0.2432
7	2.2107	0.4523	10.0890	0.0991	4.5638	0.2191
8	2.4760	0.4039	12.2997	0.0813	4.9676	0.2013
9	2.7731	0.3606	14.7757	0.0677	5.3282	0.1877
10	3.1058	0.3220	17.5487	0.0570	5.6502	0.1770
11	3.4786	0.2875	20.6546	0.0484	5.9377	0.1684
12	3.8960	0.2567	24.1331	0.0414	6.1944	0.1614
13	4.3635	0.2292	28.0291	0.0357	6.4235	0.1557
14	4.8871	0.2046	32.3926	0.0309	6.6282	0.1509
15	5.4736	0.1827	37.2797	0.0268	6.8109	0.1468
16	6.1304	0.1631	42.7533	0.0234	6.9740	0.1434
17	6.8660	0.1456	48.8837	0.0205	7.1196	0.1405
18	7.6900	0.1300	55.7497	0.0179	7.2497	0.1379
19	8.6128	0.1161	63.4397	0.0158	7.3658	0.1358
20	9.6463	0.1037	72.0524	0.0139	7.4694	0.1339
21	10.8038	0.0926	81.6987	0.0122	7.5620	0.1322
22	12.1003	0.0826	92.5026	0.0108	7.6446	0.1308
23	13.5523	0.0738	104.6029	0.0096	7.7184	0.1296
24	15.1786	0.0659	118.1553	0.0085	7.7843	0.1285
25	17.0001	0.0588	133.3339	0.0075	7.8431	0.1275
26	19.0401	0.0525	150.3340	0.0067	7.8957	0.1267
27	21.3249	0.0469	169.3740	0.0059	7.9426	0.1259
28	23.8839	0.0419	190.6989	0.0052	7.9844	0.1252
29	26.7499	0.0374	214.5828	0.0047	8.0218	0.1247
30	29.9599	0.0334	241.3327	0.0041	8.0552	0.1241
31	33.5551	0.0298	271.2926	0.0037	8.0850	0.1237
32	37.5817	0.0266	304.8478	0.0033	8.1116	0.1233
33	42.0915	0.0238	342.4295	0.0029	8.1354	0.1229
34	47.1425	0.0212	384.5210	0.0026	8.1566	0.1226
35	52.7996	0.0189	431.6636	0.0023	8.1755	0.1223
36	59.1356	0.0169	484.4632	0.0021	8.1924	0.1221
37	66.2319	0.0151	543.5988	0.0018	8.2075	0.1218
38	74.1797	0.0135	609.8306	0.0016	8.2210	0.1216
39	83.0812	0.0120	684.0103	0.0015	8.2330	0.1215
40	93.0510	0.0107	767.0916	0.0013	8.2438	0.1213
41	104.2171	0.0096	860.1425	0.0012	8.2534	0.1212
42	116.7232	0.0086	964.3596	0.0010	8.2619	0.1210
43	130.7299	0.0076	1081.0828	0.0009	8.2696	0.1209
44	146.4175	0.0068	1211.8128	0.0008	8.2764	0.1208
45	163.9876	0.0061	1358.2303	0.0007	8.2825	0.1207
46	183.6662	0.0054	1522.2179	0.0007	8.2880	0.1207
47	205.7061	0.0049	1705.8841	0.0006	8.2928	0.1206
48	230.3908	0.0043	1911.5902	0.0005	8.2972	0.1205
49	258.0377	0.0039	2141.9810	0.0005	8.3010	0.1205
50	289.0023	0.0035	2400.0188	0.0004	8.3045	0.1204
51	323.6825	0.0031	2689.0210	0.0004	8.3076	0.1204
52	362.5244	0.0028	3012.7036	0.0003	8.3103	0.1203
53	406.0274	0.0025	3375.2280	0.0003	8.3128	0.1203
54	454.7506	0.0022	3781.2554	0.0003	8.3150	0.1203
55	509.3207	0.0020	4236.0060	0.0002	8.3170	0.1202
56	570.4392	0.0018	4745.3268	0.0002	8.3187	0.1202
57	638.8919	0.0016	5315.7660	0.0002	8.3203	0.1202
58	715.5590	0.0014	5954.6580	0.0002	8.3217	0.1202
59	801.4260	0.0012	6670.2170	0.0001	8.3229	0.1201
60	897.5972	0.0011	7471.6430	0.0001	8.3240	0.1201
61	1005.3088	0.0010	8369.2402	0.0001	8.3250	0.1201
62	1125.9459	0.0009	9374.5491	0.0001	8.3259	0.1201
63	1261.0594	0.0008	10500.4950	0.0001	8.3267	0.1201
64	1412.3865	0.0007	11761.5544	0.0001	8.3274	0.1201
65	1581.8729	0.0006	13173.9410	0.0001	8.3281	0.1201
66	1771.6977	0.0006	14755.8141	0.0001	8.3286	0.1201
67	1984.3014	0.0005	16527.5117	0.0001	8.3291	0.1201
68	2222.4176	0.0004	18511.8132	0.0001	8.3296	0.1201

i = 13%

n	$\left(\frac{F}{P}\right)$	$\left(\frac{P}{F}\right)$	$\left(\frac{F}{A}\right)$	$\left(\frac{A}{F}\right)$	$\left(\frac{P}{A}\right)$	$\left(\frac{A}{P}\right)$
1	1.1300	0.8850	1.0000	1.0000	0.8850	1.1300
2	1.2769	0.7831	2.1300	0.4695	1.6681	0.5995
3	1.4429	0.6931	3.4069	0.2935	2.3612	0.4235
4	1.6305	0.6133	4.8498	0.2062	2.9745	0.3362
5	1.8424	0.5428	6.4803	0.1543	3.5172	0.2843
6	2.0820	0.4803	8.3227	0.1202	3.9975	0.2502
7	2.3526	0.4251	10.4047	0.0961	4.4226	0.2261
8	2.6584	0.3762	12.7573	0.0784	4.7988	0.2084
9	3.0040	0.3329	15.4157	0.0649	5.1317	0.1949
10	3.3946	0.2946	18.4197	0.0543	5.4262	0.1843
11	3.8359	0.2607	21.8143	0.0458	5.6869	0.1758
12	4.3345	0.2307	25.6502	0.0390	5.9176	0.1690
13	4.8980	0.2042	29.9847	0.0334	6.1218	0.1634
14	5.5348	0.1807	34.8827	0.0287	6.3025	0.1587
15	6.2543	0.1599	40.4175	0.0247	6.4624	0.1547
16	7.0673	0.1415	46.6717	0.0214	6.6039	0.1514
17	7.9861	0.1252	53.7391	0.0186	6.7291	0.1486
18	9.0243	0.1108	61.7251	0.0162	6.8399	0.1462
19	10.1974	0.0981	70.7494	0.0141	6.9380	0.1441
20	11.5231	0.0868	80.9468	0.0124	7.0248	0.1424
21	13.0211	0.0768	92.4699	0.0108	7.1015	0.1408
22	14.7138	0.0680	105.4910	0.0095	7.1695	0.1395
23	16.6266	0.0601	120.2048	0.0083	7.2297	0.1383
24	18.7881	0.0532	136.8314	0.0073	7.2829	0.1373
25	21.2305	0.0471	155.6195	0.0064	7.3300	0.1364
26	23.9905	0.0417	176.8501	0.0057	7.3717	0.1357
27	27.1093	0.0369	200.8406	0.0050	7.4086	0.1350
28	30.6335	0.0326	227.9499	0.0044	7.4412	0.1344
29	34.6158	0.0289	258.5833	0.0039	7.4701	0.1339
30	39.1159	0.0256	293.1992	0.0034	7.4957	0.1334
31	44.2010	0.0226	332.3151	0.0030	7.5183	0.1330
32	49.9471	0.0200	376.5160	0.0027	7.5383	0.1327
33	56.4402	0.0177	426.4631	0.0023	7.5560	0.1323
34	63.7774	0.0157	482.9033	0.0021	7.5717	0.1321
35	72.0685	0.0139	546.6807	0.0018	7.5856	0.1318
36	81.4374	0.0123	618.7492	0.0016	7.5979	0.1316
37	92.0243	0.0109	700.1866	0.0014	7.6087	0.1314
38	103.9874	0.0096	792.2109	0.0013	7.6183	0.1313
39	117.5058	0.0085	896.1983	0.0011	7.6268	0.1311
40	132.7815	0.0075	1013.7041	0.0010	7.6344	0.1310
41	150.0431	0.0067	1146.4856	0.0009	7.6410	0.1309
42	169.5487	0.0059	1296.5287	0.0008	7.6469	0.1308
43	191.5901	0.0052	1466.0774	0.0007	7.6522	0.1307
44	216.4968	0.0046	1657.6675	0.0006	7.6568	0.1306
45	244.6414	0.0041	1874.1643	0.0005	7.6609	0.1305
46	276.4447	0.0036	2118.8056	0.0005	7.6645	0.1305
47	312.3825	0.0032	2395.2503	0.0004	7.6677	0.1304
48	352.9923	0.0028	2707.6328	0.0004	7.6705	0.1304
49	398.8813	0.0025	3060.6251	0.0003	7.6730	0.1303
50	450.7358	0.0022	3459.5064	0.0003	7.6752	0.1303
51	509.3315	0.0020	3910.2422	0.0003	7.6772	0.1303
52	575.5446	0.0017	4419.5736	0.0002	7.6789	0.1302
53	650.3654	0.0015	4995.1182	0.0002	7.6805	0.1302
54	734.9129	0.0014	5645.4835	0.0002	7.6818	0.1302
55	830.4515	0.0012	6380.3964	0.0002	7.6830	0.1302
56	938.4102	0.0011	7210.8478	0.0001	7.6841	0.1301
57	1060.4035	0.0009	8149.2581	0.0001	7.6851	0.1301
58	1198.2560	0.0008	9209.6615	0.0001	7.6859	0.1301
59	1354.0293	0.0007	10407.9175	0.0001	7.6866	0.1301
60	1530.0531	0.0007	11761.9468	0.0001	7.6873	0.1301
61	1728.9600	0.0006	13291.9998	0.0001	7.6879	0.1301
62	1953.7248	0.0005	15020.9597	0.0001	7.6884	0.1301
63	2207.7090	0.0005	16974.6843	0.0001	7.6888	0.1301
64	2494.7112	0.0004	19182.3933	0.0001	7.6892	0.1301
65	2819.0236	0.0004	21677.1042	0.0000	7.6896	0.1300
66	3185.4966	0.0003	24496.1277	0.0000	7.6899	0.1300
67	3599.6112	0.0003	27681.6243	0.0000	7.6902	0.1300
68	4067.5606	0.0002	31281.2351	0.0000	7.6904	0.1300

i = 14%

n	$\left(\dfrac{F}{P}\right)$	$\left(\dfrac{P}{F}\right)$	$\left(\dfrac{F}{A}\right)$	$\left(\dfrac{A}{F}\right)$	$\left(\dfrac{P}{A}\right)$	$\left(\dfrac{A}{P}\right)$
1	1.1400	0.8772	1.0000	1.0000	0.8772	1.1400
2	1.2996	0.7695	2.1400	0.4673	1.6467	0.6073
3	1.4815	0.6750	3.4396	0.2907	2.3216	0.4307
4	1.6890	0.5921	4.9211	0.2032	2.9137	0.3432
5	1.9254	0.5194	6.6101	0.1513	3.4331	0.2913
6	2.1950	0.4556	8.5355	0.1172	3.8887	0.2572
7	2.5023	0.3996	10.7305	0.0932	4.2883	0.2332
8	2.8526	0.3506	13.2328	0.0756	4.6389	0.2156
9	3.2519	0.3075	16.0853	0.0622	4.9464	0.2022
10	3.7072	0.2697	19.3373	0.0517	5.2161	0.1917
11	4.2262	0.2366	23.0445	0.0434	5.4527	0.1834
12	4.8179	0.2076	27.2707	0.0367	5.6603	0.1767
13	5.4924	0.1821	32.0887	0.0312	5.8424	0.1712
14	6.2613	0.1597	37.5811	0.0266	6.0021	0.1666
15	7.1379	0.1401	43.8424	0.0228	6.1422	0.1628
16	8.1372	0.1229	50.9804	0.0196	6.2651	0.1596
17	9.2765	0.1078	59.1176	0.0169	6.3729	0.1569
18	10.5752	0.0946	68.3941	0.0146	6.4674	0.1546
19	12.0557	0.0829	78.9692	0.0127	6.5504	0.1527
20	13.7435	0.0728	91.0249	0.0110	6.6231	0.1510
21	15.6676	0.0638	104.7684	0.0095	6.6870	0.1495
22	17.8610	0.0560	120.4360	0.0083	6.7429	0.1483
23	20.3616	0.0491	138.2970	0.0072	6.7921	0.1472
24	23.2122	0.0431	158.6586	0.0063	6.8351	0.1463
25	26.4619	0.0378	181.8708	0.0055	6.8729	0.1455
26	30.1666	0.0331	208.3327	0.0048	6.9061	0.1448
27	34.3899	0.0291	238.4993	0.0042	6.9352	0.1442
28	39.2045	0.0255	272.8892	0.0037	6.9607	0.1437
29	44.6931	0.0224	312.0937	0.0032	6.9830	0.1432
30	50.9502	0.0196	356.7868	0.0028	7.0027	0.1428
31	58.0832	0.0172	407.7370	0.0025	7.0199	0.1425
32	66.2148	0.0151	465.8202	0.0021	7.0350	0.1421
33	75.4849	0.0132	532.0350	0.0019	7.0482	0.1419
34	86.0528	0.0116	607.5199	0.0016	7.0599	0.1416
35	98.1002	0.0102	693.5727	0.0014	7.0700	0.1414
36	111.8342	0.0089	791.6729	0.0013	7.0790	0.1413
37	127.4910	0.0078	903.5071	0.0011	7.0868	0.1411
38	145.3397	0.0069	1030.9981	0.0010	7.0937	0.1410
39	165.6373	0.0060	1176.3378	0.0009	7.0997	0.1409
40	188.8835	0.0053	1342.0251	0.0007	7.1050	0.1407
41	215.3272	0.0046	1530.9086	0.0007	7.1097	0.1407
42	245.4730	0.0041	1746.2359	0.0006	7.1138	0.1406
43	279.8392	0.0036	1991.7089	0.0005	7.1173	0.1405
44	319.0167	0.0031	2271.5481	0.0004	7.1205	0.1404
45	363.6791	0.0027	2590.5648	0.0004	7.1232	0.1404
46	414.5942	0.0024	2954.2439	0.0003	7.1256	0.1403
47	472.6373	0.0021	3368.8381	0.0003	7.1277	0.1403
48	538.8066	0.0019	3841.4754	0.0003	7.1296	0.1403
49	614.2395	0.0016	4380.2820	0.0002	7.1312	0.1402
50	700.2330	0.0014	4994.5215	0.0002	7.1327	0.1402
51	798.2656	0.0013	5694.7545	0.0002	7.1339	0.1402
52	910.0228	0.0011	6493.0201	0.0002	7.1350	0.1402
53	1037.4260	0.0010	7403.0429	0.0001	7.1360	0.1401
54	1182.6656	0.0008	8440.4689	0.0001	7.1368	0.1401
55	1348.2388	0.0007	9623.1345	0.0001	7.1376	0.1401
56	1536.9923	0.0007	10971.3734	0.0001	7.1382	0.1401
57	1752.1712	0.0006	12508.3657	0.0001	7.1388	0.1401
58	1997.4752	0.0005	14260.5369	0.0001	7.1393	0.1401
59	2277.1217	0.0004	16258.0121	0.0001	7.1397	0.1401
60	2595.9187	0.0004	18535.1338	0.0001	7.1401	0.1401
61	2959.3474	0.0003	21131.0525	0.0000	7.1404	0.1400
62	3373.6560	0.0003	24090.3999	0.0000	7.1407	0.1400
63	3845.9678	0.0003	27464.0559	0.0000	7.1410	0.1400
64	4384.4033	0.0002	31310.0237	0.0000	7.1412	0.1400
65	4998.2198	0.0002	35694.4272	0.0000	7.1414	0.1400
66	5697.9706	0.0002	40692.6470	0.0000	7.1416	0.1400
67	6495.6865	0.0002	46390.6177	0.0000	7.1418	0.1400
68	7405.0826	0.0001	52886.3037	0.0000	7.1419	0.1400

i = 15%

n	$\left(\dfrac{F}{P}\right)$	$\left(\dfrac{P}{F}\right)$	$\left(\dfrac{F}{A}\right)$	$\left(\dfrac{A}{F}\right)$	$\left(\dfrac{P}{A}\right)$	$\left(\dfrac{A}{P}\right)$
1	1.1500	0.8696	1.0000	1.0000	0.8696	1.1500
2	1.3225	0.7561	2.1500	0.4651	1.6257	0.6151
3	1.5209	0.6575	3.4725	0.2880	2.2832	0.4380
4	1.7490	0.5718	4.9934	0.2003	2.8550	0.3503
5	2.0114	0.4972	6.7424	0.1483	3.3522	0.2983
6	2.3131	0.4323	8.7537	0.1142	3.7845	0.2642
7	2.6600	0.3759	11.0668	0.0904	4.1604	0.2404
8	3.0590	0.3269	13.7268	0.0729	4.4873	0.2229
9	3.5179	0.2843	16.7858	0.0596	4.7716	0.2096
10	4.0456	0.2472	20.3037	0.0493	5.0188	0.1993
11	4.6524	0.2149	24.3493	0.0411	5.2337	0.1911
12	5.3503	0.1869	29.0017	0.0345	5.4206	0.1845
13	6.1528	0.1625	34.3519	0.0291	5.5831	0.1791
14	7.0757	0.1413	40.5047	0.0247	5.7245	0.1747
15	8.1371	0.1229	47.5804	0.0210	5.8474	0.1710
16	9.3576	0.1069	55.7175	0.0179	5.9542	0.1679
17	10.7613	0.0929	65.0751	0.0154	6.0472	0.1654
18	12.3755	0.0808	75.8364	0.0132	6.1280	0.1632
19	14.2318	0.0703	88.2118	0.0113	6.1982	0.1613
20	16.3665	0.0611	102.4436	0.0098	6.2593	0.1598
21	18.8215	0.0531	118.8101	0.0084	6.3125	0.1584
22	21.6447	0.0462	137.6317	0.0073	6.3587	0.1573
23	24.8915	0.0402	159.2764	0.0063	6.3988	0.1563
24	28.6252	0.0349	184.1679	0.0054	6.4338	0.1554
25	32.9190	0.0304	212.7930	0.0047	6.4641	0.1547
26	37.8568	0.0264	245.7120	0.0041	6.4906	0.1541
27	43.5353	0.0230	283.5688	0.0035	6.5135	0.1535
28	50.0656	0.0200	327.1041	0.0031	6.5335	0.1531
29	57.5755	0.0174	377.1698	0.0027	6.5509	0.1527
30	66.2118	0.0151	434.7452	0.0023	6.5660	0.1523
31	76.1435	0.0131	500.9570	0.0020	6.5791	0.1520
32	87.5651	0.0114	577.1005	0.0017	6.5905	0.1517
33	100.6998	0.0099	664.6656	0.0015	6.6005	0.1515
34	115.8048	0.0086	765.3655	0.0013	6.6091	0.1513
35	133.1755	0.0075	881.1703	0.0011	6.6166	0.1511
36	153.1519	0.0065	1014.3459	0.0010	6.6231	0.1510
37	176.1247	0.0057	1167.4978	0.0009	6.6288	0.1509
38	202.5434	0.0049	1343.6224	0.0007	6.6338	0.1507
39	232.9249	0.0043	1546.1658	0.0006	6.6380	0.1506
40	267.8636	0.0037	1779.0907	0.0006	6.6418	0.1506
41	308.0431	0.0032	2046.9543	0.0005	6.6450	0.1505
42	354.2496	0.0028	2354.9974	0.0004	6.6478	0.1504
43	407.3871	0.0025	2709.2471	0.0004	6.6503	0.1504
44	468.4951	0.0021	3116.6342	0.0003	6.6524	0.1503
45	538.7694	0.0019	3585.1293	0.0003	6.6543	0.1503
46	619.5848	0.0016	4123.8987	0.0002	6.6559	0.1502
47	712.5225	0.0014	4743.4835	0.0002	6.6573	0.1502
48	819.4009	0.0012	5456.0061	0.0002	6.6585	0.1502
49	942.3111	0.0011	6275.4070	0.0002	6.6596	0.1502
50	1083.6577	0.0009	7217.7181	0.0001	6.6605	0.1501
51	1246.2064	0.0008	8301.3759	0.0001	6.6613	0.1501
52	1433.1374	0.0007	9547.5823	0.0001	6.6620	0.1501
53	1648.1080	0.0006	10980.7197	0.0001	6.6626	0.1501
54	1895.3242	0.0005	12628.8278	0.0001	6.6631	0.1501
55	2179.6228	0.0005	14524.1520	0.0001	6.6636	0.1501
56	2506.5663	0.0004	16703.7749	0.0001	6.6640	0.1501
57	2882.5512	0.0003	19210.3413	0.0001	6.6644	0.1501
58	3314.9339	0.0003	22092.8926	0.0000	6.6647	0.1500
59	3812.1740	0.0003	25407.8264	0.0000	6.6649	0.1500
60	4384.0001	0.0002	29220.0007	0.0000	6.6651	0.1500
61	5041.6002	0.0002	33604.0010	0.0000	6.6653	0.1500
62	5797.8402	0.0002	38645.6011	0.0000	6.6655	0.1500
63	6667.5163	0.0001	44443.4419	0.0000	6.6657	0.1500
64	7667.6437	0.0001	51110.9580	0.0000	6.6658	0.1500
65	8817.7904	0.0001	58778.6021	0.0000	6.6659	0.1500
66	10140.4590	0.0001	67596.3926	0.0000	6.6660	0.1500
67	11661.5278	0.0001	77736.8525	0.0000	6.6661	0.1500
68	13410.7571	0.0001	89398.3809	0.0000	6.6662	0.1500

$$i = 16\%$$

n	$\left(\dfrac{F}{P}\right)$	$\left(\dfrac{P}{F}\right)$	$\left(\dfrac{F}{A}\right)$	$\left(\dfrac{A}{F}\right)$	$\left(\dfrac{P}{A}\right)$	$\left(\dfrac{A}{P}\right)$
1	1.1600	0.8621	1.0000	1.0000	0.8621	1.1600
2	1.3456	0.7432	2.1600	0.4630	1.6052	0.6230
3	1.5609	0.6407	3.5056	0.2853	2.2459	0.4453
4	1.8106	0.5523	5.0665	0.1974	2.7982	0.3574
5	2.1003	0.4761	6.8771	0.1454	3.2743	0.3054
6	2.4364	0.4104	8.9775	0.1114	3.6847	0.2714
7	2.8262	0.3538	11.4139	0.0876	4.0386	0.2476
8	3.2784	0.3050	14.2401	0.0702	4.3436	0.2302
9	3.8030	0.2630	17.5185	0.0571	4.6065	0.2171
10	4.4114	0.2267	21.3215	0.0469	4.8332	0.2069
11	5.1173	0.1954	25.7329	0.0389	5.0286	0.1989
12	5.9360	0.1685	30.8502	0.0324	5.1971	0.1924
13	6.8858	0.1452	36.7862	0.0272	5.3423	0.1872
14	7.9875	0.1252	43.6720	0.0229	5.4675	0.1829
15	9.2655	0.1079	51.6595	0.0194	5.5755	0.1794
16	10.7480	0.0930	60.9250	0.0164	5.6685	0.1764
17	12.4677	0.0802	71.6730	0.0140	5.7487	0.1740
18	14.4625	0.0691	84.1407	0.0119	5.8178	0.1719
19	16.7765	0.0596	98.6032	0.0101	5.8775	0.1701
20	19.4608	0.0514	115.3797	0.0087	5.9288	0.1687
21	22.5745	0.0443	134.8405	0.0074	5.9731	0.1674
22	26.1864	0.0382	157.4150	0.0064	6.0113	0.1664
23	30.3762	0.0329	183.6014	0.0054	6.0442	0.1654
24	35.2364	0.0284	213.9776	0.0047	6.0726	0.1647
25	40.8742	0.0245	249.2140	0.0040	6.0971	0.1640
26	47.4141	0.0211	290.0882	0.0034	6.1182	0.1634
27	55.0004	0.0182	337.5024	0.0030	6.1364	0.1630
28	63.8004	0.0157	392.5027	0.0025	6.1520	0.1625
29	74.0085	0.0135	456.3032	0.0022	6.1656	0.1622
30	85.8499	0.0116	530.3117	0.0019	6.1772	0.1619
31	99.5858	0.0100	616.1615	0.0016	6.1872	0.1616
32	115.5196	0.0087	715.7474	0.0014	6.1959	0.1614
33	134.0027	0.0075	831.2670	0.0012	6.2034	0.1612
34	155.4431	0.0064	965.2697	0.0010	6.2098	0.1610
35	180.3141	0.0055	1120.7128	0.0009	6.2153	0.1609
36	209.1643	0.0048	1301.0269	0.0008	6.2201	0.1608
37	242.6306	0.0041	1510.1912	0.0007	6.2242	0.1607
38	281.4515	0.0036	1752.8218	0.0006	6.2278	0.1606
39	326.4837	0.0031	2034.2732	0.0005	6.2309	0.1605
40	378.7211	0.0026	2360.7570	0.0004	6.2335	0.1604
41	439.3165	0.0023	2739.4781	0.0004	6.2358	0.1604
42	509.6071	0.0020	3178.7945	0.0003	6.2377	0.1603
43	591.1443	0.0017	3688.4016	0.0003	6.2394	0.1603
44	685.7273	0.0015	4279.5459	0.0002	6.2409	0.1602
45	795.4437	0.0013	4965.2732	0.0002	6.2421	0.1602
46	922.7147	0.0011	5760.7169	0.0002	6.2432	0.1602
47	1070.3491	0.0009	6683.4316	0.0001	6.2442	0.1601
48	1241.6049	0.0008	7753.7806	0.0001	6.2450	0.1601
49	1440.2617	0.0007	8995.3855	0.0001	6.2457	0.1601
50	1670.7036	0.0006	10435.6471	0.0001	6.2463	0.1601
51	1938.0161	0.0005	12106.3507	0.0001	6.2468	0.1601
52	2248.0987	0.0004	14044.3667	0.0001	6.2472	0.1601
53	2607.7945	0.0004	16292.4653	0.0001	6.2476	0.1601
54	3025.0416	0.0003	18900.2598	0.0001	6.2479	0.1601
55	3509.0482	0.0003	21925.3013	0.0000	6.2482	0.1600
56	4070.4959	0.0002	25434.3494	0.0000	6.2485	0.1600
57	4721.7752	0.0002	29504.8452	0.0000	6.2487	0.1600
58	5477.2593	0.0002	34226.6201	0.0000	6.2489	0.1600
59	6353.6207	0.0002	39703.8794	0.0000	6.2490	0.1600
60	7370.2000	0.0001	46057.5000	0.0000	6.2492	0.1600
61	8549.4320	0.0001	53427.6997	0.0000	6.2493	0.1600
62	9917.3411	0.0001	61977.1313	0.0000	6.2494	0.1600
63	11504.1156	0.0001	71894.4727	0.0000	6.2495	0.1600
64	13344.7740	0.0001	83398.5879	0.0000	6.2495	0.1600
65	15479.9379	0.0001	96743.3613	0.0000	6.2496	0.1600
66	17956.7278	0.0001	112223.2988	0.0000	6.2497	0.1600
67	20829.8042	0.0000	130180.0264	0.0000	6.2497	0.1600
68	24162.5728	0.0000	151009.8301	0.0000	6.2497	0.1600

$$i = 17\%$$

n	$\left(\frac{F}{P}\right)$	$\left(\frac{P}{F}\right)$	$\left(\frac{F}{A}\right)$	$\left(\frac{A}{F}\right)$	$\left(\frac{P}{A}\right)$	$\left(\frac{A}{P}\right)$
1	1.1700	0.8547	1.0000	1.0000	0.8547	1.1700
2	1.3689	0.7305	2.1700	0.4608	1.5852	0.6308
3	1.6016	0.6244	3.5389	0.2826	2.2096	0.4526
4	1.8739	0.5337	5.1405	0.1945	2.7432	0.3645
5	2.1924	0.4561	7.0144	0.1426	3.1993	0.3126
6	2.5652	0.3898	9.2068	0.1086	3.5892	0.2786
7	3.0012	0.3332	11.7720	0.0849	3.9224	0.2549
8	3.5115	0.2848	14.7733	0.0677	4.2072	0.2377
9	4.1084	0.2434	18.2847	0.0547	4.4506	0.2247
10	4.8068	0.2080	22.3931	0.0447	4.6586	0.2147
11	5.6240	0.1778	27.1999	0.0368	4.8364	0.2068
12	6.5801	0.1520	32.8239	0.0305	4.9884	0.2005
13	7.6987	0.1299	39.4040	0.0254	5.1183	0.1954
14	9.0075	0.1110	47.1027	0.0212	5.2293	0.1912
15	10.5387	0.0949	56.1101	0.0178	5.3242	0.1878
16	12.3303	0.0811	66.6488	0.0150	5.4053	0.1850
17	14.4265	0.0693	78.9792	0.0127	5.4746	0.1827
18	16.8790	0.0592	93.4056	0.0107	5.5339	0.1807
19	19.7484	0.0506	110.2846	0.0091	5.5845	0.1791
20	23.1056	0.0433	130.0329	0.0077	5.6278	0.1777
21	27.0336	0.0370	153.1385	0.0065	5.6648	0.1765
22	31.6293	0.0316	180.1721	0.0056	5.6964	0.1756
23	37.0062	0.0270	211.8013	0.0047	5.7234	0.1747
24	43.2973	0.0231	248.8076	0.0040	5.7465	0.1740
25	50.6578	0.0197	292.1049	0.0034	5.7662	0.1734
26	59.2697	0.0169	342.7627	0.0029	5.7831	0.1729
27	69.3455	0.0144	402.0324	0.0025	5.7975	0.1725
28	81.1342	0.0123	471.3779	0.0021	5.8099	0.1721
29	94.9271	0.0105	552.5121	0.0018	5.8204	0.1718
30	111.0647	0.0090	647.4391	0.0015	5.8294	0.1715
31	129.9456	0.0077	758.5038	0.0013	5.8371	0.1713
32	152.0364	0.0066	888.4495	0.0011	5.8437	0.1711
33	177.8826	0.0056	1040.4859	0.0010	5.8493	0.1710
34	208.1226	0.0048	1218.3685	0.0008	5.8541	0.1708
35	243.5035	0.0041	1426.4911	0.0007	5.8582	0.1707
36	284.8991	0.0035	1669.9946	0.0006	5.8617	0.1706
37	333.3319	0.0030	1954.8937	0.0005	5.8647	0.1705
38	389.9984	0.0026	2288.2256	0.0004	5.8673	0.1704
39	456.2981	0.0022	2678.2240	0.0004	5.8695	0.1704
40	533.8687	0.0019	3134.5220	0.0003	5.8713	0.1703
41	624.6264	0.0016	3668.3908	0.0003	5.8729	0.1703
42	730.8129	0.0014	4293.0172	0.0002	5.8743	0.1702
43	855.0511	0.0012	5023.8301	0.0002	5.8755	0.1702
44	1000.4098	0.0010	5878.8813	0.0002	5.8765	0.1702
45	1170.4795	0.0009	6879.2911	0.0001	5.8773	0.1701
46	1369.4610	0.0007	8049.7706	0.0001	5.8781	0.1701
47	1602.2694	0.0006	9419.2317	0.0001	5.8787	0.1701
48	1874.6552	0.0005	11021.5011	0.0001	5.8792	0.1701
49	2193.3466	0.0005	12896.1563	0.0001	5.8797	0.1701
50	2566.2155	0.0004	15089.5028	0.0001	5.8801	0.1701
51	3002.4721	0.0003	17655.7183	0.0001	5.8804	0.1701
52	3512.8924	0.0003	20658.1904	0.0000	5.8807	0.1700
53	4110.0841	0.0002	24171.0830	0.0000	5.8809	0.1700
54	4808.7984	0.0002	28281.1670	0.0000	5.8811	0.1700
55	5626.2941	0.0002	33089.9653	0.0000	5.8813	0.1700
56	6582.7642	0.0002	38716.2598	0.0000	5.8815	0.1700
57	7701.8340	0.0001	45299.0239	0.0000	5.8816	0.1700
58	9011.1459	0.0001	53000.8579	0.0000	5.8817	0.1700
59	10543.0406	0.0001	62012.0039	0.0000	5.8818	0.1700
60	12335.3577	0.0001	72555.0449	0.0000	5.8819	0.1700
61	14432.3684	0.0001	84890.4023	0.0000	5.8819	0.1700
62	16835.8711	0.0001	99322.7715	0.0000	5.8820	0.1700
63	19756.4692	0.0001	116208.6426	0.0000	5.8821	0.1700
64	23115.0691	0.0000	135965.1113	0.0000	5.8821	0.1700
65	27044.6309	0.0000	159080.1816	0.0000	5.8821	0.1700
66	31642.2180	0.0000	186124.8125	0.0000	5.8822	0.1700
67	37021.3950	0.0000	217767.0313	0.0000	5.8822	0.1700
68	43315.0322	0.0000	254788.4258	0.0000	5.8822	0.1700

			i = 18%			
n	$\left(\dfrac{F}{P}\right)$	$\left(\dfrac{P}{F}\right)$	$\left(\dfrac{F}{A}\right)$	$\left(\dfrac{A}{F}\right)$	$\left(\dfrac{P}{A}\right)$	$\left(\dfrac{A}{P}\right)$
1	1.1800	0.8475	1.0000	1.0000	0.8475	1.1800
2	1.3924	0.7182	2.1800	0.4587	1.5656	0.6387
3	1.6430	0.6086	3.5724	0.2799	2.1743	0.4599
4	1.9388	0.5158	5.2154	0.1917	2.6901	0.3717
5	2.2878	0.4371	7.1542	0.1398	3.1272	0.3198
6	2.6996	0.3704	9.4420	0.1059	3.4976	0.2859
7	3.1855	0.3139	12.1415	0.0824	3.8115	0.2624
8	3.7589	0.2660	15.3270	0.0652	4.0776	0.2452
9	4.4355	0.2255	19.0859	0.0524	4.3030	0.2324
10	5.2338	0.1911	23.5213	0.0425	4.4941	0.2225
11	6.1759	0.1619	28.7551	0.0348	4.6560	0.2148
12	7.2876	0.1372	34.9311	0.0286	4.7932	0.2086
13	8.5994	0.1163	42.2187	0.0237	4.9095	0.2037
14	10.1472	0.0985	50.8180	0.0197	5.0081	0.1997
15	11.9737	0.0835	60.9653	0.0164	5.0916	0.1964
16	14.1290	0.0708	72.9390	0.0137	5.1624	0.1937
17	16.6722	0.0600	87.0680	0.0115	5.2223	0.1915
18	19.6733	0.0508	103.7403	0.0096	5.2732	0.1896
19	23.2144	0.0431	123.4135	0.0081	5.3162	0.1881
20	27.3930	0.0365	146.6280	0.0068	5.3527	0.1868
21	32.3238	0.0309	174.0210	0.0057	5.3837	0.1857
22	38.1421	0.0262	206.3448	0.0048	5.4099	0.1848
23	45.0076	0.0222	244.4869	0.0041	5.4321	0.1841
24	53.1090	0.0188	289.4945	0.0035	5.4509	0.1835
25	62.6686	0.0160	342.6035	0.0029	5.4669	0.1829
26	73.9490	0.0135	405.2722	0.0025	5.4804	0.1825
27	87.2598	0.0115	479.2212	0.0021	5.4919	0.1821
28	102.9666	0.0097	566.4810	0.0018	5.5016	0.1818
29	121.5006	0.0082	669.4476	0.0015	5.5098	0.1815
30	143.3707	0.0070	790.9481	0.0013	5.5168	0.1813
31	169.1774	0.0059	934.3188	0.0011	5.5227	0.1811
32	199.6293	0.0050	1103.4962	0.0009	5.5277	0.1809
33	235.5626	0.0042	1303.1255	0.0008	5.5320	0.1808
34	277.9639	0.0036	1538.6881	0.0006	5.5356	0.1806
35	327.9974	0.0030	1816.6520	0.0006	5.5386	0.1806
36	387.0369	0.0026	2144.6494	0.0005	5.5412	0.1805
37	456.7035	0.0022	2531.6863	0.0004	5.5434	0.1804
38	538.9102	0.0019	2988.3898	0.0003	5.5452	0.1803
39	635.9140	0.0016	3527.3000	0.0003	5.5468	0.1803
40	750.3785	0.0013	4163.2141	0.0002	5.5482	0.1802
41	885.4467	0.0011	4913.5926	0.0002	5.5493	0.1802
42	1044.8271	0.0010	5799.0393	0.0002	5.5502	0.1802
43	1232.8960	0.0008	6843.8664	0.0001	5.5510	0.1801
44	1454.8172	0.0007	8076.7625	0.0001	5.5517	0.1801
45	1716.6843	0.0006	9531.5797	0.0001	5.5523	0.1801
46	2025.6875	0.0005	11248.2642	0.0001	5.5528	0.1801
47	2390.3113	0.0004	13273.9518	0.0001	5.5532	0.1801
48	2820.5674	0.0004	15664.2632	0.0001	5.5536	0.1801
49	3328.2695	0.0003	18484.8306	0.0001	5.5539	0.1801
50	3927.3581	0.0003	21813.1003	0.0000	5.5541	0.1800
51	4634.2825	0.0002	25740.4585	0.0000	5.5544	0.1800
52	5468.4534	0.0002	30374.7412	0.0000	5.5545	0.1800
53	6452.7751	0.0002	35843.1948	0.0000	5.5547	0.1800
54	7614.2747	0.0001	42295.9702	0.0000	5.5548	0.1800
55	8984.8441	0.0001	49910.2451	0.0000	5.5549	0.1800
56	10602.1161	0.0001	58895.0898	0.0000	5.5550	0.1800
57	12510.4971	0.0001	69497.2061	0.0000	5.5551	0.1800
58	14762.3867	0.0001	82007.7041	0.0000	5.5552	0.1800
59	17419.6165	0.0001	96770.0908	0.0000	5.5552	0.1800
60	20555.1475	0.0000	114189.7080	0.0000	5.5553	0.1800
61	24255.0742	0.0000	134744.8574	0.0000	5.5553	0.1800
62	28620.9875	0.0000	158999.9316	0.0000	5.5554	0.1800
63	33772.7656	0.0000	187620.9199	0.0000	5.5554	0.1800
64	39851.8638	0.0000	221393.6875	0.0000	5.5554	0.1800
65	47025.1992	0.0000	261245.5527	0.0000	5.5555	0.1800
66	55489.7358	0.0000	308270.7539	0.0000	5.5555	0.1800
67	65477.8887	0.0000	363760.4922	0.0000	5.5555	0.1800
68	77263.9092	0.0000	429238.3828	0.0000	5.5555	0.1800

i = 19%

n	$\left(\frac{F}{P}\right)$	$\left(\frac{P}{F}\right)$	$\left(\frac{F}{A}\right)$	$\left(\frac{A}{F}\right)$	$\left(\frac{P}{A}\right)$	$\left(\frac{A}{P}\right)$
1	1.1900	0.8403	1.0000	1.0000	0.8403	1.1900
2	1.4161	0.7062	2.1900	0.4566	1.5465	0.6466
3	1.6852	0.5934	3.6061	0.2773	2.1399	0.4673
4	2.0053	0.4987	5.2913	0.1890	2.6386	0.3790
5	2.3864	0.4190	7.2966	0.1371	3.0576	0.3271
6	2.8398	0.3521	9.6830	0.1033	3.4098	0.2933
7	3.3793	0.2959	12.5227	0.0799	3.7057	0.2699
8	4.0214	0.2487	15.9020	0.0629	3.9544	0.2529
9	4.7854	0.2090	19.9234	0.0502	4.1633	0.2402
10	5.6947	0.1756	24.7089	0.0405	4.3389	0.2305
11	6.7767	0.1476	30.4035	0.0329	4.4865	0.2229
12	8.0642	0.1240	37.1802	0.0269	4.6105	0.2169
13	9.5964	0.1042	45.2445	0.0221	4.7147	0.2121
14	11.4198	0.0876	54.8409	0.0182	4.8023	0.2082
15	13.5895	0.0736	66.2607	0.0151	4.8759	0.2051
16	16.1715	0.0618	79.8502	0.0125	4.9377	0.2025
17	19.2441	0.0520	96.0217	0.0104	4.9897	0.2004
18	22.9005	0.0437	115.2659	0.0087	5.0333	0.1987
19	27.2516	0.0367	138.1664	0.0072	5.0700	0.1972
20	32.4294	0.0308	165.4180	0.0060	5.1009	0.1960
21	38.5910	0.0259	197.8474	0.0051	5.1268	0.1951
22	45.9233	0.0218	236.4384	0.0042	5.1486	0.1942
23	54.6487	0.0183	282.3618	0.0035	5.1668	0.1935
24	65.0320	0.0154	337.0105	0.0030	5.1822	0.1930
25	77.3881	0.0129	402.0425	0.0025	5.1951	0.1925
26	92.0918	0.0109	479.4305	0.0021	5.2060	0.1921
27	109.5892	0.0091	571.5223	0.0017	5.2151	0.1917
28	130.4112	0.0077	681.1116	0.0015	5.2228	0.1915
29	155.1893	0.0064	811.5228	0.0012	5.2292	0.1912
30	184.6753	0.0054	966.7121	0.0010	5.2347	0.1910
31	219.7636	0.0046	1151.3874	0.0009	5.2392	0.1909
32	261.5187	0.0038	1371.1510	0.0007	5.2430	0.1907
33	311.2072	0.0032	1632.6697	0.0006	5.2462	0.1906
34	370.3366	0.0027	1943.8770	0.0005	5.2489	0.1905
35	440.7006	0.0023	2314.2136	0.0004	5.2512	0.1904
36	524.4337	0.0019	2754.9141	0.0004	5.2531	0.1904
37	624.0761	0.0016	3279.3478	0.0003	5.2547	0.1903
38	742.6505	0.0013	3903.4239	0.0003	5.2561	0.1903
39	883.7541	0.0011	4646.0744	0.0002	5.2572	0.1902
40	1051.6674	0.0010	5529.8286	0.0002	5.2582	0.1902
41	1251.4842	0.0008	6581.4960	0.0002	5.2590	0.1902
42	1489.2662	0.0007	7832.9802	0.0001	5.2596	0.1901
43	1772.2268	0.0006	9322.2465	0.0001	5.2602	0.1901
44	2108.9499	0.0005	11094.4731	0.0001	5.2607	0.1901
45	2509.6504	0.0004	13203.4231	0.0001	5.2611	0.1901
46	2986.4839	0.0003	15713.0735	0.0001	5.2614	0.1901
47	3553.9159	0.0003	18699.5574	0.0001	5.2617	0.1901
48	4229.1599	0.0002	22253.4731	0.0000	5.2619	0.1900
49	5032.7003	0.0002	26482.6331	0.0000	5.2621	0.1900
50	5988.9133	0.0002	31515.3333	0.0000	5.2623	0.1900
51	7126.8068	0.0001	37504.2466	0.0000	5.2624	0.1900
52	8480.9001	0.0001	44631.0532	0.0000	5.2625	0.1900
53	10092.2711	0.0001	53111.9531	0.0000	5.2626	0.1900
54	12009.8026	0.0001	63204.2241	0.0000	5.2627	0.1900
55	14291.6650	0.0001	75214.0264	0.0000	5.2628	0.1900
56	17007.0813	0.0001	89505.6914	0.0000	5.2628	0.1900
57	20238.4268	0.0000	106512.7725	0.0000	5.2629	0.1900
58	24083.7278	0.0000	126751.1992	0.0000	5.2629	0.1900
59	28659.6360	0.0000	150834.9277	0.0000	5.2630	0.1900
60	34104.9668	0.0000	179494.5625	0.0000	5.2630	0.1900

REFERENCES

Chapter 1

1-1 *1980 Annual Report to Congress* (DOE Report DOE/EIA-0173(80)3SUM, Vol. 3)

Chapter 2

2-1 Edward Dean: Introduction to the 1975 Berkeley Summer Study (*Energy & Buildings*, Vol. 1, No. l, May 1977).

2-2 *Energy Conservation in New Building Design* (ANSI/ASHRAE/IES 90A-1980, 90B-1975, 90C-1977; ASHRAE, Atlanta, Ga., 1980).

2-3 C. Lentz: RUF and Energy Conservation in New Construction (*ASHRAE Journal*, June 1978, p. 35).

Chapter 3

3-1 *ASHRAE Handbook, 1977 Fundamentals* (ASHRAE, Atlanta).

3-2 *ASHRAE Handbook, 1976 Systems* (ASHRAE, Atlanta, Ga.).

3-3 *Engineering Weather Data* (Air Force Manual 88-29, 1978).

3-4 *Technical Guidelines for Energy Conservation* (NBSIR 77-1238, U.S. Dept. of Commerce, Washington, D.C., 1977).

Chapter 5

5-1 James C. Webster, Randolph R. Sweich: A Design Procedure for Providing Optimum Illumination (*Specifying Engineer*, October 1978, p. 94).

5-2 H. Richard Blackwell: Energy conservation by selective lighting standards, graded in terms of task and observer characteristics (*Lighting Design & Application*, Jan. 1976, p.17).

5-3 *IES Lighting Handbook, Fifth Edition* (Illumination Engineering Society, New York, N.Y., 1972).

5-4 Francis Clark: Evaluating lighting system performance to save energy (*Lighting Design & Application*, August 1977).

5-5 *Westinghouse Lighting Handbook* (January 1976).

5-6 James F. Finn: Energy efficient lighting — a management guide (*Lighting Design & Application*, September 1977).

5-7 K. Chen, M.C. Unglert, R.L. Malafa: Energy saving lighting for industrial applications (*Conference Record, IAS Annual Meeting*, IEE Industry Applications Society).

5-8 *IES Recommended Practice of Daylighting* (Illumination Engineering Society, New York, N.Y., 1962).

5-9 Joseph B. Murdoch: A procedure for calculating the potential savings in lighting energy from the use of skylights (International Conference of Illumination Engineering Society, 1976).

5-10 IES Committee on Lighting and Air Conditioning: Lighting Systems as Heat Sources (*Illuminating Engineering*, March 1966, p. 130).

5-11 James F. Finn: Efficient application of lighting energy–a luminaire air heat-transfer evaluation (*Lighting Design & Application*, January 1976, p. 40).

5-12 D.K. Ross: Electric Lighting Controls (*Energy and Buildings*, Vol. 1, No. 1, May 1977).

5-13 Harry B. Zackrison: Interior lighting and energy consciousness (*Lighting Design & Application*, April 1978).

5-14 D.K. Ross: Energy Conservation Applied to Task Lighting (*IEEE Transactions on Industry Applications*, Jan./Feb. 1976).

Chapter 6

6-1 *Site Energy Handbook*, Vol. I, Methodology for Energy Survey & Appraisal (Energy Research and Development Administration, U.S. Dept. of Commerce, Washington, D.C.).

6-2 K.S. Bajaj, Robert G. LaMont: An Energy Analysis Program for Electrical Distribution Systems (*Consulting Engineer*, May 1979).

6-3 Donald Beeman: *Industrial Power Systems Handbook* (McGraw-Hill, 1955).

6-4 *ABC of Large AC Motors and Control* (Electric Machinery Mfg. Co.).

6-5 *A Guide to Power Factor Correction for the Plant Engineer* (Sprague Electric Co.).

6-6 Irwin Lazar: The Engineering Basics of Power Factor Improvement (*Specifying Engineer*, April 1975).

6-7 Paul Lindhorst: High Efficiency Motors–A New Way to Save Energy (*Specifying Engineer*, October, 1975).

6-8 H.E. Albright: Evaluating the economics of power transformer efficiencies (*Plant Engineering*, June 1975).

6-9 *Norpak Demand Control* (Square D Company).

6-10 Robert Oliverson: Energy Management Equipment (*Specifying Engineer*, July 1977).

Chapter 7

7-1 H.G. Thuesen, W.J. Fabrycky, G.J. Thuesen: *Engineering Economy, 4th Ed.* (Prentice-Hall, Inc., 1971).

7-2 George A. Taylor: *Managerial and Engineering Economy* (D. Van Nostrand Co., 1964).

7-3 James L. Riggs: *Economic Models for Engineers and Managers* (McGraw-Hill Inc., 1968).

7-4 Craig Lentz: Initial vs. Life Cycle Cost, the Economics of Conservation (*Consulting Engineer*, October 1976).

Chapter 8

8-1Quentin Looney: Energy Audits and Information Systems (Energy Management Seminar Proceedings, IEEE Industry Applications Society, Chicago, Ill., October, 1976).

8-2*Site Energy Handbook*, Vol. I, II, III, Methodology for Energy Survey & Appraisal (Energy Research and Development Administration, U.S. Dept. of Commerce, Washington, D.C.).

8-3Gary Vanderweil: Savings Begin with an Energy Audit (*Consulting Engineer*, November 1976).

8-4*Industrial Fuel Conservation through Energy Management* (Exxon Report, 1976).

8-5James L. Coggins: An Effective Energy Audit: The Flow Sheet Approach (*ASHRAE Journal*, June 1976).

8-6K.S. Bajaj, T. Singh: Systematic Approach to Plant Energy Conservation (*Specifying Engineer*, August 1977, p. 76).

Chapter 9

9-1Joseph B. Olivieri: A Consultant Looks at . . . Heat recovery systems (*Air Conditioning, Heating & Refrigeration News*, Jan. 25, 1971, p. 23; Feb. 22, 1971, p. 24; March 8, 1971, p. 18; April 19, 1971, p. 53; Aug. 16, 1971, p. 22; Nov. 15, 1971, p. 26; Jan. 17, 1972, p. 17).

10-1*Conserve Energy by Design* (The Trane Co., LaCrosse, Wis., 1974).

Chapter 10

10-1*Conserve Energy by Design* (The Trane Co., LaCrosse, Wis., 1974).

10-2R.G. Nevins and F.H. Rohles: The nature of thermal comfort for sedentary man (*ASHRAE Transactions*, Vol. 77, Part I, 1971).

10-3*Technical Guidelines for Energy Conservation* (NBSIR 77-1238, U.S. Dept. of Commerce, Washington, D.C., 1977).

10-4*Methods for Testing Air Cleaning Devices* (ASHRAE Standard 52-68).

10-5Smith, Hinchman & Grylls Associates, Inc. (Detroit, MI, 1977).

Chapter 11

11-1S.Y.S. Chen: Dissecting Computer Programs (*Heating/Piping Air Conditioning*, Sept. 1975, p. 41 and Oct. 1975, p. 59).

11-2L.C. Spielvogel: Comparison of Energy Analysis Programs (*ASHRAE Journal*, January 1975, p. 45).

11-3J.M. Ayers: Predicting Building Energy Requirement(*Energy & Buildings*, Vol. 1, No. 1, May 1977, pp. 11-18).

11-4*Technical Guidelines for Energy Conservation* (NBSIR 77-1238, U.S. Dept. of Commerce, Washington, D.C., 1977).

11-5E. Stamper: Cross Checking Energy Analysis Procedure and Standardizing Weather Input (*ASHRAE Transactions*, 1976, Part I, p. 323).

11-6*NECAP Engineering Manual* (September 1975, p. 5-126).

11-7*Procedure for Determining Heating and Cooling Loads for Computerizing Energy Calculations* (ASHRAE, February 1975).

11-8W.F. Stoccne: *Procedures for Simulating the Performance of Components and Systems for Energy Calculations, 3rd Ed.* (ASHRAE, 1974).

Chapter 12

12-1R.C. Jordan, Ed.: *Applications of Solar Energy for Heating and Cooling Buildings* (ASHRAE Publication GRP 170, 1977).

12-2G.O.G. Lof, S. Karakari, and C.C. Smith: Comparative Performance of Solar Heating with Air and Liquid Systems (*Proceedings of Joint Conference, America Section, International Solar Energy Society*, Vol. 3, 1976, p. 186).

12-3W.A. Beckman, S.A. Klein, and J.A. Duffie: *Solar Heating Design by the f-Chart Method* (John Wiley, New York, 1977).

12-4S.A. Klein, W.A. Beckman, and J.A. Duffie: A design procedure for solar heating systems (*Solar Energy*, Vol. 18, No. 2, 1976, pp. 113-126).

12-5ITT Fluid Handling Div.: *Solar Heating Systems Design Manual* (ITT, Morton Grove, Ill., 1976).

12-6Smith, Hinchman & Grylls Associates, Inc. (Detroit, Michigan, 1977).

12-7*Technical Guidelines for Energy Conservation* (NBSIR 77-1238, U.S. Dept. of Commerce, Washington, D.C., 1977).